Multifunctional Concrete Technology

Feng Naiqian

Lu Jin Ping

Peng Gai-Fei

Tsinghua University
American Concrete Institute - Singapore Chapter
Beijing Jiaotong University

Published by **Materials Research Forum LLC**
Millersville, PA 17551, USA

Published as part of the book series
Materials Research Foundations
Volume 127 (2022)
ISSN 2471-8890 (Print)
ISSN 2471-8904 (Online)

Print ISBN 978-1-64490-198-4
ePDF ISBN 978-1-64490-199-1

Distributed worldwide by

Materials Research Forum LLC
105 Springdale Lane
Millersville, PA 17551
USA
http://www.mrforum.com

Printed in the United States of America
10 9 8 7 6 5 4 3 2 1

Table of Contents

Foreword

With the development and progress of human civilization, concrete materials and application technology are also continuously improved. From the Dadiwan site in Gansu Province, China, a ground material similar to ancient Roman volcanic ash concrete has been excavated. This ground material from about 5,000 years ago has surprising durability. Several companies and research units in Japan took samples from the Dadiwan site in China and carried out a systematic microstructure analysis. Using the basic principle of the long life of concrete, they developed the "EIEN" (forever) concrete technology, produced formwork for concrete structures and, for marine engineering. The bridge piers were wrapped with this permanent formwork, also known as Ten-thousand years concrete. The Dadiwan site is also the oldest concrete in the world today.

In the 1970s, in order to improve the durability of offshore reinforced concrete platforms, Norway added silica fume into the concrete. As a result, it not only improved the durability of the concrete, but also increased the strength of the concrete and improved the fluidity of the concrete. This kind of concrete cannot be measured by a certain performance and was later named as high-performance concrete.

On the basis of the development of mineral powder technology and superplasticizer technology, ultra-high-performance concrete was further developed from high-performance concrete. Japan has used ultra-high-performance concrete with a strength of 200 MPa in the bottom column of super high-rise buildings. In the United States, ultra-high performance concrete with a strength of 250 MPa has been commercialized. In Europe, ultra-high performance concrete with a strength of 200-250 MPa has been used in bridges and special structures. A bridge panel in Quebec, Canada, used ultra-high-performance concrete with a strength of 250 MPa.

In China, for the super high-rise building of the East Tower in Guangzhou, ultra-high performance concrete with a strength of 150MPa was used. This concrete has multiple functions: self-compacting, self-curing, low hydration heat, low shrinkage, high slump retention and high durability. As the function of a certain aspect cannot fully reflect the characteristics of this concrete, this type of concrete is named as multifunctional concrete. In the East Tower project, the ultra-high pumping test was carried out, and the concrete was directly pumped from the ground to a height of 510 m, and beams, slabs, columns and shear walls were cast with high quality and there was no cracks found on the concrete surface after demolding.

Norwegian concrete experts point out: "Ultra-high performance concrete is a breakthrough in concrete technology".

With the multifunctional concrete technology the strength of concrete can reach up to 150 MPa. Low strength concrete such as C30 concrete can also have a variety of functions. Multifunctional concrete has the same raw material as ordinary concrete, and can also be produced from industrial waste. In addition to ordinary concrete, there are also lightweight aggregates multifunctional concrete, etc. The multifunctional concrete technology is the result of Professor Feng Naiqian's research and development based on the needs of three landmark super high-rise building projects of Guangzhou's West Tower, East Tower and Shenzhen Kingkey Building. Professor Feng was invited to participate in the expert committee of the three projects. With the support of Chairman Mr. Ye Haowen, the project leader, he developed C100 ultra-high performance concrete (UHPC) and C100 ultra-high performance self-compacting concrete. These two kinds of concrete were pumped to 420 m high, and the steel tube concrete was cast. Then, combined with Kingkey Building, C120UHPC was developed. Fresh concrete could keep flowable for 3 hours. After 1.5

hours from Zhengqiang concrete mixing station in Bao'an, it was transported to the construction site of Caiwuwei and pumped to a height of more than 500 m for the casting of beam-slab-columns. In the East Tower project, the C80 ultra-high performance concrete shear wall on the first floor was cast. After demoulding, it was found that there were cracks on the wall. For some walls, there were more than 120 cracks on the same wall. This was caused by the autogenous shrinkage of the concrete. With the support of the East Tower project, in the laboratory of Tianda Concrete Company, the research on inhibiting the self-shrinkage cracking of ultra-high performance concrete was carried out. The performance of multi-functional concrete with low to high strength was studied from raw materials, small simulation test, shear wall simulation test, and actual size structure test, etc. From the traditional raw materials of concrete to the resource application of industrial waste residues, multi-functional concrete with the application of paper-making white mud and industrial waste was developed. C130 multifunctional concrete was pumped to a height of 510 m in the Guangzhou East Tower project, and the concrete of beams, slabs, columns and shear walls was casted. This project won three gold medals at the China (Shanghai) International Invention Expo, and one Gold Medal at the Geneva Invention Expo held in the Silicon Valley of the United States.

This book is divided into 22 chapters. Chapter 1 describes the development of concrete materials and technologies. Chapter 2 describes how powder technology and water reducing agent technology promote concrete technology to the development and application of high-performance, ultra-high performance. Chapter 3 shows the development and production application of high performance concrete. Chapter 4 covers the development and application of ultra high performance concrete. Chapter 5 covers coarse and fine aggregate of ultra high-performance concrete materials. Chapter 6 reviews the development and application of new superplasticizers. Chapter 7 describes concrete shrinkage and cracking and application of shrinkage reducing agents. Chapter 8 introduces the use of anti-rust agents for steel bars in concrete. Chapter 9 gives the development and experimental application of C l⁻ solidifying agent in concrete. Chapter 10 covers microbead ultrafine powder and its function. Chapter 11 covers the properties of natural zeolite ultrafine powder and application in concrete. Chapter 12 explores the minerals properties and applications of slag ultrafine powder. Chapter 13 reviews the application of silica fume in concrete. Chapter 14 reviews the application of fly ash in concrete. Chapter 15 shows the development and engineering application of multifunctional concrete technology. Chapter 16 covers high-strength and ultra-high-strength multifunctional concrete. Chapter 17 covers the C120 development and application of MPC. Chapter 18 shows the development and performance testing of MPC with a strength of 150MPa. Chapter 19 reviews research and development and application of MPC with C30, C60 strength grades with papermaking white mud. Chapter 20 covers the ultra-high pumping technology for ultra-high strength MPC. Chapter 21 covers lightweight aggregate multifunctional concrete and its applications. Chapter 22 covers electrochemical protection technology of multifunctional concrete structures.

This book discusses multi-functional concrete technology from various aspects. It can be used as a reference for teachers and students in colleges and universities, and can also be used for reference by engineers and technicians.

Also, best efforts for correctness have been made by the author there may be some errors, the author would appreciate any feedback for correction.

During the research and testing of multi-functional concrete, the China Construction Fourth Engineering Bureau and Guangzhou Tianda Concrete Co., Ltd, Civil Engineering Company, Shenzhen Construction Engineering Group, Shenzhen Zhengqiang Investment Company, and

Zoomlion and Sany Heavy Industry, etc. provided strong support and assistance and the author would like to express the heartfelt thanks. Professor Takeshi Mukai, Yoshio Kasai and S. Nishibayashi were the author's mentor and friends when he studied in Japan, and they gave the author great help and guidance. Heartfelt thanks to them too.

The manuscript of the whole book was written by Feng Naiqian in Chinese and the English manuscript was translated by ACI Singapore Chapter President Lu Jinping with the support of ACI Singapore Chapter Board members Dr Wong Sook Fun, Dr Geng Guoqing, Dr Qian Shunzhi, Ms Li Wei, Dr Tan Jun Yew, Dr Tao Neng Fu and Mr Jiang Shan. Professor Peng Gai-Fei proofread and corrected Chapter 1 and Chapter 3-13 on the basis of the translation.

Professor Feng Naiqian, Beijing

Chapter 1

Review of the Development of Concrete Materials and Technology

Concrete is currently the most used material for construction in the world, with an annual production of about one ton per capita, and the world's annual production of concrete is about 6.5-7 billion tons. China accounts for more than half of the total production.

When we discuss on the development of modern concrete technology, we should note that, from the invention of cement in 1824 and to the emergence of concrete in 1835, the history of cement concrete is less than 200 years. Concrete has undergone deep-level high-tech development: from ancient concrete, ancient Roman volcanic ash concrete, to ordinary concrete, high performance and ultra-high performance concrete, to multifunctional concrete, etc.

Currently ultra-high performance concrete is a cutting-edge advance in concrete technology.

1.1 The oldest concrete in the world

The concrete in primitive society is the oldest concrete in the world. The Dadiwan Ancient Cultural Site in Qin'an County, Tianshui City, Gansu Province, China, has a total area of 1.1 million m^2. The age might be as early as 7800 years ago, if not, at least 4800 years ago. It had 3000 years of cultural continuity. There is a large-scale building belonging to the Yangshao cultural period, with three doors and a verandah (Figure 1.1). The site area is 270 m^2, and the indoor area is 150 m^2. Built on the ground, the main room of the F901 palace-style building with an area of 130 m^2, all floor materials were made of loess and sand, similar to the modern cement floor (Figure 1.2).

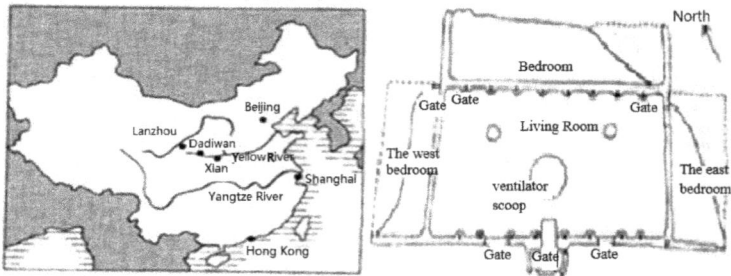

(a) Location of Dadiwan Ancient Cultural Site (b) Sanmen Kaifang site area is 270 m^2

Figure 1.1 The big building with three-door opening and eaves in the late period of Yangshao Culture.

Figure 1.2 The floor in the main room of the F901 palace-style building, similar to a modern cement concrete floor.

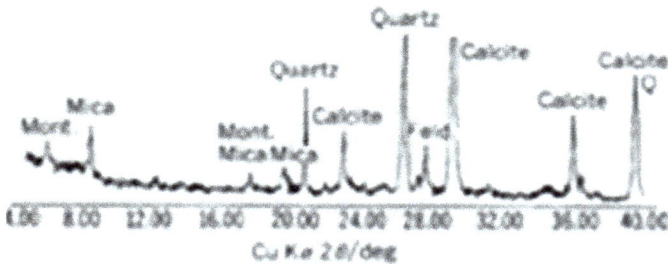

Figure 1.3 XRD of Loess Stone powder.

In a book "Lightweight concrete" edited by Professor Yoshio Kasai of Japan, it was stated: "The world's oldest lightweight concrete is in the site of a large-scale residential complex in Dadiwan, Qin'an County, Gansu Province, China." The researchers from the Department of Industrial Engineering, Nihon University and Japan Kashima Co., Ltd., took samples from Dadiwan and performed XRD (Figure 1.3) and SEM analyses. It was further pointed out that the floor of the Dadiwan residential group site was built by using a material composed of a hydraulic cementitious material produced by calcining a loess stone containing calcium carbonate and clay, and aggregate, also fired with the same loess stone to become an artificial lightweight aggregate with countless pores. The density of the latter is 1.68 g/cm^3, the particle size is about 20 mm, and the packing ratio is 67%. The compressive strength of the cube is 10.78 MPa. This structural material has maintained sufficient strength after 5000 years." (Figure 1.4)

Materials Research Forum LLC

https://doi.org/10.21741/9781644901991

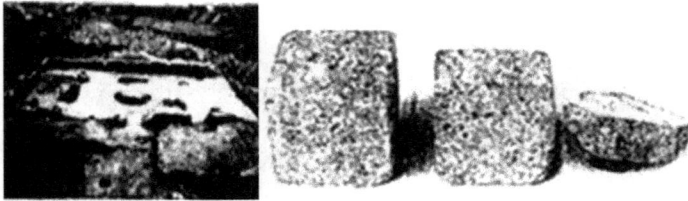

(a) Loess stone floor   (b) Ancient concrete samples

Figure 1.4 Loess stone floor site and specimens taken from site

The loess was heated at different temperatures, and the weight loss is shown in Table 1.1. It can be seen that, after long-term carbonization of the floor concrete, most of the hydrates were carbonized and became calcium carbonate.

Table1.1. The weight loss of loess stone at different temperature

110 °C dry to constant weight	Weight loss 0.85 %
1000 °C heating 1h	27.1 %
550 °C-1000 °C heating	25.8 %
$CaCO_3$ Content	60 %

1.2 Enlightenment from ancient concrete

1.2.1 Comparison of chemical composition between cement and loess stone

Comparison of chemical composition of the loess stone and Portland cement: The chemical composition of the loess stone in the loess concrete material is shown in Table 1.2, and the polarized light microscope observation is shown in Figure 1.5. The XRD pattern of the ancient concrete sample at 800 °C, 900 °C and 1000 °C after calcination is shown in Figure 1.6.

Table 1.2 Comparison of chemical composition of loess stone and Portland cement

Sample No.	SiO_2	Al_2O_3	Fe_2O_3	CaO	MgO	SO_3	Na_2O	K_2O	TiO_2	P_2O_5	MnO
00	35.5	8.7	5.0	45.2	2.1	0.1	-	2.3	0.6	-	0.4
01	21.8	4.7	2.8	67.6	1.2	0.1	-	1.5	0.4	-	-
Average	20.9	5.3	4.4	65.2	1.8	0.2	-	1.1	0.4	-	0.2
Cement	21.2	5.4	2.7	64.7	2.0	1.9	0.28	0.56	0.3	0.22	0.09

Note: Na_2O in No. 00 and No. 01 samples have not been detected. Other components such as SiO_2, Al_2O_3, CaO, MgO and etc., in the loess stone are similar to those of Portland cement.

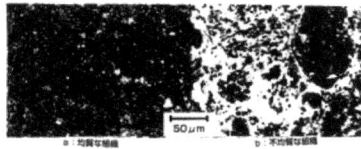

Figure 1.5 Polarized light microscope spectrum of loess stone.

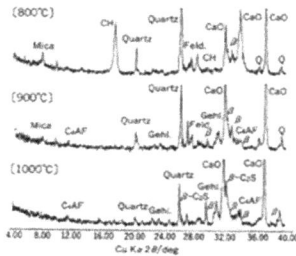

Figure 1.6 XRD patterns of ancient concrete samples calcined at 800 °C, 900 °C, and 1000 °C.

It can be found from the XRD patterns of ancient concrete samples calcined at 800 °C, 900 °C and 1000 °C for 8 h: (1) When heated at 800 °C, calcite and montmorillonite decomposed to produce hydraulic mineral β-C_2S. In addition, at this temperature, CaO generated by the decomposition of montmorillonite is highly active. During the cooling process in the atmosphere, it absorbed water vapor and generated $Ca(OH)_2$. (2) A large amount of quartz exists in the sample fired at 900 °C; The characteristic peaks of muscoviter and feldspar can also be seen; CaO and the hydraulic mineral β-C_2S can be seen at the same time, and occasionally the characteristic peak of C_4AF can be found; (3) The characteristic peaks of muscovite and feldspar disappeared when the sample was calcined at 1000 °C ; At this temperature, a large amount of calcite and quartz reacted to form β-C_2S. From the ancient concrete samples, through the calcining conditions reproduced, the same mineral composition as that of the present Portland cement is obtained.

1.2.2 XRD patterns of ancient concrete samples

The sample calcined at 1000 °C was mixed with water at water/solid ratio of 0.5 to produce paste. After molding, it was cured for 1 month and 6 months, respectively, and then analyzed by XRD, as shown in Figure 1.7.

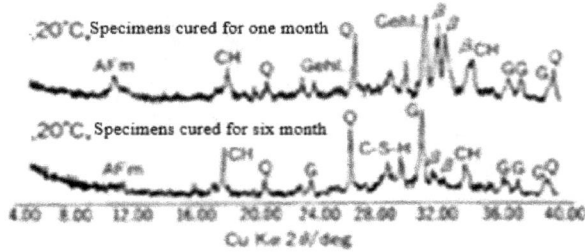

Figure 1.7 XRD pattern of the specimens of loess stone powder fired at 1000 °C mixed with water.

In addition to the characteristic peaks of β-C_2S, quartz, the characteristic peaks of AFm, CH and C-S-H can also be observed; there is indeed a hydrate of β-C_2S. This proves that the loess stone powder has undergone hydration reaction.

1.2.3 Strength of the calcined Loess stone powder concrete

The strength of loess stone powder concrete with different calcination temperature and different curing age is shown in Figure 1.8.

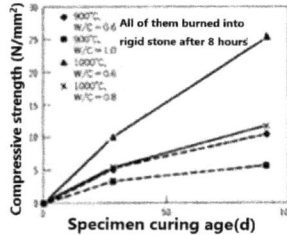

Figure 1.8 Strength of loess power concrete.

The strength of concretes made with the loess stone power calcined at different temperature and cured at constant temperature for 8 hours, is shown in Table 1.3.

Table 1.3 Strength of concrete specimens made with calcined loess stone power.

Calcined Temperature (°C)	Curing hours (h)	Water/cement ratio(W/C)	Compressive strength (N/mm²)	
			28d	91d
900	8	0.6	6	10.5
1000	8	0.6	10	25.4

Note: The concrete strength of the sample collected from the floor is 10 N/mm².

1.2.4   Conclusion

It can be seen from the analyses of the above materials: The main mineral composition of the loess stone concrete 5000 years ago found in Dadiwan of China is similar to β-C_2S, which can bond sand and other particles together, to become concrete. The concrete has been fully carbonized to form $CaCO_3$, which has an extremely long service life.

1.3   Development of concrete materials with a life span of ten thousand years

Japan's Kashima Corporation Electric Chemical Industry and Ishikawa Island Building Materials Industry, etc., made use of the chemical principles of this oldest concrete material to develop concrete with a ten-thousand-year life. They called this material EIEN (For Ever) and such a concrete has been used in the ultra-high durability concrete structure. (Figure 1.9).

(a) The permanent formwork made of concrete simulated to the oldest concrete in Dadiwan

(b) The setup of simulation   (c) Bridge pile made from concrete with permanent formwork

Figure 1.9 Technique and applications of the permanent formwork made of concrete simulated to the oldest concrete in Dadiwa.

The performance of the formwork simulated from the oldest concrete in Dadiwan: 1) Shrinkage: 1/2 of ordinary concrete; 2) Chloride penetration: 1/8 of ordinary concrete; 3) Neutralization depth: 0 mm (20 °C, 60 % RH, CO_2 5 %, 1 year).

1.4 Volcanic ash concrete in ancient Rome

The concrete material invented in ancient Rome is one of the main engineering materials in human construction and life. About 2000 years ago, the ancient Romans skilfully mixed the volcanic ash, lime and marble fragments which were rich in Rome, with water to become a mixture which can automatically hydrate and harden into concrete. The ancient Romans used this material to build ancient Roman buildings, such as the Colosseum, churches, and Pantheon. It was once glorious in history, as shown in Figure 1.10. Ancient Roman buildings made of volcanic ash concrete in ancient Rome were heroic, and prosperous for a while. In 1084 AD, the Germans invaded Rome, Rome was looted, and many buildings were destroyed by the war. Only remnants are left for future generations to see. The power of volcanic ash concrete should surprise future generations!

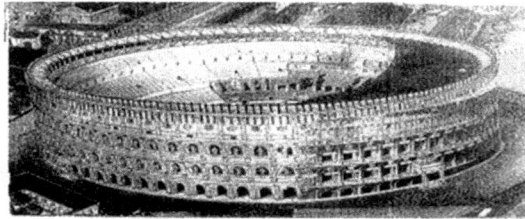

Figure 1.10 Restoration Ancient Roman Colosseum (72-80 AD).

1.5 Normal concrete

The development of human civilization based on the afterglow of ancient Roman architecture helps cement concrete to be invented. In 1824, the British Joseph Aspdin obtained the patent right to manufacture Portland cement. As the colour of this hardened cement is similar to the colour of British Portland stone, so it is named as Portland cement. The emergence of Portland cement laid the material foundation for the establishment of ordinary concrete. The cement material used to make concrete has undergone a qualitative change. The production technology of cement concrete has developed rapidly, the amount of concrete has also increased sharply, and the scope of its application has been expanding. So far, it has become the most used artificial stone in the world.

After the Second World War, cement concrete was used to rebuild the world. Not only the old buildings were restored, but new magnificent buildings also sprung up like mushrooms. This heralded the arrival of the cement concrete civilization era.

1.6 The first full concrete house

In 1835, the Roman cement manufacturer, John Bazley-White built the first full-concrete house with concrete walls, concrete roof tiles, concrete window purlins and concrete decorations. And even in the garden, the statue of God was also built with concrete. Afterwards, many full-concrete houses were built according to his model. However, the floor slabs at that time were still wooden or steel structures. The first full-concrete house built by White is shown in Figure 1.11.

Figure 1.11 The first all-concrete house built by White.

1.7   Emergence of reinforced concrete, prestressed reinforced concrete and fiber concrete

In 1850, the French Lambot built a small cement boat with the method of adding steel mesh to cement. This is the earliest reinforced concrete structure. Later, steel bars were used to reinforce concrete to compensate the low tensile and flexural strength of the concrete. It has greatly promoted the application of concrete to various engineering structures. In 1887, M. Koenen published a calculation method for reinforced concrete. After the concrete is reinforced, it can be used to produce members to bear bending or tension. But it did not solve the problem of easy cracking of concrete. In 1928, E. Freyssinet in France proposed a theory of shrinkage and creep of concrete. By using high-strength steel wires, and pre-stressed anchors he invented, he has developed the foundation of the application of pre-stressed concrete technology in concrete engineering. Pre-stressing concrete with tensile steel bars can make the concrete members resist tension and prevent cracks under loading, especially for application of high-strength concrete where the pre-stressing method is the most effective.

The emergence of pre-stressed reinforced concrete is a breakthrough in concrete technology. It has excellent performance in large-span buildings, high-rise and super high-rise buildings, earthquake-resistant structures, and structures with anti-cracking and internal pressure resistance. This greatly expands the application of concrete.

In order to reduce the brittleness of concrete and increase the ductility, the theory of dispersed reinforcement was proposed. In 1940, L. Nervi in Italy proposed a steel mesh cement reinforcement material to make reinforced concrete with homogeneous properties. Thereafter, large-span reinforced concrete buildings and thin-shell structures were built. Later, the concept of fiber as reinforcement was proposed by using fibers as dispersed reinforcement in concrete, which greatly improved the crack resistance of concrete and increased the ductility of concrete.

1.8   The calculation of concrete strength and the theory of water-cement ratio

In 1918, D.A. Abrams published the famous water-cement ratio theory for calculating the strength of concrete. The strength (R) of concrete is inversely proportional to the water-cement ratio (W/C) and the prerequisite for durability is based on the reduction of W/C. It has laid a theoretical foundation for ordinary concrete. The law of water-cement ratio (W/C) still plays a guiding role in the preparation of high-performance concrete and ultra-high-performance concrete. In 1875, Willian Lascellss applied external pressure to the formed concrete members through a steel mould, so that the mixing water in the concrete overflows through the pores of the formwork, reducing the water-cement ratio of the concrete from the original 31 % to 22 %, thus obtaining a high-strength concrete with a compressive strength of 110 MPa. This has already provided an example for the law of water-cement ratio (W/C).

1.9 Research on the performance of ordinary concrete

Advanced materials science shows that, when the composition of the material is determined, its various properties are closely related to its internal structure. Hardened concrete is a multi-phase, porous composite material. It has a highly inhomogeneous and complex internal structure. On the macro level, concrete can be regarded as a two-phase material composed of aggregate particles (coarse aggregate and sand) dispersed in the cement matrix, as shown in Figure 1.12. On the micro level, as shown in Figure 1.13, not only is the distribution of the two phases of the aggregate and the cement matrix inhomogeneous, but the two phases themselves are also not homogeneous. In the microstructure of the cement matrix, there are hydration products, unhydrated cement particles, pores, etc., and the distribution is extremely inhomogeneous.

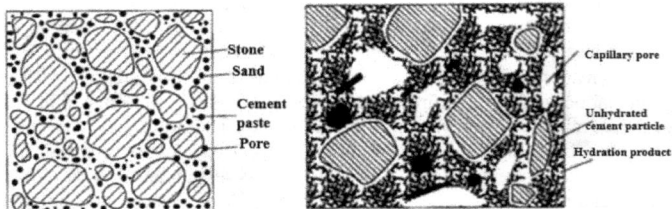

Figure 1.12 Macro structure of concrete paste.

Figure 1.13 Micro structure of hardened cement

Although the concrete structure is highly inhomogeneous and unstable, it is difficult to establish a structure-performance relationship model from the perspective of materials science. However, a deeper understanding of the important structural characteristics of each component of concrete is fundamental and meaningful to control the performance of concrete.

1.9.1 The structure of the aggregate phase

In ordinary concrete, the strength of the aggregate is higher than that of the other two phases, so the aggregate has no direct effect on the strength of the concrete. However, with larger particle size, the more the needle- and flake-shaped particles, the lower the strength of concrete. Because the water film accumulated on the surface of these kinds of aggregate is thicker, the interfacial transition zone tends to become weaker and more likely to cause micro-cracks (Figure 1.14a). If the composition of the concrete is unreasonable, it will cause segregation and stratification due to excessive water consumption and integrity of the concrete will be greatly affected (Figure 1-14b). Due to the change of cement paste to aggregate ratio in concrete, 3 states of concrete macro structures will be developed (Figure 1.14c). These will have an impact on the mechanical properties of concrete.

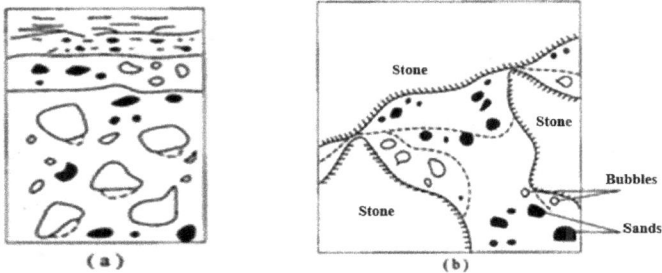

(a) State of externally layered concrete structure

(b) State of internally layered concrete structure

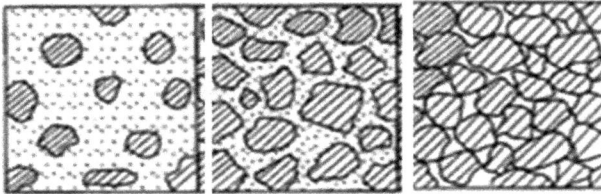

(c) Concrete macrostructure status

Figure 1.14 The effect of quantity and quality of coarse aggregate on concrete structure.

1.9.2 The structure of hardened cement paste

Hardened cement paste is a complex structure system composed of solid, liquid and gas phases. It changes with time, temperature and humidity. The hardened cement paste in a concrete structural system is the structural system formed after cement hydration. The solids, pores and water in the cement paste play key roles in the properties of the hardened cement paste and even the performance of the concrete.

1) There are four main solid phases in the hardened cement paste observed under the scanning electron microscope: hydrated calcium silicate gel phase (CSH phase), calcium hydroxide phase (CH phase), hydrated calcium sulfoaluminate phase and unhydrated cement clinker particles.

   i. Hydrated calcium silicate gel phase (CSH phase): It occupies about 50-75 % by volume of the cement paste, which plays an important role in the performance of the cement paste. The CSH phase has no fixed composition; it is a gel with C/S varying from 1.5 to 2.0, and H/S varies greatly. The morphology of the CSH phase often varies between fibrous and network. The exact structure of CSH is not yet known, but some scholars have established a variety of structural models to explain the properties of hardened cement paste. Various models have one thing in common, that the CSH phase has a very high specific surface area and surface energy. Therefore, changing the surface energy or increasing the ionic covalent bonds between particles can change the physical and mechanical properties of the hardened cement paste; the common points of these models emphasize the role of water in the structure and performance of the CSH phase.

ii.   Calcium hydroxide phase (CH phase): about 10 %-25 % of the volume of the cement paste, with a fixed chemical composition—$Ca(OH)_2$. It is a well-crystallized hexagonal column crystal. Due to its small specific surface area, and layered structure, cleavage will appear, resulting in the weakening of the mechanical properties of the cement stone.

iii.   Hydrated calcium sulfoaluminate phase: about 10 %-15 % of the volume of the paste. It plays a small role in the performance of cement stone, but has a greater impact on the corrosion of sulfate on cement concrete. In the early stage of cement hydration, ettringite crystals formed, later transformed into hexagonal plate-shaped calcium sulfoaluminate.

iv.   Unhydrated cement clinker particle phase: about <5 % of the volume of the slurry; because the size of the cement particles is about 1-50 μm, with the hydration process, the small particles will quickly hydrate, while the large particles will hydrate for a long time. It can still be found in hardened cement paste after a long time.

2)   Porosity in hardened cement paste: It is an important structural component. It has a great influence on the physical and mechanical properties of cement stone. The characteristics of the pore structure include two parameters, porosity and the pore size distribution. In addition, there are pore morphology, connectivity, openness, etc. Figure 1.15 shows a general size range of solid phase and pores in hardened cement paste.

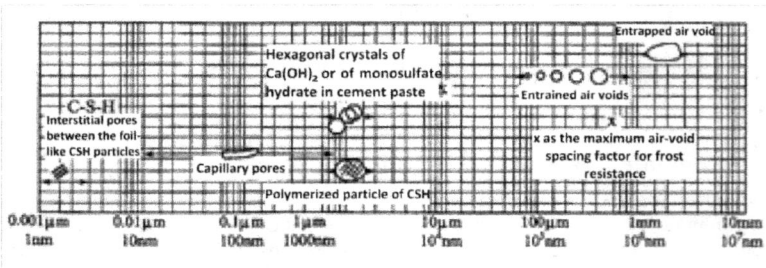

Figure 1.15 Schematic diagram of the general size range of solid phase and pores in hardened cement paste.

The pores in hardened cement paste can be divided into two categories: capillary pores and gel pores according to their size and properties, as shown in Table 1.4. The hydration process of cement can be regarded as the space originally occupied by cement and water was replaced gradually by hydration products. And those spaces that are not occupied by cement or hydration products will form capillary pores. The size and volume of the pores mainly depend on the water-cement ratio and the degree of cement hydration (Figure 1.16). In the well hydrated paste with low water-cement ratio, the pore size in the early stage of hydration is about 10 μm, and in the later stage, the size of the pores is mostly about 0.05 μm. Under a scanning electron microscope, the

Multifunctional Concrete Technology

Materials Research Foundations **127** (2022)

Materials Research Forum LLC

https://doi.org/10.21741/9781644901991

capillary pores can be seen, but the gel pores cannot be distinguished. Gel pores are contained in the volume occupied by CSH and can be regarded as part of CSH.

In addition, some pores will be embedded during the mixing process of cement paste, of which the size range is about 5-200 μm. Sometimes it can be as large as 3 mm. This type of pores has a harmful effect on hardened cement paste.

Table 1.4 Classification of pores in cement stone and its effect on performance.

Name	Description	Pore size/nm	The effect of water	on the performance of cement stone
Capillary pores	Macropores	50-10	Free water	Strength, permeability
	Medium pores	10-50	Causes medium surface tension	Strength, permeability
Gel pores	Small capillary pores (gel pores)	2.5-10	Causes strong surface tension	Shrinks at 50 % relative humidity
		0.5-2.5	Strong adsorption, does not form a concave lunar surface	Shrinkage, creep
		≤0.5	Structural water contained in the molecular	Shrinkage, creep

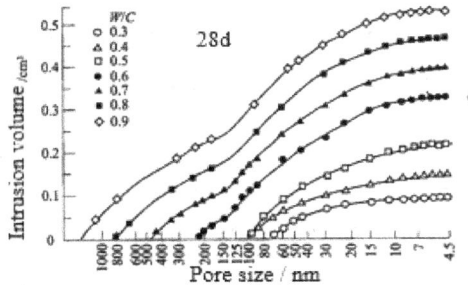

Figure 1.16 Pore size distribution in hardened cement paste at different W/C ratios determined by using mercury intrusion method.

It can be seen that reducing capillary porosity is a necessary condition for cement concrete to achieve high performance and ultra-high performance. In a laboratory in the UK, the capillary pores in hardened cement paste were reduced to 2 %, which achieved compressive strength of 665 MPa (Figure 1.17).

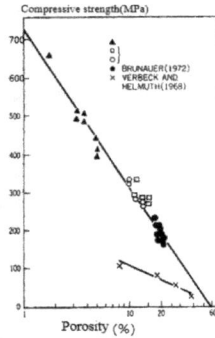

Figure 1.17 The relationship between strength and porosity of hardened cement paste (Teamura, Sakai).

3) Water in hardened cement paste

According to the difficulty of reducing water from hardened cement paste, water can be divided into the following four types, as shown in Figure 1.18.

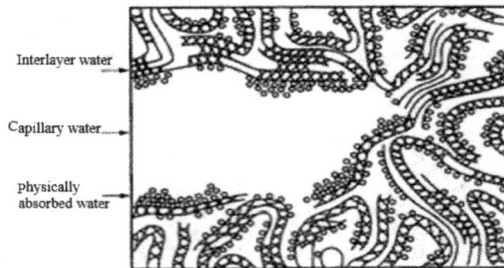

Figure 1.18 Types of water related to C-S-H gel.

(1)   Chemically bound water: It is a part of hydration products. Only when the hydration product is decomposed by heating, this water will be released. (2) Interlayer water: It is located between the CSH layers and firmly fixed by hydrogen bonds. It will only be lost when it is strongly dried (relative humidity <11 %). It has a great impact on shrinkage and creep. (3) Adsorbed water: The water is physically adsorbed on the surface of the cement paste and it can form a multi-molecular layer adsorption. When the relative humidity is less than or equal to 30 %, most of this part of water is lost; it has a greater impact on shrinkage and creep; (4) Capillary water: A kind of gravity water is divided into two categories. The part in the pores at size > 5nm can be regarded as free water. Loss in this part of water will not cause volume deformation of hardened cement paste; The

13

other part in the pores at size <5 nm has stronger capillary action. If this water is lost, the hardened cement paste will shrink.

The water content of hardened cement paste also affects its strength. The compressive strength of water-saturated hardened cement paste is about 10 % lower than that of dry cement stone.

1.9.3   The structure of interfacial transition zone between aggregate and cement stone

The structure of interfacial transition zone has a great influence on the performance of concrete. For example, concrete is brittle when it fails in tension, but it is elastic-plastic when it fails in compression. In the uniaxial compression test, each component of the concrete remains elastic until failure, but after being compounded into concrete, it shows inelastic behaviour. Why is the compressive strength of concrete an order of magnitude higher than the tensile strength? Why is mortar stronger than concrete even if the same materials are used? Why does the increase in the size of the coarse aggregate reduce the strength of the concrete? And so on, it can be said that all problems are caused by the interfacial transition zone.

1) Structure of interfacial transition zone: During the casting and molding process of concrete, before it sets, due to the different density and particle size of the constituent materials, different degrees of segregation may occur. For example, the water in concrete, because the density is lower than that of cement and aggregate, gradually floats from the bottom to the upper surface. When it encounters the coarse aggregate, it accumulates near the surface under the coarse aggregate to form water pockets and form internal stratification after hardening. The water rises to the surface to form an outer layer. Due to internal and external delamination, the concrete structure is more inhomogeneous (see Figure 1.14). Due to the internal stratification, the concrete has an anisotropic characteristic, which shows that the tensile strength in the vertical direction is lower than that in the horizontal direction.

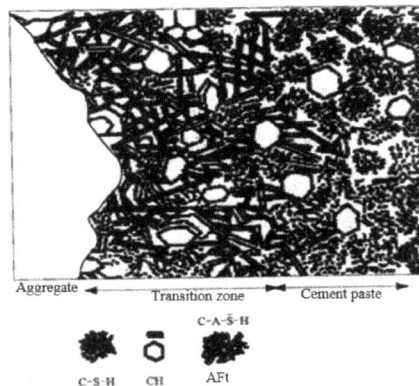

Figure 1.19 Cement paste, aggregate and interfacial transition zone (ITZ).

Due to internal stratification, the water-cement ratio in the cement paste near the lower surface of the coarse aggregate is large, and the crystal of the crystalline product formed by cement hydration is also larger; the pores of the hardened cement paste at the interfacial transition zone are larger than the pores in the bulk cement paste. The formation of calcium hydroxide crystal orientation layer makes the C axis perpendicular to the surface of the coarse aggregate. As the hydration continues, the smaller crystals of poorly crystallized CSH, CH and ettringite will fill in the skeleton composed of the larger crystalline ettringite and CH crystals. A schematic diagram of hardened cement paste, aggregate and interfacial transition zone in concrete is shown in Figure 1.19.

2) The strength of interfacial transition zone: The strength of the interfacial transition zone is generally lower than the strength of the bulk hardened cement paste, as shown in Figure 1.20. It is the weakest link in the three-phase composition of concrete.

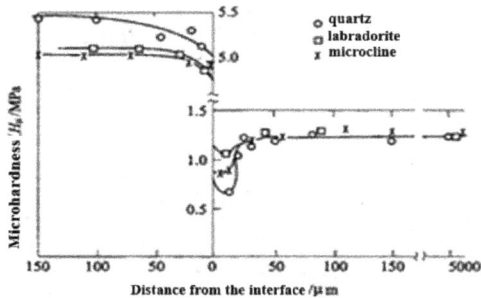

Figure 1.20 Changes in the microscopic hardness of interfacial transition zone.

The main reasons for the low strength of interfacial transition zone are: (1) The water/cement ratio in interfacial transition zone is large, and its pore volume and pore size are larger than the bulk hardened cement paste or the mortar matrix; (2) There are more and large calcium hydroxide crystals orientated with low strength at the interfacial transition zone; (3) The existence of micro-cracks in the interfacial transition zone is the main factor for the decrease in strength.

3) The effect of the interfacial transition zone on concrete performance: In hardened concrete, the interfacial transition zone is the weakest link in strength. Because its strength is lower than the bulk hardened cement paste phase and aggregate phase in hardened concrete, the interfacial transition zone phase is the weakest link which can determine the upper limit value of concrete strength. There are micro-cracks in the interfacial transition zone before the concrete is loaded. After loading, cracks will propagate due to the weak interfacial transition zone, which makes the concrete exhibit inelastic behavior during the compression process. While in the tensile process, cracks in the interfacial transition zone propagate more rapidly than when under pressure. Therefore, the tensile strength of concrete is much lower than the compressive strength, and results in brittle failure.

The interfacial transition zone also affects the elasticity and rigidity of concrete. Due to the weak bonding of the interfacial transition zone, the stress cannot be transferred well, so the rigidity of the concrete is somewhat low. When the concrete is exposed to fire or high temperature, the micro-cracks in the interfacial transition zone propagate dramatically, so that the elastic modulus of the concrete decreases faster than the compressive strength. The characteristics of the interfacial transition zone also affect the durability of the concrete. Due to the existence of cracks in the interfacial transition zone, the impermeability is lower than that of cement paste or cement mortar; and it also has an adverse effect on the corrosion of the steel bars. Except for the interfacial transition zone, the concrete structure is highly heterogeneous. Figure 1.21 presents the relationship between the structure and the strength of the concrete at a given cross-section; it illustrates the heterogeneity of the concrete structure and performance.

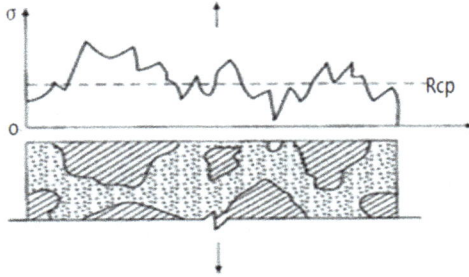

Figure 1.21 Structure-related strength distribution at a given concrete cross-section.

1.10 The key technology of ordinary concrete performance research

The study and research on ordinary concrete is continuously progressed and improved. The key technologies for the research and solution of ordinary concrete performance are summarized in Figure 1.22.

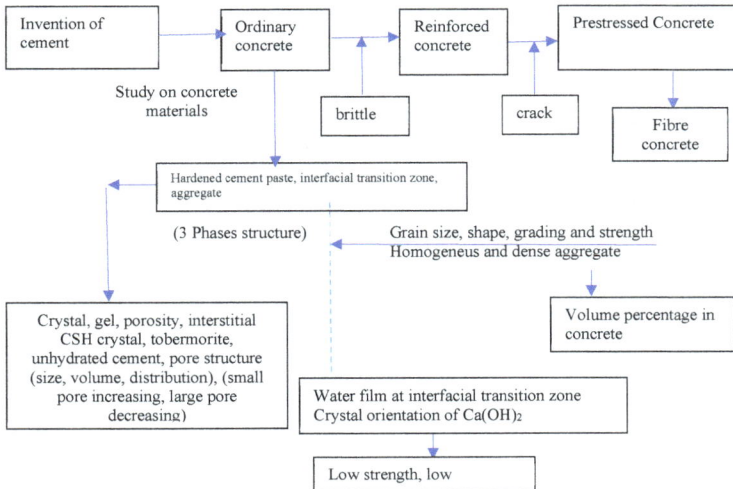

Figure 1.22 The key problems solved in the research of ordinary concrete performance.

1.11 The Characteristics of the development of concrete materials in the 20th century

After the invention of Portland cement, from the development and application of ordinary concrete during the past less than 200 years of historical development, the following basic trends can be summarized:

1) Strength improvement: The most typical example is that, in Norway in the 1970s, the strength of concrete was 50 MPa and the slump was 120 mm; the strength of 1990s' concrete was 100 MPa. Ultra-high performance concrete with compressive strength of 250 MPa and 200 MPa has been applied in the US and Japan, respectively. The strength grade of concrete commonly used in China was also increased from C20 to C30, C40. Ultra-high performance concrete with compressive strength of 150 MPa has been applied in China.

2) Improved workability: The concrete mixture has been developed from dry mix, semi-dry mix to plastic, fluid concrete and self-compacting concrete. A multi-function concrete characterized by self-compacting, self-curing, highly plastic-keeping and high durability, etc., has been developed in China.

3) Extending the working life of concrete structures: The offshore platform for oil production in Norway is 370 m high and reaches about 300 m underwater. In order to improve durability, silica fume is added to the concrete. Thus, both the densification and durability of the concrete is improved, so that the service life of the concrete is effectively prolonged. The service life of concrete structures has always been emphasised worldwide. For the construction of Otaru Port in Japan, specimens are retained for more than 100 years. The structural durability of the Tsingdao bay bridge in China has also been designed to have a service life of 100 years.

4) Resource saving, energy saving and economic benefits: In 1981, the ACI 226 committee began to study the application of fly ash (FA) and blast-furnace slag (BFS) in concrete. This has improved the durability of concrete and effectively utilized the industrial by-products. Industrial waste and construction waste became useful resources for concrete. Mr. Shen Dansheng in China has contributed to FA research, promotion and application of industrial by-products and the improvement of the durability of concrete.

5) The law of water/cement ratio: The law of water/cement ratio was proposed in 1918 - the strength (R) of concrete is inversely proportional to the water-cement ratio (W/C) and the prerequisite for durability is also to reduce W/C as the main approach. This has laid a theoretical foundation for ordinary concrete! The law of water-cement ratio still plays a guiding role in the preparation of HPC and UHPC.

Chapter 2

Powder Technology & Water Reducing Technology in Promoting Concrete Technology for the Development of High Performance, Ultra-High Performance Concrete

In the late 19th century, research on aggregates, interfaces and cement stones as raw ingredients for concrete had progressively transformed into studies on powder technology in this area of application.

From the 1970s to 1990s, high performance concrete was developed in Norway by mixing silica powder with concrete to improve performance. Subsequently, ultrafine powder, spherical cement, specialty cement and others as well as cement with strength of 200 MPa were introduced in Japan in recent years.

In the United States, kaolin clay ultrafine powder had been utilized. The development of natural zeolite ultrafine powder in China had greatly enhanced the growth of high performance and ultra-high performance concrete. The development and application of high-efficiency water-reducing agent can make the water-to-binder ratio (W/B) of concrete materials lower than 20%. Originally, when naphthalene-based water-reducing agent was used to prepare C60 strength grade concrete, the viscosity was high and the construction was difficult. While the production and application of sulfamate and polycarboxylate superplasticizers made the W/B of concrete below 0.15, and the combination of powder technology and superplasticizer technology can make the strength of concrete ≥150 MPa.

2.1 Function and effect of ultrafine powder on HPC and UHPC

The function of mineral ultrafine powder in HPC and UHPC is first demonstrated in the pore filling and densification effect as shown in Figure 2.1.

The average particle size of Portland cement is 10.4 μm while the average particle size of fly ash is 10.09 μm. Both Portland cement and fly ash have pore volumes of nearly similar percentages!

However, if silica fume with average particle size of 0.1 μm is used to mix with Portland cement, the pore volume of the two powders combined is the lowest at about 15%, when the volume of silica powder is 30% and the volume of Portland cement is 70%. If the average particle size of fly ash is 0.95 μm and the volume of 70% Portland cement is combined with 30% of fly ash, the pore volume of the combined powders is the lowest at about 20%. In other words, when the particle size ratio of the two powders is 1/10-1/20, the pore volume of the powders is the lowest and the density is the highest when the ratio is 70%-30%.

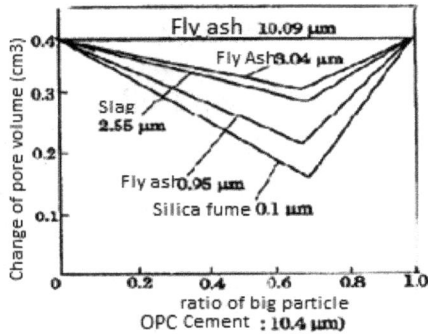

Figure 2.1 Variation of pore volume in blended cement with powders of different particle sizes.

Thus, the flowability of cement paste is the best under similar conditions as the flowability of cement paste in conventional concrete. If the slurry of similar flowability is obtained, the water demand of combined powders is the lowest while the strength of hardened cement stone is the highest; and the resulting concrete would exhibit the highest strength and the greatest durability!

The filling effect of ultrafine powder in cement slurry can also be explained from Figure 2.2.

The slurry structure of Portland cement after adding water is shown in Figure 2.2 (a), a large amount of free water is bound in the flocculation structure formed by the cement particles, the fluidity is poor, the pores are large after hardening, and the performance is not good. After the free water is discharged, the cement particle gap is low, and it has higher strength and durability after hardening, as shown in Figure 2.2(b). From Figure 2.2(c), it can be seen that the ultrafine powder is filled between the cement particles. After hardening, the compactness, strength and durability of the cement stone are improved. This is also the effect of the double addition of superplasticizer and ultrafine powder in HPC and UHPC. That means, the double addition technology of superplasticizer and ultrafine powder makes the cement stone obtain the dual effect of fluidization and densification.

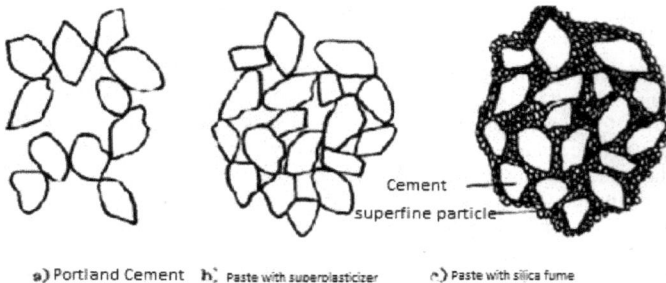

a) Portland Cement   b) Paste with superplasticizer   c) Paste with silica fume

Figure 2.2 Effect of double addition of superplasticizer and ultrafine powder.

2.2 Flowability effect of high-range water reducing agent and ultrafine powder

2.2.1 Flowability of cement and ultrafine powder mixtures at different proportions

The water content of standard cement paste is 27%, with the amount of naphthalene-based high-range water reducing agent at 1.0%, and the setting of the cement paste by 32 mm. Cement was replaced with different types of ultrafine powder at the same amount, while other variables were unchanged. Data regarding the setting of these cement pastes are presented in Table 2.1. The specific surface areas of various ultrafine powders are shown in Table 2.2.

Table 2.1 Setting of cement pastes containing different types of ultrafine powder.

Cement:powder	95:5	90:10	85:15	80:20	75:25
GGBS	33.5	34	35	37	38
Phosphorus slag powder	34	35.5	36	38	38.5
Zeolite powder	31	30	26	20	16.5
Silica fume	31	21	14.5	/	/

Table 2.1 shows that:

(1) after phosphorus slag and slag ultrafine powder are added to the cement paste, with the pores and flocculated structures filled with water between the cement particles, water is squeezed out resulting in the dilution of the paste and increase in flowability;

(2) in addition, zeolite ultrafine powder readily absorbs water due to its porous crystalline particles, such that water absorption is more dominant than water removal, thus reducing the flowability of the paste; and

(3) silica powder has a large specific surface area and large amount of water can be adsorbed onto the surface, which significantly increases the consistency of the cement paste.

Table 2.2 Specific surface areas of various ultrafine powders.

Name	Surface area(cm^2/g)	Property	Name	Surface area(cm^2/g)	Property
Phosphorus slag powder	6820	glassy	Phosphorus slag powder (2)	8560	glassy
GGBS	6820	glassy	GGBS(2)	8560	glassy
Zeolite	6820	Porous crystal	Zeolite(2)	8560	Porous crystal
Silica fume	200000	glassy			

Different types of ultrafine powder have their unique characteristics of high surface energy; ultrafine powder produced from the grinding of vitreous glass results in many broken bonds during grinding, highlighting its characteristics of high surface energy.

Water-absorption and surface adsorption of cement particles can also form flocculated cement paste to some extent, and the water adsorption and surface adsorption of some ultrafine powders (e.g. natural zeolite powders) may reduce or eliminate the densification effects of these powders.

2.2.2   Effect of superplasticizer dosage on slurry flowability

There are four types of slurry in the experimental protocol: (1) cement; (2) cement + 20% BFS, (3) cement + 20% PS, (4) cement + 10% NZ. BFS-superfine mineral powder, PS-superfine phosphorus slag powder, NZ-superfine natural zeolite powder. The content of superplasticizer NF is from 0.4%-0.9% and the results are shown in Table 2.3.

Table 2.3 Effect of superplasticizer dosage (flow value mm).

NF Dosage%	0.4	0.5	0.6	0.7	0.8	0.9
Cement	129	138	155	190	235	236
C+20%BFS	125	136	185	230	265	267
C+20%PS	132	170	215	250	270	272
C+10%NZ	/	No flow	130	195	237	237

It can be seen from Table 2.3 that:

(1) for slag superfine powder, when the content of superplasticizer NF is 0.4%, the fluidity of C+20% BFS is lower than that of the reference slurry; when the content of NF is 0.5%, the flow value of these two slurries are equal; when the NF content is 0.6%, the fluidity of C+20%BFS is higher than that of the reference slurry. When the NF content is 0.8%-0.9%, the fluidity of the slurry is also higher than that of the reference slurry, the flow value reaches the maximum, but remains basically unchanged.

(2) For superfine phosphorus slag powder, when the content of superplasticizer NF is 0.4%, the flow value of the slurry is larger than that of the reference slurry and also larger than that of the mineral powder slurry. When the content of NF is 0.8%-0.9%, the flow value of the slurry is also in a steady-state (the flow value is 270-272 mm).

(3) For natural zeolite ultrafine powder, when the content of superplasticizer NF is 0.4-0.5%, the slurry does not flow, and when the content of NF is 0.8%-0.9%, the flow value of the slurry reaches 237 mm, slightly higher than the reference slurry, but also in a steady state.

This shows that for different superfine powders, in order to overcome the adsorption phenomenon of superfine powders, there is a minimum dosage of superplasticizer NF. When the dosage of superplasticizer NF reaches a certain value, the slurry flow value tends to a certain stable value, and then with the increase of dosage of superplasticizer NF, the flow value does not increase.

For example, if the NF content in the table is 0.8%-0.9%, the flow value of the slurry is basically unchanged. That is to say, the adsorption capacity of cement and ultrafine powder particles to superplasticizer has a limit value. When the dosage of water reducing agent increases again, the Zeta potential of the particle surface after adsorbing the water reducing agent does not increase, and the fluidity does not increase.

2.2.3   Influence of co-blending of superplasticizer and ultrafine powder on slurry fluidity

The mixtures of cement with various types of ultrafine powder at different proportions, W/B=29%, incorporating naphthalene-based high-range water reducing agent (NF) 0.9%, and with nominal flowability of 240 mm are presented in Figure 2.3.

It can be seen from Figure 2.3 that under the same superplasticizer content of 0.9%, the fluidity of the phosphorus slag superfine powder and slag superfine powder increases with the increase of the replacement amount of cement. When the replacement amount of natural zeolite ultrafine powder to cement is 5%, the fluidity of the slurry is slightly higher than that of the reference, and with the increase of the replacement amount of cement, the fluidity of the slurry is lower than that of the reference slurry. The effect of composite powder with zeolite powder and mineral powder, and composite powder with zeolite powder and silica powder on slurry fluidity is similar to that of natural zeolite powder.

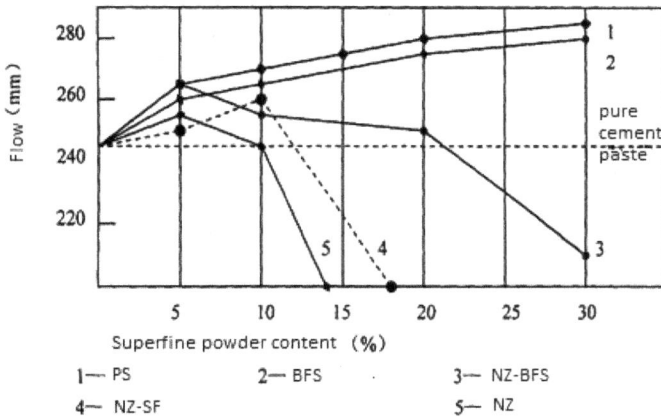

Figure 2.3 Flowability of cement paste with different types and proportions of ultrafine powder.

2.2.4 Understanding of slurry mechanism by co-blending of mineral ultrafine powder and superplasticizer

Due to the different properties of the powder, the effect of increasing the fluidity of the slurry is different. Both slag and phosphorus slag ultrafine powder are glassy powders. They do not absorb water. However, the surface of the particles can also adsorb superplasticizers to form an electric double layer, and the electrostatic repulsion can disperse the ultrafine powder particles and the cement particles. When the flocculation structure of the cement particles is stretched, the bound free water is released, and the fluidity of the slurry increases. On the other hand, the particle size of the ultrafine powder is about 1/2 of that of the cement particle, and it is filled in the micropores of the cement powder. After the cement particles and the ultrafine powder particles adsorb the superplasticizer, double electricity (Zeta potential) is formed on the surface of the particles. The superposition of the electric double layer potentials of the two results in a relatively large repulsive force. Moreover, the force field of the ultrafine powder particles in the pores of the cement particles is similar to the wedge action, and the cement particles are easier to disperse. As shown in Figure 2.4.

Phosphorus slag and slag ultrafine powder belong to the dispersing ultrafine powder; while natural zeolite ultrafine powder and silica fume belong to the filling ultrafine powder.

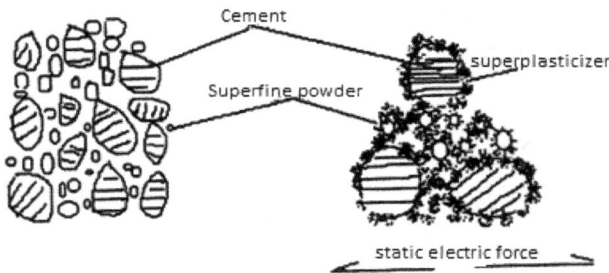

Figure 2.4 Effect of NF ultrafine powder on mixture flowability.

2.3 Porosity and Strength of Hardened Cement Paste

Hardened cement paste is a porous material with capillary and gel pores, as outlined in Table 2.4.

Table 2.4. Effect of pore size distribution on the properties of hardened cement paste.

Name	Diameter	Classification	Influence on the performance of cement stone
capillary pores	10-0.05μ	Big capillary pores	Strength, permeability
capillary pores	500-25Å	Middle capillary pores	Strength, permeability and shrinkage
gel pores	100-25Å	Small gel pores	Shrinkage under RH%＜50%
gel pores	25-5Å	Micro pores	Shrinkage, creep
gel pores	＞5Å	interlayer pores	Shrinkage, creep

It can be seen that capillary pores have a great influence on the macroscopic mechanical properties and permeability resistance, while gel pores have a great influence on dry shrinkage. To achieve high performance and ultra-high performance in concrete, the volume of capillary pores must first be reduced. In cement stone, after adding mineral ultrafine powder, the change of pore structure is shown in Figure 2.5.

It can be seen from Figure 2.5 that at the age of 7 d, the reference specimens have higher macropores $\geq$ 100 nm, while the specimens containing natural zeolite ultrafine powder have higher content of small pores less than 100 nm than the reference specimens. At the age of 28 d, the sample containing natural zeolite ultrafine powder has much higher content of small pores less than 100 nm than that of the reference sample; while for the large pores $\geq$100 nm, the sample containing natural zeolite ultrafine powder is much lower than the reference sample.

Note: W/C=0.35 20% composite superfine powder replaces the cement specimen and the benchmark specimen in an equivalent amount.

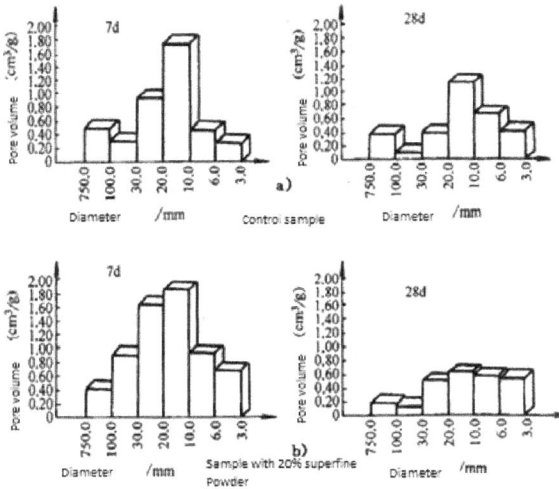

Figure 2.5 Changes in the pore structure of cement stone after co-blending of composite ultrafine powder and superplasticizer (Feng Naiqian).

Tests have shown that the large content of macropores $\geq$ 100 nm is detrimental to the strength, impermeability and durability of concrete; the high content of small pores smaller than 100 nm is beneficial to the strength, impermeability and durability of the concrete.

The pore structure of cement stone in concrete is closely related to the water-cement ratio of concrete as shown in Figure 2.6. When the W/C=0.2 of the cement paste, the cement particles are not fully hydrated, and there are some unhydrated cement particles in the hardened cement stone. When W/C=0.6, the cement particles are fully hydrated. But there is still excess free water in the cement stone. After the water evaporates, it becomes a capillary and remains, which reduces impermeability and durability. Only when W/C=0.4 (there is also literatures with proposal of W/C=0.38-0.42), as shown in Figure 2.7, the cement particles are all hydrated, and there is neither capillary water nor unhydrated cement particles, the concrete achieved good impermeability.

The W/C of ultra-high performance concrete is generally below 25%. After replacing cement with a part of mineral ultra-fine powder, superplasticizer is added at the same time to ensure the fluidity of concrete and obtain a low water-to-binder ratio (W/B), to obtain high-strength or ultra-high-strength concrete.

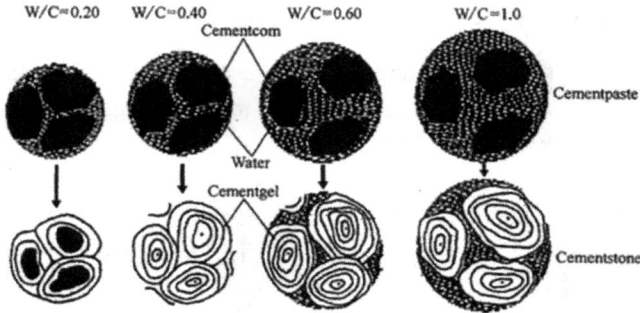

Figure 2.6 Water-cement ratio and cement-stone structure (Zhang Xinhua).

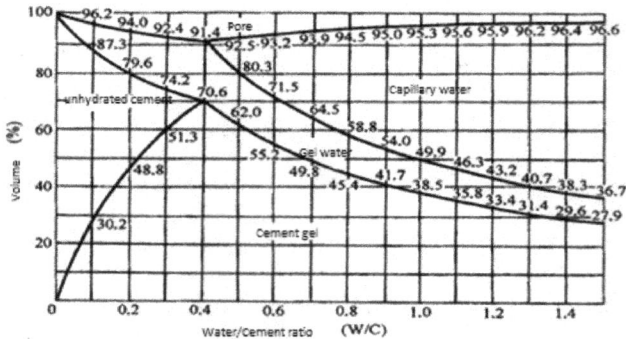

Figure 2.7 Relationship between the composition of cement stone and water-cement ratio (100% hydration of cement).

2.4   Co-Blending of superfine powder and superplasticizer improves the interface structure between cement stone and coarse aggregate

Concrete is composed of three phases of cement stone, aggregate and interface transition zone. The concentration and orientation of $Ca(OH)_2$ crystals in the interface transition zone is the main reason for the low strength and durability of concrete as shown in Figure 2.8. Therefore, in order to obtain ultra-high performance of concrete, the interface structure must be improved.

Suppressing the formation of the transition zone is also a key technology that must be solved in the development of concrete to HPC and UHPC.

The interface transition area between coarse aggregate A and cement stone in Fig. 2.8 is enlarged, as shown in Fig. 2.9

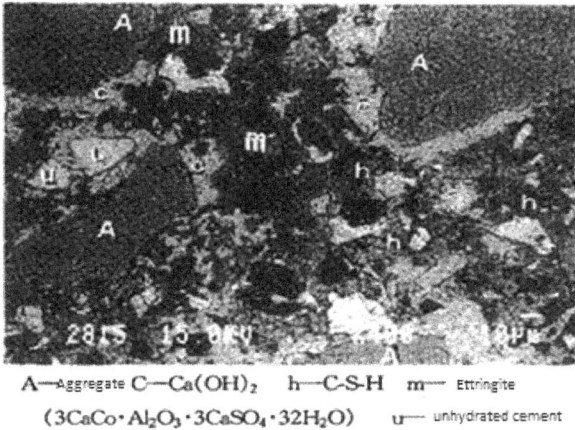

A—Aggregate C—Ca(OH)$_2$ h—C-S-H   m— Ettringite
($3CaCo \cdot Al_2O_3 \cdot 3CaSO_4 \cdot 32H_2O$)   u— unhydrated cement

Figure 2.8 EPMA of hardened concrete polished surface (Uchida et al.).

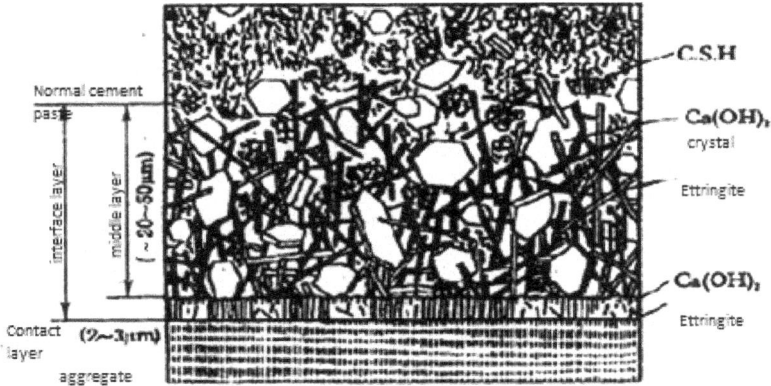

Figure 2.9 Microstructure model of the cement stone-aggregate interface (Morino).

Khagat, Aitein et al. replaced cement with 15% silica fume to prepare concrete with W/B=0.33, the CH crystalline content in the interface transition zone decreased, and the crystalline orientation of CH also decreased, and the porosity decreased, as shown in Figure 2.10.

In Figure 2.10, after the reference concrete is freshly formed, due to bleeding, water gaps are formed under the coarse aggregates, and the cement particles at the interface are also insufficient. In the transition zone after coagulation and hardening, there are CH phases, CSH gels, and due to the evaporation of pore water at the interface, a large number of pores and some needle-like hydrates are left. After the silica fume is added, the silica fume particles fill the space around the coarse aggregate, and there is no water gap on the bottom surface under the aggregate. In the transition zone after hardening, the porosity is very low, and there is no problem of enrichment and orientation of the CH phase. The inhomogeneity of the transition zone is eliminated. The improvement of the structure makes the performance of the concrete improved.

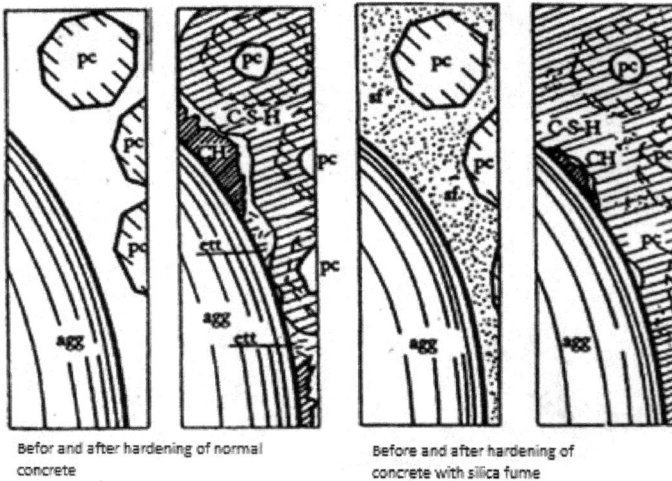

Befor and after hardening of normal concrete — Before and after hardening of concrete with silica fume

Figure 2.10 Interface transition zone between reference concrete and concrete containing 15% silica fume.

It can be concluded that superfine powder technology and superplasticizer technology can be the technical and material basis for the development of concrete towards high performance and ultra-high performance.

2.5 Modeling of cement-based materials

In order to improve the performance of cement concrete, a global study on cement-based materials has been conducted. A systematic research on the high strength and high performance characteristics of these materials is carried out, and two models are presented.

2.5.1   MDF (macro-defect free cement)

Studies on MDF were first proposed by Birchall and co-authored by the British Chemical Industry Corporation and Oxford University in 1979. Subsequently, these research works were also conducted in the United States, Japan and Sweden.

Using ordinary Portland cement or aluminate cement (90-99%), water-soluble resin (4-7%), with water to ash ratio of ≤20%; high-efficiency shearing and mixing equipment; as well as hot pressing and molding technology, the MDF characteristics were obtained with compressive strength of 300 MPa, flexural strength of 150 MPa and impact resistance of 50 GPa.

Preparation process: (Cement + PVA + Admixture) → Concrete → Shearing and Mixing →Hot Pressing →Curing → Finished Product

The main objective of this model is to fill the pores within cement particles with water-soluble resin; and at the same time, the water-soluble resin particles would bind the cement particles together to obtain a high-density, high-strength cementitious structure. However, due to the complexity of this manufacturing process, it is not as practical.

2.5.2   DSP (density system containing homogeneously arranged ultrafine particles)

As documented in Bache's patent, DSP was introduced based on the development and application of silica powder in countries such as Sweden, Norway and Iceland. The Japan Electrochemical Industries first introduced this technology to their country, Japan. The basic composition comprises: cement, silica powder, polycarboxylic acid high-range water reducing agent.

The DSP schematics and performance are shown in Figure 2.4, illustrating the densified filler system which comprises ordinary Portland cement + ultrafine silica powder + high-range water reducing agent and water.

If the particle size of the ultrafine powder is 1/10-1/100 of cement particle size, the micro-pore filling effect can be achieved. In this case, the mixing water is minimized, and the best flowability is obtained by incorporating a high-range water reducing agent. This can be easily adopted in the built environment, hence the applications of HPC and UHPC that we see today. The compressive strength of these HPC and UHPC materials can range from 180 to 300 MPa and their flexural strength from 20 to 40 MPa.

Ultra-high performance concrete such as ECC can be prepared by mixing SF and quartz sand fiber based on the DSP model, while ultra-high performance concrete such as HPC and UHPC can also be achieved by mixing sand and gravel based on the DSP model.

In conclusion, the above briefly expounds the development of HPC and UHPC based on materials and theoretical knowledge as well as technical requirements and objective demands.

Different scholars in the industry use different processes to obtain cement stones with different pore contents and strengths. Among them, the cement stone obtained by British scholars has the highest strength, with a compressive strength of 665 MPa, and the pore content in the cement stone is only 2%.

Chapter 3

Development and Production Application of High Performance Concrete

3.1 Introduction

It should be pointed out that high-performance concrete originated from Norway. In the construction of drilling platforms in the North Sea Oilfield in Norway, in order to improve the durability of concrete, silica fume was added to concrete. As a result, not only the durability and strength were improved, but also the workability was significantly improved. The performance of this type of concrete cannot be evaluated from only one aspect. This kind of concrete was proposed as high-performance concrete (HPC). In 1987, the first international conference on high-strength and high-performance concrete (HS/HPC) was held in Stavanger, Norway. There were 235 participants from 24 countries. Although there were only 56 papers at the time, the proceedings had 688 pages. At that time, the concrete strength in Norway has grown from 50 MPa in 1970 to 100 MPa in 1990.

In Europe, high-strength concrete and high-performance concrete are closely related, so that "high-strength and high-performance concrete (HS/HPC)" is usually used. The concrete technology with addition of silica fume has been gradually developed In Norway and introduced to other countries.

In China, in the 1970s, in the construction of the protective door of the Beijing subway, C60 high-fluidity and high-strength concrete was used. In order to meet the construction requirements, the concrete was kept plastic for more than 3 hours. Since 1992, many universities, scientific research institutes and engineering units in China have begun to study high-strength and high-performance concrete. The China Engineering Construction Standardization Association Standard CECS207--2006 "Technical Regulations for High Performance Concrete Application" was developed by The High-Performance Concrete Committee and published in 2006.

In 1981, the ACI 226 committee began to study the application of FA and BFS to concrete. It improved the durability of concrete and promoted the effective application of industrial by-products. In 1986, concrete with a compressive strength of 12,000 psi was commercialized in the United States. In the same year, Tokyo University of Japan has carried out the research of SCC. In 1998, the Japanese Civil Engineering Society officially promulgated a technical standard of SCC, and classified SCC also as HPC. Since then, making SCC entered the commercial application prospect.

In 1993, ACI established the HPC Committee, and made a preliminary definition of HPC: "High-performance concrete (HPC) is concrete that has been designed to be more durable and, if necessary, stronger than conventional concrete. HPC mixtures are composed of essentially the same materials as conventional concrete mixtures, but the proportions are designed, or engineered, to provide the strength and durability needed for the structural and environmental requirements of the project. High-strength concrete is defined as having a specified compressive strength of 8000 psi (55 MPa) or greater. The value of 8000 psi (55 MPa) was selected because it represented a strength level at which special care is required for production and testing of the concrete and at which special structural design requirements may be needed.

In 1996, RILEM also established the SCC and HPC committees.

Norway held the first international conference on high-strength and high-performance concrete (HS/HPC) in 1987. After that, the HS/HPC international conference was held every three years. The 10th international conference was held in Beijing in September 2014.

The important feature of high performance concrete is its high durability. In terms of composition materials, in addition to the use of superplasticizers, the application of inorganic powder (ultrafine powder) is another important feature. Now the research on HPC is mainly aimed at high strength and high performance.

3.2   Technical standards for HS/HPC in different countries

The technical standards of HPC in different countries have their own characteristics, and even the definitions of HPC by different technical associations in the same country may be quite different.

The most representative one is the definition put forward jointly by the National Institute of Standards and Technology (NIST) and the American Institute of Concrete (ACI) in May 1990: " High-performance concrete is concrete having desired properties and uniformity that cannot be obtained routinely using only traditional constituents and normal mixing, placing, and curing practices. As examples these properties may include: 1. Ease of placement and compaction without segregation; 2. Enhanced long-term mechanical properties; 3. High early-age strength; 4. High toughness; 5. Volume stability; 6. Long life in severe environments." HPC is especially suitable for high-rise buildings, bridges and building structures exposed to harsh environments! In 1998, ACI further pointed out: "When certain characteristics of concrete are formulated for a specific use and environment, this is high-performance concrete!"

The authors of this book would like to further clarify that, for marine engineering concrete structures, the concrete in the splashing zone needs to withstand the splashing effect of sea water, the effect of dry and wet changes, the effect of freeze-thaw cycles, and the penetration and diffusion of sea salt, etc. This requires the concrete to have high strength, high durability, and impact resistance, etc. The concrete that meets these performance requirements is high-performance concrete!

In 1998, Japan Civil Engineering Society formally promulgated the SCC specification, and also classified SCC as HPC. This has made SCC applicable to commercial applications, and integrate SCC technology with international standards. Furthermore, concrete researchers in various countries also chose SCC as major research topics.

In 1993, China Civil Engineering Society established the High Strength Concrete Committee under the Prestressed Reinforced Concrete Branch. In 1998, the High performance Concrete (HPC) Committee was established by Cement Concrete Branch of Chinese Ceramic Society.

Now, many countries have published technical standards for high-performance concrete, which include specifications for design and construction of HPC. In addition, there are a large number of HPC literature, covering HPC engineering technology and providing relevant guidance. Therefore the technical capability of the relevant professionals has been greatly enhanced!

3.3   High-strength concrete (HSC) and high-performance concrete (HPC)

In general, high-performance concrete (HSC) has high strength, but high-strength concrete does not necessarily have high performance. The most important indicator of high performance is high durability. The strength and durability of concrete are not related to each other. The strength of concrete can be obtained by appropriately reducing the water/binder ratio. The durability of concrete and concrete structures can be obtained by adding appropriate types of ultra-fine powder in addition to appropriately reducing the water/binder ratio, as shown in Table 3.1.

Table 3.1 Comparison of composition and performance between HSC and HPC.

No.	W/C (W/B)	Cement	Slag powder	Sand	Aggregate	Water	Water reducing agent (%)	Compressive strength (Mpa)	Chloride diffusion ($10^{-9}cm^2/s$)	RCPT (column)
1	0.28	600	/	650	1050	168	1.30	78.5	10.024	1536
2	0.28	360	240	650	1050	168	0.7	79.6	3.899	281

NO.1 is high-strength concrete, while NO.2 is high-performance concrete. The chloride ion resistance coefficient of HSC is 1/2.57 of HPC. The RCPT of HPC is 1/5.5 of HSC. When evaluating its durability according to the diffusion coefficient of chloride ions, the service life of high-performance concrete is about 3 times of that of high-strength cement.

The internal and external layered structures of NO.1 (HSC) concrete are shown in Figure 3.1, and the microstructure is shown in Figure 3.2. After the concrete is cast, during the formation of the structure, because the free water has a relatively small specific gravity, it floats upward from the bottom layer to form capillaries at different sizes in the concrete. When the free water rises and encounters the obstacle of a coarse aggregate particle, an internal layer is formed under the aggregate particle. When the water arrives at the surface layer of the structural element, a bleeding layer may be formed at the surface, which is the outer layer. The inner layer, the outer layer and the capillaries are all caused by the movement of water in the concrete. Therefore HSC (NO.1) (High-strength concrete with pure cement) has reduced strength, reduced impermeability, and reduced durability. But HPC (NO.2) (high performance concrete containing ultra-fine slag powder) does not have the above-mentioned interfacial pores, and the interfacial structure is densified.

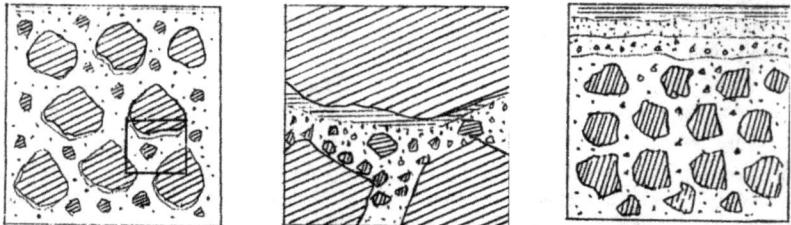

(a) Inner layered structure of ordinary HSC

(b) Enlarged view of inner layered structure

(c) Outer layered structure

Figure 3.1 Internal and external layered structures formed in HSC.

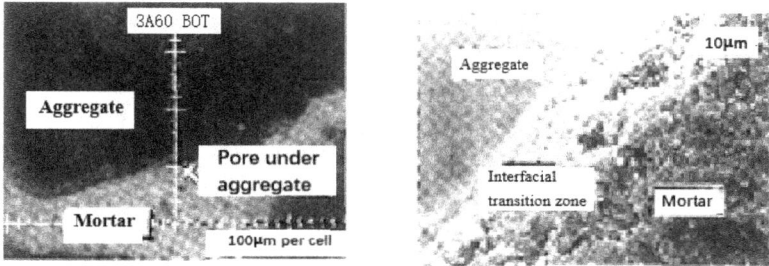

(a) Example of pores near the bottom
surface of an aggregate

(b) Example of transition layer

Figure 3.2 Microstructure of the internal layer (Nagataki).

Professor S.P. Shah of the United States proposed: "Although high-strength concrete (HSC) has high strength, it does not have the required comprehensive durability." HPC includes both mechanical properties and some other properties, such as filling, impermeability, non-segregation, corrosion resistance, and volume stability, etc.

In Norway, in the 1970's, the average strength of concrete was only 50 MPa, the slump was 120 mm, and the cost of construction labour was very high. Thanks to the 20 years' development and application of HPC, the average strength of concrete reached 100 MPa, and the slump reached 270 mm. Pumping construction and self-compacting concrete are some successful results of these achievements. The main factors for such achievements are summarized in Figure 3.3.

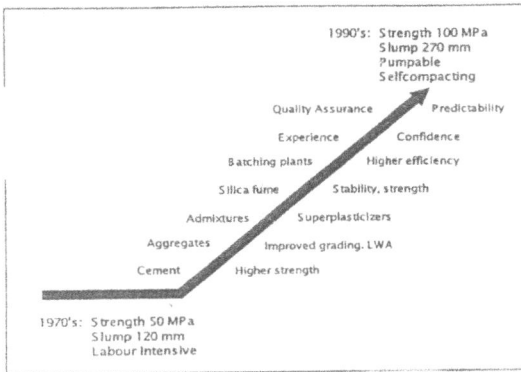

Figure 3.3 Norway's successful technical approach to develop high-performance concrete.

3.4 Preparation and application examples of HS/HPC

3.4.1 HS/HPC with natural zeolite powder

In an HPC test, Japan's Otani stone and Futsui zeolite were processed into ultra-fine powders, and the replacement amount of cement was 10 %. Under the same preparation conditions, the changes in concrete performance were evaluated. The mix design of HPC is shown in Table 3.2. The test results are shown in Table 3.3.

Table 3.2 Mix Design of HC/HPC made with natural zeolite powder.

Cement + addition		W/C+Z (%)	water (kg/m3)	Sand ratio (%)	Admixture (kg/m3)	A(kg/m3)			
type	Size (μm)					Cement	Zeolite	Sand	Aggregate
1	8.0	35	175	32	5.0	500	---	560	1207
2	6.7	35	175	32	5.9	450	50	560	1207
2	6.2	35	175	32	5.0	450	50	560	1207
2	5.6	35	175	32	5.0	459	50	560	1207
3	6.8	35	175	32	5.0	450	50	560	1207
3	6.4	35	175	32	5.0	450	50	560	1207
3	5.6	35	175	32	5.0	450	50	560	1207

Table 3.3 Performance of HS/HPC with natural zeolite powder.

Type	Slump (cm)	Air content (%)	Compressive strength (MPa)		
			3d	7d	28d
Cement	20	3.1	40.375	50.575	60.18
Otani	17.5	3.5	41.65	55.675	66.215
Otani	17.0	3.5	43.605	56.61	66.215
Otani	17.0	3.4	43.605	58.14	68
Futsui	17.0	3.5	41.225	44.625	64.345
Futsui	17.0	3.3	41.225	55.675	64.94
Futsui	17.0	3.2	44.37	54.57	63.155

It can be seen from Table 3.2 and 3.3 that the concrete strength at 3 d and 7 d are higher after replacing cement with 10% Otani stone and Futsui stone ultrafine powder, and the compressive strength of 28 d is about 10% higher than that of the control concrete. The active SiO_2 and active Al_2O_3 content of Otani stone were 12.08% and 8.39% respectively, which participated in the pozzolanic reaction as a part of the hydration in concrete and promoted the improvement of concrete strength.

It can be seen from the microscopic analysis, in Figure 3.4. Figure 3.4 (a) shows hardened cement paste incorporating 20% Otani stone ultrafine powder. The characteristic peak of $Ca(OH)_2$ in the figure disappeared. This is due to the reaction between Otani stone ultrafine powder and $Ca(OH)_2$. Figure 3.4 (b) is the reference cement paste, where the characteristic peak of $Ca(OH)_2$ in the figure is very obvious. On the other hand, it can also be seen from the diagram of the interfacial transition zone of concrete that the mass ratio of SiO_2/CaO in the interfacial transition zone containing Otani

Materials Research Forum LLC

https://doi.org/10.21741/9781644901991

stone was higher than that of the control concrete. The concentration of Ca $(OH)_2$ was reduced. Therefore, the performance of HS/HPC containing 10-15% of natural zeolite powder is improved.

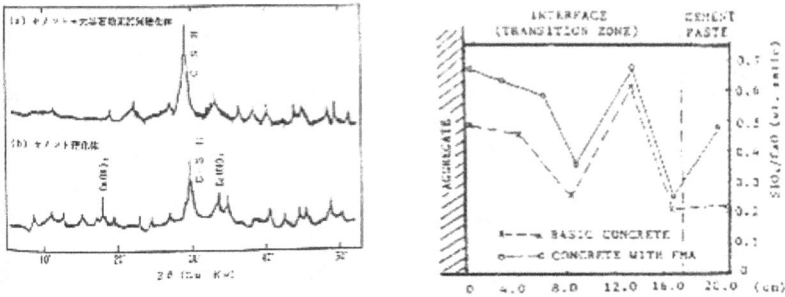

(a) XRD of reference concrete and Otani concrete

(b) Structure of transition zone

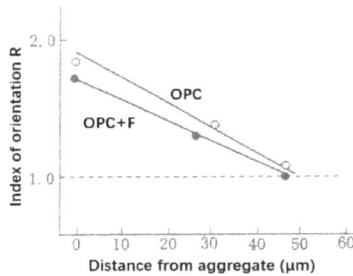

(c) Ca $(OH)_2$ orientation index of the interfacial transition layer

Figure 3.4 Microstructure of reference concrete and Otani concrete.

The lower the Ca $(OH)_2$ orientation index of the interfacial transition zone is, the more beneficial it is to the strength. After adding 10% natural zeolite ultrafine powder, the orientation index decreases, so the strength is higher than the reference concrete.

3.4.2   Application of HS/HPC in subway protective door

Around 1965, about 600 m^3 of HPC with strength grade of C60 was needed for construction of a subway protective door. The fresh concrete required a plastic retention of more than 3 h. Natural zeolite powder and superplasticizer was mixed to prepare balls with particle size of 5-10 mm. The balls were added to the concrete with the dosage of 1.5-2.0% of the cement amount. The superplasticizer was slowly released from the balls, and the slump loss could be controlled for more than 2 h. This product was entitled the fluidizing agent carrier (CFA). The mix proportion of the concrete mix is shown in Table 3.4, and the compressive strength is shown in Table 3.5. The slump changes with time are shown in Figure 3.5.

Table 3.4 Mix design of slump controlled HPC.

NO.	W/C	C	NZ	Water	Sand	Aggregate	Retarder	NF	CFA
1	0.37	500	--	185	600	1200	1.25	2.5	0
2	0.37	450	50	185	600	1200		2.5	2.0%

Table 3.5 Strength development of concrete.

NO	W/C	Compressive Strength (MPa)			
		3d	7d	28d	90d
1	0.37	39.2	51.8	70.5	74.2
2	0.37	44.2	55.8	75.3	81.2

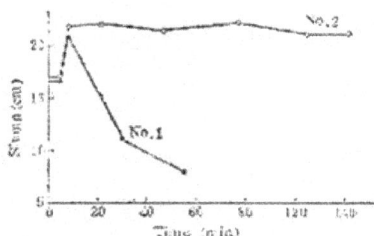

Figure 3.5 Slump loos of CFA controlled HPC.

No.1 was the reference concrete, without the carrier fluidizing agent CFA, the slump loss was fast. No.2 was the concrete with 2.0% of CFA of the cementing material content, and the slump could be controlled for more than 2 h without change. This enabled the concrete construction to be very easy.

3.4.3 HS/HPC prepared with ultrafine slag powder and silica fume

HS/HPC was prepared with ordinary Portland cement of grade 42.5, silica fume and ultrafine GGBS powder (specific surface area 7000 cm^2/g), W/C=0.30. The mix design is given in Table 3.6. The compressive strength at 3 d, 28 d and 90 d are shown in Table 3.7.

Table 3.6 Mix design and properties of fresh concrete.

No	Cement	Silica fume	GGBS	Sand	Aggregate	Water	Admixture	Slump	Slump flow
1	500	--	--	645	1055	150	1.3%	21 cm	48 cm
2	450	50	--	645	1055	150	1.3%	21 cm	44 cm
3	350	--	150	645	1055	150	1.3%	23 cm	60 cm
4	300	50	150	645	1055	150	1.3%	22 cm	55 cm

Table 3.7 Compressive strength of HS/HPC.

NO	Compressive Strength (MPa)			Flexural Strength (MPa)			Tensile Split Strength (MPa)		
Age (days)	3	28	90	3	28	90	3	28	90
1	51.8	66.1	78.0	6.09	7.20	7.21	3.91	4.46	5.65
2	54.3	89.3	99.1	7.62	10.14	11.4	4.06	6.16	7.22
3	64.5	91.6	97.8	7.11	9.96	11.1	4.20	6.09	6.29
4	57.5	113.7	113.4	7.50	11.64	12.36	4.14	6.09	6.59

The above results proved that the strength and workability of HS/HPC were the best when the cement was replaced by 10% silica fume and 30 % GGBS. The resistance to acid attack was also improved.

3.5 The further development and application of HS/HPC, the problems that need to be further solved

3.5.1 Autogenous shrinkage and cracking

In the constituent materials of HS/HPC, cement represents a high amount while water has relatively low amount. After the initial setting of the concrete, the cement undergoes further hydration, absorbing the capillary water in the concrete, consuming the capillary water, and turning the capillary pores into a near-vacuum state. The tensile stress will be created by capillary shrinkage, and when this tensile stress exceeds the tensile strength of the concrete, the concrete will crack. This kind of cracking is caused by loss in water in capillary pores and capillary tension under the condition of no medium exchange between the concrete and the surrounding environment, so it is called autogenous shrinkage cracking. As shown in Figure 3.6, HPC at strength grade of C80 was used for the first-story shear wall of an ultra-high-rise building. After demoulding, all the concrete of the shear wall cracked, and some walls cracked severely and had more than 120 cracks. Autogenous shrinkage cracking becomes one of the technical difficulties for HPC to be solved.

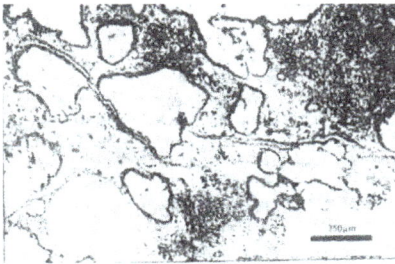

(a) Polished surface of HS/HPC

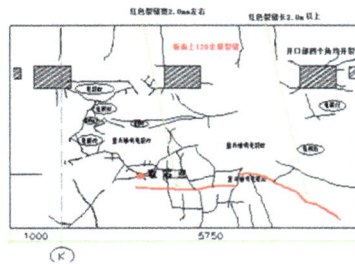

(b) Auto-shrinkage cracking of C80HPC shear wall

Figure 3.6 Auto-shrinkage cracking of HS/HPC.

3.5.2   Swelling and cracking

Another technical difficulty in the development and application of HS/HPC is swelling and cracking. When HS/HPC is under water for a long time, due to the high amount of cement and low W/C, when it is under water for a long time, the water may diffuse through and penetrate the concrete. The unhydrated cement residues will absorb water, hydrate and cause swelling. This swelling may cause cracking of HS/HPC as shown in Figure 3.7.

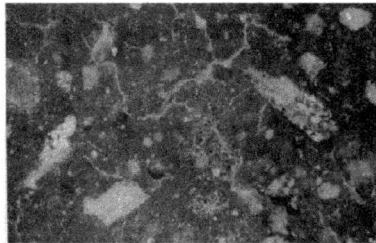

Figure 3.7 Swelling cracking of HC/HPC soaked in water for long time.

3.5.3   Brittleness

The descending section of the stress-strain relationship curve of HS/HPC is relatively steep, reflecting its greater brittleness, as shown in Figure 3.8. The increased brittleness of concrete will cause problems for engineering structures, especially in those with seismic requirements.

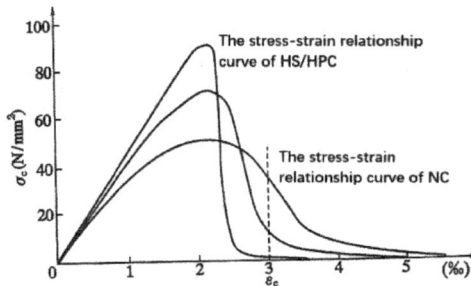

Figure 3.8 The stress-strain curve of ordinary concrete and HS/HPC.

The autogenous shrinkage cracking, swelling cracking, and high brittleness of HS/HPC are all caused by factors such as low W/C, high cement amount, and high compressive strength. These three problems of ultra-high performance concrete are even worse than those of HS/HPC. How to solve these problems will be further elaborated in the section of UHPC.

Chapter 4

Research and Application of Ultra-High Performance Concrete (UHPC)

UHPC is a high-performance concrete with ultra-high strength. In China, high-performance concrete with strength over 100 MPa is called ultra-high-performance concrete (UHPC). In Japan, high-performance concrete with strength grade over C80 is called ultra-high-performance concrete (UHPC). In the United States, high-performance concrete with strength over 150 MPa is called ultra-high-performance concrete (UHPC). It can be seen that, due to the level of national concrete technology, the strength level of ultra-high-performance concrete is slightly different. At the present stage, the strength threshold of ultra-high performance concrete may be 100 MPa.

4.1 The superiority of UHPC application in structures

4.1.1 Technical and economic effects

The effects of applications of HPC and UHPC in structures are shown in Table 4.1.

Table 4.1 The effect of HPC and UHPC in structures.

Concrete strength	Effect
60 MPa replacing 30-40 MPa	Reduce 40% concrete volume, 39% steel and 20-35% cost
80 MPa replacing 40 MPa	Reduce 30% volume and self-weight
150 MPa replacing 60-80 MPa	Reduce cross-section area of column, increase span of beam, increase usable areas.

4.1.2 Changes in the structural system

If UHPC is used to replace ordinary concrete, a framed structure with thick columns and short beams can be changed to a tubular structure with thin columns and long beams, as shown in Figure 4.1.

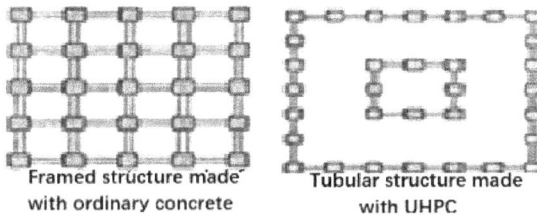

Framed structure made with ordinary concrete

Tubular structure made with UHPC

Figure 4.1 Transition from a framed structure with thick columns and short beams to a tubular structure with thin columns and long beams.

As a result, the cross-section of the column becomes smaller, the span of the beam increases, and the available area and space increase.

4.1.3   The effect of UHPC with a strength of 150 MPa instead of HPC with a strength of 60 to 80 MPa

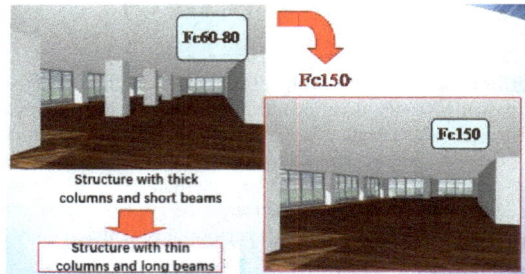

(a) Schematic illustration of using Fc 150 MPa replacing Fc 60-80 MPa

(b) Construction site

Figure 4.2 Concrete structure with F_c 150 MPa instead of F_c 60-80 MPa.

The structural system has changed from thick columns and short beams to thin columns and long beams, increasing the usable area and space, as shown in Figure 4.2.

4.1.4   Comparison of structure self-weight under the same bending moment

For a given bending moment as 150 kN-m, the mass per meter of I-beam and prestressed UHPC may be the same, which is 50 kg/m. But the mass per meter of a prestressed member with ordinary concrete exceeds 210 kg/m, which is more than 4 times of the former, as shown in figure 4.3. A prestressed UHPC structure has a longer service life and lower maintenance costs than those of a steel structure. For columns and shear walls in ultra-high-rise buildings, UHPC may be one of the most economical materials to carry compressive loading, and more usable area can be obtained.

UHPC concrete technology is either the direction or a breakthrough progress in the development of concrete technology.

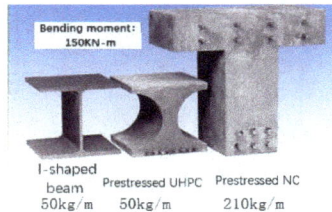

Figure 4.3 Comparison of the masses of structures under the same bending moment

4.2   UHPC applications in engineering

4.2.1   The applications of UHPC in Europe

In Germany, UHPC of C105 was firstly used in ultra-high-rise buildings in Frankfurt. UHPC of C115 was used for construction of cast-in-place and prefabricated components. UHPC was also used to build a pedestrian bridge in Kassel (Figure 4.4). In French, UHPC was used for a toll station (Figure 4.5).

Figure 4.4 UHPC pedestrian bridge built in Germany.

In Denmark, UHPC was used to produce prefabricated stairs (Figure 4.6). In the Netherlands, prefabricated panels of UHPC were used for protective doors in ocean caves (Figure 4.7).

Figure 4.5 UHPC toll station in French.

Figure 4.6 Staircase by UHPC in Denmark.

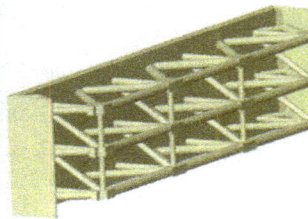

(a) Production of UHPC panels   (b) UHPC protection door

Figure 4.7 Protective door of UHPC prefabricated panel (protective door 45m wide x 15m high) in the Netherlands.

4.2.2   The applications of UHPC in North America

In the United States, in 1987, Seattle Two-Union Square adopted UHPC with strength of 131 MPa. In 2008, UHPC with strength of 250 MPa was produced and installed. UHPC with strength of 250 MPa has become commercialized and available in the market.

200 MPa UHPC was used for the footbridge deck of a pedestrian bridge in Quebec, Canada.

4.2.3   The applications of UHPC in Japan

1) The application of HS/HPC in an ultra-high-rise buildings in Japan is shown in Figure 4.8. Ultra-high-strength concrete with strength over 150 MPp was produced in a ready-mixed concrete factory. Based on the experiments, it was used in ultra-high-rise building structures. The cementitious materials used in this concrete are ordinary cement, slag, gypsum and silica fume. All the test results of slump, slump flow value, air content and compressive strength met the requirements of 150 MPa concrete. Based on these tests, production and construction applications were carried out. In a 53-story ultra-high-rise building, UHPC with strength of 180 MPa was applied (Figure 4.9).

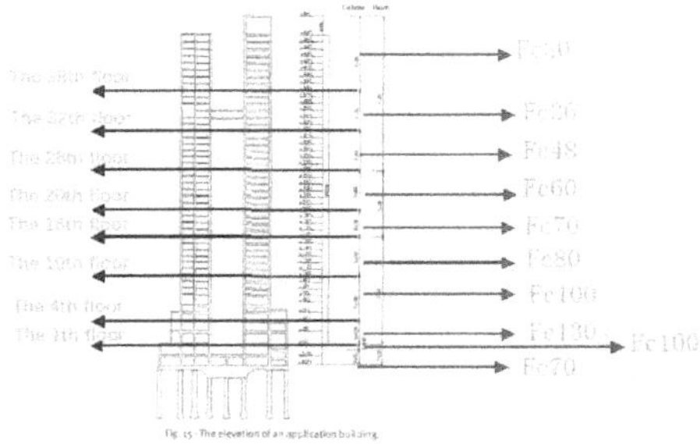

(a) HPC and UHPC used in different floors of an ultra-high-rise building

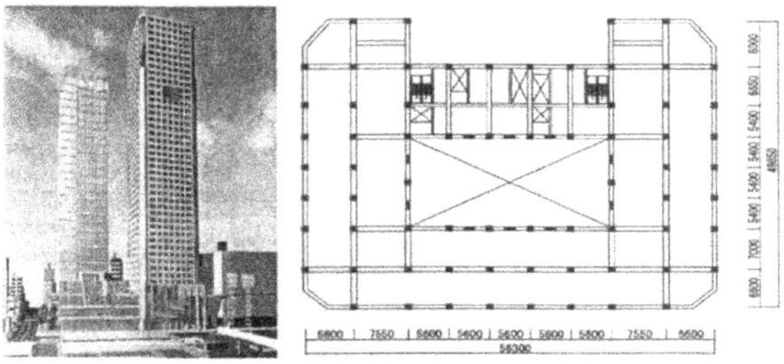

(b) The plane and elevation of an ultra- high-rise building with HPC and UHPC

Figure 4.8 HPC and UHPC used in an ultra-high-rise building in Japan.

Figure 4.9 UHPC with strength of 180 MPa used in an ultra-high-rise building.

2) The application of engineered composite materials in Japan for ultra-high-rise and large-span residential buildings: engineered cementitious composites (ECC for short) was jointly researched and developed by Victor. C. Li and Kanda in Japan. The PVA-ECC was successfully developed by using polyvinyl alcohol fibre. ECC is also a type of UHPC, with no coarse aggregate in the constituent materials. It is produced by mixing paste, sand and fibre. It

has high bending strength and deformation ability. This composite material was applied to earthquake-resistant building structures (Figure 4.10).

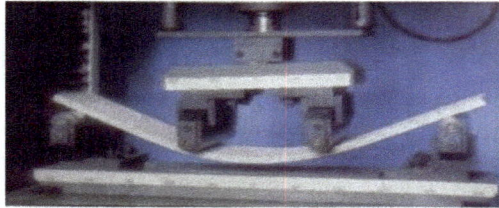

(a) Bending test of ECC material

(b) Earthquake-resistant building with ECC materials

(The new construction method derived from this technology can provide earthquake-resistance, high-strength and comfortable living space)

(c) Earthquake-resistant large span structural building with ECC material

Figure 4.10 Earthquake-resistant large span structure with ECC material.

Moreover, in Japan UHPC has also been successfully used in many construction projects in different periods, such as:

- From 1993 to 1999, UHPC with strength of 100 MPa was used in super high-rise residential buildings.
- In the 1990s, 105 MPa was used for the Japan Center Office.
- In 2001, 200 MPa UHPC was used for the Sakata Mirai bridge.
- In 2004, 130 MPa UHPC was used for ultra-high-rise residential buildings
- Japan's Pacific Cement Group has developed a special cement with 200 MPa strength. This cement can be used for production of ultra-high performance concrete with 80-120 MPa (up to 150 MPa).

4.2.4   The research and applications of UHPC in China

In China, an HPC technical standards has been developed by China Engineering Technology Standards Association. According to this standard, HPC must be designed for durability in addition to ensuring a designed strength. In China, the C60 high strength and high performance concrete has been widely used. C80 high-performance concrete has been often used in production of pipe piles. Here are some examples: Beijing airport expansion project, Guangzhou Baiyin Square steel tube concrete, Shenzhen Antuoshan C80 high-strength high-performance concrete, Changsha ultra-fine fly ash powder C80 high-performance Concrete, and Shenyang C100 high-performance concrete, etc. The highest strength for high-performance concrete achieved in laboratory is approximately 150 MPa.

1) Guangzhou West Tower Project (437 m high, as shown in Figure 4.11): In this project, both C100 UHPC and C100 UHP-SCC were used for steel-tube-confined concrete. The mix proportions are shown in Table 4.2. The compressive strength at different ages is shown in Table 4.3. The concrete was pumped to a height of 421 m.

Table 4.2 Mix designs of UHPC and C100UHP-SCC (kg/m³).

W/B	C	BFS	SF	CFA	NZP	S	G	W	HWRA
20%	500	190	60	-	-	750	900	150	18
22%	450	190	60	14	28	750	900	154	15.4

Table 4.3 Compressive strength of C100 UHPC and C00 UHP-SCC (MPa).

Type of concrete	3d	7d	28d	56d
UHPC	87	108.7	130.8	130.8
UHP-SCC	97.2	106.5	117.3	118.0

Figure 4.11 Guanzhou West Tower under construction.

2) C120 UHPC was developed for the Shenzhen Jingji Building: The C120 ultra-high-performance concrete was developed based on the needs of the ultra-high-rise reinforced concrete structure of the Jingji Building and the research on high-performance ultra-high-performance concrete both internationally and in China. Table 4.4 shows the mix proportion for C120 ultra-high performance concrete. The target strength at 28d was more than 125 MPa without shrinkage or with slight shrinkage only (shrinkage value below 4/10000). The concrete; prepared must be of high toughness, good fire resistance, excellent durability, good construction performance, excellent plastic retention performance of fresh concrete, and convenience for ultra-high pumping, etc.

Table 4.4 Mix design of C120 high performance concrete for the Jingji Building.

W/B	C	MB	SF	S	G	W	Compound plasticizer
0.18	500	170	80	700	1000	130	4.0%

Concrete strength: 3 d – 100 MPa; 7d - 108 MPa; 28 d - average strength of 13 groups of specimens, 135 MPa; Standard deviation 5.56 MPa.

The construction quality inspection centre of Shenzhen has tested the strength of 4 groups of C120 concrete, which reached 105%, 103%, 105% and 114% of the standard values, respectively.

Jingji C120 concrete was pumped to a height of 521 m when the project was about to be completed, and the volume of such concrete for this project was about 100 m^3.

3) UHPC developed for the Guangzhou East Tower Building project

The UHPC with a strength of 150 MPa developed for the East Tower Building project was pumped from the ground to a height of 510 m in the fall of 2015. Shear walls, beams, slabs, columns and other components were cast with this concrete. And the concrete has vibration-free, self-compacting, self-curing and other features. After demoulding, no cracks were found on the surface of the structure. The mix proportion of UHPC for the East Tower Building project for ultra-high pumping is shown in Table 4.5. The properties of fresh concrete are shown in Table 4.6. The compressive strength of concrete at different ages is shown in Table 4.7. The surface of the demoulded member of the concrete by ultra-high pumping is shown in Figure 4.12. No cracking caused by autogenous shrinkage has been found on the surface.

Table 4.5 The mix design of UHPC (kg/m³).

Cement 52.5R	Micro powder	Silica fume	Zeolite powder	Sand	Aggregate	Water	Admixture	CFA	Expansion agent
600	200	100	20	750	850	130	(2.4%)	20	10

Table 4.6 Properties of fresh concrete.

properties time	Flow time (s)	Slump (mm)	Slump flow (mm)	Raising of U Channel (mm/s)	workability
initial	2.88	280	770*770	340mm/14s	good
2h	4.06	275	740*750	330mm/30s	good

Table 4.7 Compressive strength at different age (MPa).

Age(day)	1	3	7	28	56
Strength	104	125	136	147	150

(C120 green multi-functional concrete development and ultra-high (500 meters) pumping
Sponsor: China State Construction Corporation Guangzhou CHOW Tai Fook Financial Center General Contracting Project Department
June 24, 2014)

(a) East Tower   (b) Casing of UHPC with 150 MPa

(c) Floor surface after demolding   (d) Shear wall surface after demolding

Figure 4.12 Ultra-high pumping of UHPC and the structure elements of East Tower.

4.3   Technical approach of UHP-SCC and UHPC

The development of ultra-high-performance concrete should take into account high-quality raw materials, including the grading, particle size and shape of sand and crushed stone; the water reducing rate and slump retention of water reducing (plasticizer); the combination of cement and ultra-fine mineral powder, and an appropriate water content and mix proportion, etc.

4.3.1   Mix design of C100 UHPC

UHPC is prepared with binding materials based on the existing cement, mixed with silica fume, microbeads, ultra-fine mineral powder. According to the experience of Japan, C100 UHPC can be prepared with the unit water content of 150 kg, W/B of 0.20, the amount of cementitious material of 750 kg/m^3, cement content of 500 kg/m^3. The mix proportion is shown in Table 4.8.

Table 4.8 Mix design of C100 UHP-SCC (kg/m3)

Cement	Micro-beads	Silica fume	Zeolite powder	Sand	Aggregate	Water	Admixture	CFA	W/B
500	150	70	30	800	900	150	1.5 %	1.5%	0.2

With such mix proportion, the actual W/C of 0.30, the water/cement ratio is relatively high, and the addition of natural zeolite ultrafine powder at a content of 30 kg/m^3 can be used as a concrete self-curing agent and viscosity modifier. Therefore, the concrete has good flowability without bleeding and segregation. CFA can make the fresh concrete keep slump. Such UHP- SCC will have good flowability, good slump retention and self-compacting ability. It ensures ultra-high pumping and self-compacting. The combination of cement, microbeads and silica fume with W/B of 0.20 can ensure sufficient strength development.

4.3.2   Fresh behaviour of C100 UHPC

Based on the existing cement, through the optimisation of particle size grading of the powders, the flowability of the paste is improved, the strength is increased, but the viscosity is reduced. Such prepared UHPC and UHP-SCC are more scientific, economical, with lower carbon emission, and energy and resource saving. As shown in Table 4.9, the effect of cementitious materials on the performance of fresh concrete will be identified firstly, and then the composition and quantity of cementitious materials will be determined. The mix proportion of concrete will then be decided. The performance of fresh concrete with different combination of binding materials is shown in Table 4.10.

Table 4.9 Mix proportion of UHP (kg/m^3).

Cement	Micro-beads	GGBS	Silica fume	Zeolite powder	Water	Sand	Aggregate	admixture	CFA
600	190	(190)	90	20	130	720	880	5.4	9

OPC, C-52.5; Microbeads (MB), specific surface area 8000-10000 cm^2/g; Mineral powder (BFS) specific surface area 8000 cm^2/g; Zeolite powder (NZ) specific surface area 6000 cm^2/g;

Materials Research Forum LLC
https://doi.org/10.21741/9781644901991

Polycarboxylic superplasticizer, 40% solid content; the carrier fluidizing agent (CFA) is made of NZ powder + superplasticizer.

Table 4.10 Properties of fresh concrete with different binding materials.

Combined binding materials	Slump(mm)	Slump flow (mm)	Flow time(sec)
C + BFS	265	650/630	18
C + BFS + SF	275	660/670	7
C + MB + SF	285	680/690	5

It can be seen that the combination of cementitious materials including cement, microbeads and silica fume has the best flowability. This is due to the micro-filling effect of the powder, the porosity is reduced, and the flowability is increased under the same water content and the same dosage of water reducing agent. In particular, the microbeads are spherical and vitreous, and require lower water, as shown in Figure 4.13.

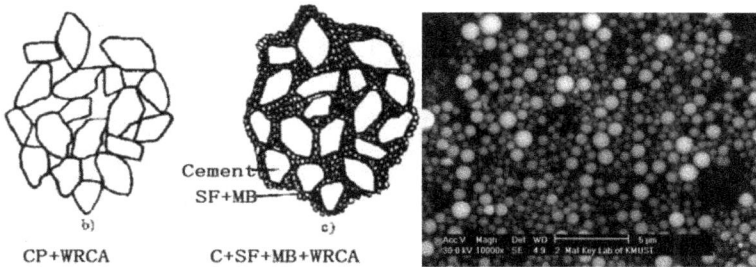

(a) Filling effect of super fine powder (b) SEM of microbeads

Figure 4.13 SEM of microbeads and filling effect.

4.3.3 Spherical cement

The direction of UHPC technology is to reduce the micro pores in the paste of binding materials by using a minimum of water content for the same workability based on the particle shape, size and grading of the cementitious material. Japan Pacific Cement Corporation has developed a new type of cement with strength of 200 MPa for the application of ultra-high-performance concrete of C150-C180. Spherical cement is shown in Figure 4.14.

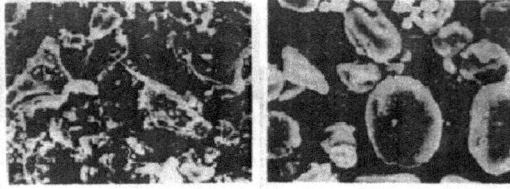

Figure 4.14 SEM of ordinary cement and spherical cement.

The evaluation of spherical cement is based on the sphericity of its particles. The sphericity is calculated as follows:

The sphericity of the particle = the diameter of the circle with the same projection area of the particle / the smallest diameter of outer circle of the projection area

The sphericity of ordinary Portland cement is 0.67, while that of spherical cement is 0.85. The sphericity of cement is close to that of a real sphere (sphericity=1).

1) The changes in cement properties caused by spheroidization: The comparison of workability between spherical cement (SC) and ordinary Portland cement (0PC) is shown in Figure 4.15. Whether paste or mortar is tested, the workability with SC is significantly higher than OPC.

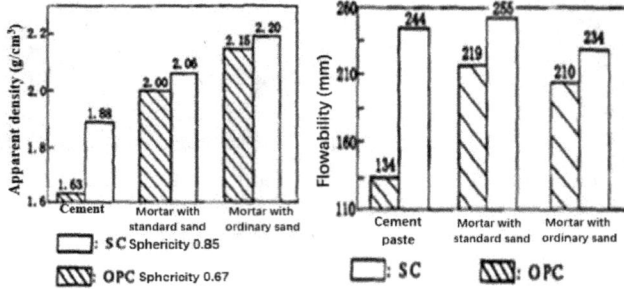

(a) Bulk density of SC and OPC (b) Workability of SC and OPC

Figure 4.15 Comparison between spherical cement (SC) and ordinary Portland Cement (OPC).

2) Changes in surface zeta potential: The elements distribution on the surface of a spherical cement particle detected by X-ray is shown in Figure 4.16 and its zeta potential is shown in Figure 4.17.

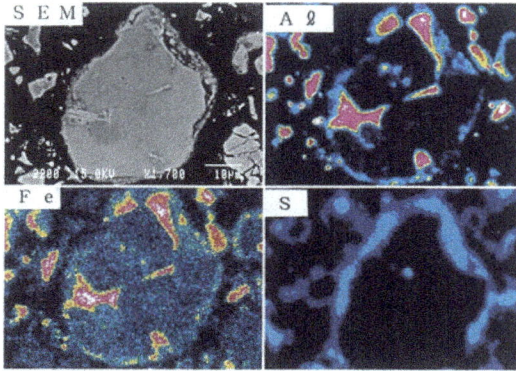

Figure 4.16 Surface analysis of spherical cement by X-ray and SEM.

Table 4.11 Zeta potential on the surface of spherical cement particles and powder properties.

	Zet Potential (mV)	Average diameter (μm)	Blaine (cm^2/g)	B.E.T (cm^2/g)	Bulk density (g/cm^3)	Dense bulk density (g/cm^3)
OPC	+1.13	13.5	3270	9694	0.98	1.86
SC	+3.42	10.1	2480	5026	1.19	1.91

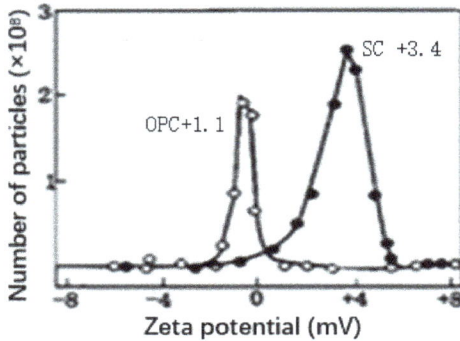

Figure 4.17 Zeta potential of spherical cement and ordinary cement.

Table 4.11 and Figure 4.17 show that the zeta potential of SC is about 3 times of that of OPC, which is caused by the electrostatic repulsion between particles after spheroidization.

3) Workability test:

- Flowability of paste: Under the same water-cement ratio, the flow value of the SC paste is about 100 mm higher than that of OPC. As shown in Figure 4.18.

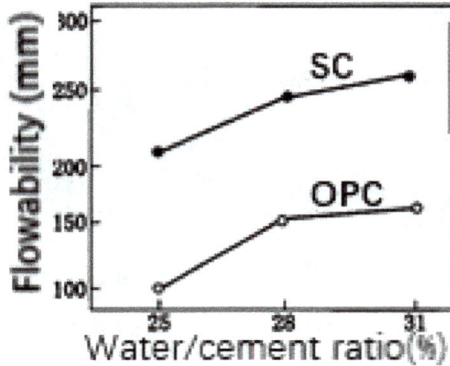

Figure 4.18 The flow value of SC and OPC slurry.

- Flowability of mortar: The flow value of SC mortar is greatly improved than that of OPC mortar, and the change of the former with time is small, as shown in Figure 4.19.

(a) Flow value of OPC Mortar: 177mm   (b)Flow value of SC mortar: 277mm

Figure 4.19 Comparison of mortar flowability (W/C=0.55, C:S=1:2).

- Flowability of concrete: Under the same water/cement ratio, the slump of SC concrete increases (maximum 22 cm). The slump of OPC concrete with w/c of 0.25 is about 5 cm, but the slump of SC concrete is about 20 cm. It shows good flowability of SC concrete. On the other hand, if the slump of SC and OPC concrete are the same, the water consumption of SC concrete can be reduced by 9-30% compared with that of OPC concrete. Using SC to prepare high-performance and ultra-high-performance concrete, it has outstanding advantages to improve the flowability of fresh concrete.

- The strength of concrete with spherical cement and ordinary cement (Figure 4.20).

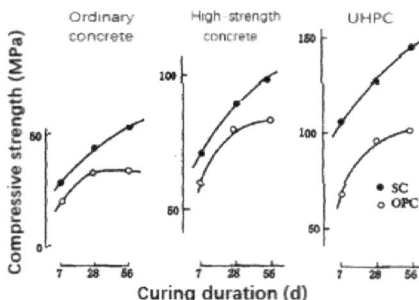

Figure 4.20 Strength of spherical cement and ordinary cement concrete.

Whether ordinary concrete, HS/HPC, or UHPC is prepared, the strength of the concrete with spherical cement is greatly improved than that with OPC. The strength at 56 d was over 150 MPa.

4.3.4 Graded cement (improve particle-size distribution and reduce porosity)

On the basis of DSP materials, graded cement has been developed. The features of it are as follows: (1) Adjust the particle size distribution of this cement in order to increase the packing density. (2) Increase the particle size of cement particles, and the particle size distribution moves to the coarse direction. (3) Add ultra-fine powder to obtain a filling as dense as possible. The cement paste has good fluidity, high early strength, low heat of hydration, and slow heat release of hydration. High performance concrete can be prepared with resource-saving and energy-saving.

1) Raw materials of graded cement as shown in Table 4.12.

Table 4.12 Raw materials of graded cement.

Raw materials	Symbol	Density	Specific surface area	Average grain size
Portland cement	N	3.17	3400 cm^2/g	19.57 μ m
Coarse cement	O	3.17	600 cm^2/g	90.74 μ m
Lime stone powder	W	2.71	18000 cm^2/g	6.04 μ m
Silica fume	S	2.26	200000 cm^2/g	0.2 μ m
Fly ash	F	2.18	4320 cm^2/g	17.38 μ m
GGBS	K	2.92	6260 cm^2/g	7.86 μ m

Maximum grain size: 90.47 μm, the minimum grain size: 0.2 μm.

2) The composition of graded cement: The graded cement can be blended with powders of different fineness and cement according to a certain proportion. From Table 4.13, it can be seen that the particle size range of graded cement has been expanded.

Table 4.13 Composition of graded cement.

NO	Symbol	(N) Cement	(O) Coarse powder	(W) Lime Stone	(S) Silica Fume	(F) Fly Ash	(K) GGBS
1	N7W30	70		30			
2	N7W10	70	20	10			
3	N7S10	70	20		10		
4	N7F10	70	20			10	
5	N7K10	70	20				10
6	N	100					

The average particle size of cement is 19.57 μm, while the size of coarse powder is extended to 90.74 μm. Ultra-fine silica fume has an average particle size of 0.2 μm, which makes the overall particle size of graded cement becomes finer. The particle size distribution of cement, composed of particles of different sizes, becomes optimum, so the powder compactness of the cementitious material is improved.

3) Properties of graded cement

• Flowability of cement paste: The measured flow value of the paste prepared according to the composition in Table 4.13 is shown in Figure 4.21.

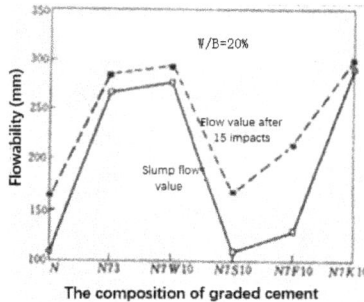

Figure 4.21 Flow value of graded cement slurry.

- Hydration heat of graded cement: As shown in Figure 4.22.

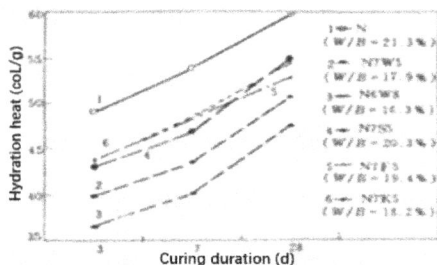

Figure 4.22 The hydration heat of graded cement.

The highest heat of hydration was detected for N-pure cement, which reached 209.3 J/g at 3 d, and 251.2 J/g at 28 d. N6W8 (Portland cement 60%, coarse powder 32%, limestone powder 8%) had the lowest heat of hydration, followed by N7W5 (Portland cement 70%, coarse powder 25%, limestone powder 5%). The lowest heat of hydration of N6W8 at 3 d age was 150.7 J/g, and 196.7 J/g at 28 d. In the graded cement containing coarse powder and limestone powder, the heat of hydration was greatly reduced.

- Strength of graded cement

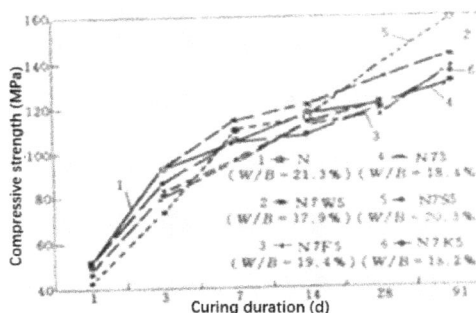

Figure 4.23 Strength of graded cement.

It can be seen from the strength of graded cement shown in Figure 4.23 that, with the same flowability, the water/powder ratio (W/P) was different for different graded cements with different compositions. As Portland cement (code N) required a large amount of water, the 28-day strength was about 120 MPa, lower compared to those of other graded cements, while N7S5 (Portland cement 70%, coarse powder 25%, silica fume 5%), and N7W5 (Portland cement 70%, coarse powder 25% lime stone 5%) had higher strength of 140 Map at 28 d.

From the three aspects of paste such as flowability, heat of hydration and compressive strength, the comprehensive evaluation can conclude that the composition of graded cement N7W5 or N7S5 is a better choice for preparation of HPC and UHPC.

4) Concrete of graded cement

Graded cement is prepared by blending of powders of different fineness and different properties. Due to the matching of each type of powders, graded cement can achieve high performance. Next, we will prepare HPC and UHPC with graded cement.

- Raw materials, powder blending and concrete mix design

The relevant properties of graded cement, sand and coarse aggregate are shown in Table 4.14. The composition of graded cement is shown in Table 4.15, and the mix proportion of graded cement concrete is shown in 4.16.

Table 4.14 Raw materials and properties.

Material	Density		Surface area (cm^2/g)	Average grain size (μm)	Fines modulus
Portland cement	3.17		3400	19.57	
Coarse cement	3.17		600	90.74	
Lime stone	2.71		18000	6.04	
Silica fume	2.26		200000		
Fine aggregate	2.58				2.73
Coarse aggregate	2.62				6.63

Table 4.15 Composition of graded cement.

Symbol	Portland cement	Coarse cement	Lime stone	Silica fume	Filling ratio
A	100	0			0.5
B	70	30			0.57
C	70	20	10		0.55
D	70	20		10	0.55
E	70	20	5	5	0.55

As to the five series of cements, i.e. A, B, C, D, and E, different water/cement ratios were used. Totally 15 series of concrete were prepared as shown in Table 4.16.

In Table 4.16:

i. A25 and A30 were ordinary Portland cement concretes, with a water/cement ratio of 0.25 and 0.30 respectively.

ii. B20, B25, and B30B were three series of graded cement concretes with a water/cement ratio of 0.20, 0.25 and 0.30 respectively.

iii. C17.5, C20, C25, C30 were four series of graded cement concrete, with a water/cement ratio of 0.175, 0.20, 0.25 and 0.30 respectively.

iv. D20, D25, D30D were three series of graded cement concrete, with a water/cement ratio of 0.20, 0.25 and 0.30 respectively.

v. E17.5, E20, E25, and E30 were four series of graded cement concrete, with a water/cement ratio of 0.175, 0.20, 0.25 and 0.30 respectively.

Table 4.16 Mix proportion of graded cement concrete

Concrete	W/C (%)	s/a (%)	Weight (kg/m3)				Chemical admixture (B x%)		Air content (%)
			Water	Cement	Sand	Aggregate	Superplasticizer	Air content adjuster	
A25	25.0	39.6	165	660	618	950	1.8	0.008	2.0
A30	30.0	43.8	155	533	736	950	1.4	0.006	2.1
B20	20.0	33.8	165	825	482	950	2.4	0.015	2.4
B25	25.0	40.7	160	640	647	950	1.4	0.007	2.5
B30	30.0	44.7	155	517	762	950	1.4	0.006	2.2
C17.5	17.5	30.2	160	914	409	950	3.5	0.018	1.7
C20	20.0	36.3	155	776	537	950	2.4	0.012	1.1
C25	25.0	42.6	150	600	700	950	1.8	0.008	1.9
C30	30.0	46.3	145	483	812	950	1.5	0.006	2.5
D20	20.0	30.6	170	850	417	950	3.2	0018	2.5
D25	25.0	38.7	165	660	594	950	2.3	0.012	2.8
D30	30.0	43.2	160	533	718	950	2.0	0.015	2.5
E17.5	17.5	27	165	943	348	950	4.0	0.028	1.9
E20	20.0	34.2	160	800	490	950	3.0	0.018	1.5
E25	25.0	41.1	155	620	658	950	2.0	0.014	2.7
E30	30.0	45.3	150	500	781	950	2.0	0.014	2.9

- The performance of each series of graded cement concrete

The performance of each series of graded cement concrete can be compared through the tests under the same WC, as shown in Table 4.17. It can be seen from the test results of concrete in Table 4.17:

i. When the W/C of each group of graded cement concrete was 0.25 and 0.30, the slump of the concrete ranged from 25 cm to 28 cm, where C25 had the largest slump up to 28 cm. The slump flow ranged from 59 cm to 71 cm, where the slump flow of C25 is the largest up to 71 cm. The yield value was also the lowest for C25, which was 15.9 Pa;

ii. When the W/C was 25%, D25 had the highest compressive strength at 28 d and 91 d, which were 107 MPa and 127 MPa respectively;

iii. The graded cement concrete E20 had a slump flow value of 70 cm and strength of 143.0 MPa at 91 d. This is a typical high-flow ultra-high performance concrete;

iv. Under the same slump, the W/C of graded cement concrete is about 5% lower than the reference Portland cement concrete.

Table 4.17 Performance of graded cement concrete.

Concrete	W/B (%)	Concrete		Mortar		Compressive Strength (MPa)			
		Slump (cm)	Slump Flow (cm)	Viscosity (s)	Yield (Pa)	3d	7d	28d	91d
A25	25.0	26.0	61	16.1	67.5	76.5	92.7	105.3	118.3
A30	30.0	26.5	65	11.9	44.2	63.6	77.8	92.6	103.0
B20	20.0	27.0	63	24.6	91.1	77.2	93.7	110.1	117.2
B25	25.0	25.0	68	10.8	27.8	64.8	80.6	94.7	110.9
B30	30.0	26.5	64	11.9	38.3	45.6	64.5	78.6	92.4
C17.5	17.5	26.5	60	54.0	161.8	86.0	96.3	108.3	121.3
C20	20.0	-	74	18.5	17.6	86.7	95.4	109.8	122.9
C25	25.0	28.0	71	13.6	15.9	63.2	77.6	91.2	100.0
C30	30.0	25.5	59	8.9	43.5	55.0	69.7	84.2	97.2
D20	20.0	27.0	64	38.1	111.7	76.8	95.8	124.5	135.3
D25	25.0	27.5	66	17.2	41.9	58.8	80.5	107.9	127.1
D30	30.0	27	64	13.9	40.5	51.2	71.6	101.0	109.4
E17.5	17.5	27.0	62	59.7	158.3	82.6	99.1	125	133.7
E20	20.0	-	70	26.6	28.6	88.0	102.9	124.8	143.1
E25	25.0	-	68	16.5	27.4	65.9	84.5	106.8	117.1
E30	30.0	27.0	66	13.1	24.7	53.6	70.9	96.6	113.3

According to the author's point of view, UHPC can be divided into two categories:

1) Cementitious materials + fibre + organic adhesive + water →mixing and casting →UHPC products, such as ECC; without coarse but with fine aggregates: If fine aggregate is added, various prefabricated boards and other products of UHPC with strength over 200 MPa can be produced by mixing, vibration, compaction and other processes.

2) Cementitious materials + sand + coarse aggregate + water, etc., to obtain concrete with strength over 150 MPa, which can be used for structural components such as shear-bearing walls, beams and columns.

Since UHPC with coarse and fine aggregates is the topic of this book, aggregates will be presented in another chapter.

Multifunctional Concrete Technology   Materials Research Forum LLC
Materials Research Foundations **127** (2022)   https://doi.org/10.21741/9781644901991

Chapter 5

Coarse and fine Aggregate for UHPC

5.1   Introduction

Aggregate is the skeleton of concrete. It accounts for about 70% of the total volume of concrete and is a main component of concrete.

Concrete aggregates are classified according to the size. Fine aggregate is with a particle size range of 0.15-5 mm, such as natural sand, sea sand and stone chips, etc. Coarse aggregate is with the particle size range of 5 mm or more, to 150 mm, such as pebbles, crushed stone and crushed pebbles, etc. In addition, in order to recycle resources, there are recycled aggregates. Recycled coarse and fine aggregate is produced by crushing of construction waste that will be demolished, such as concrete blocks, bricks, etc. Recycled aggregate is also within the scope of this chapter.

Aggregates in concrete have important technical, economic and environmental protection functions. The correct choice of aggregates, which can meet the requirements of relevant technical standards, is the basis for preparation of high-performance or ultra-high-performance concrete.

In ordinary concrete, the strength of general aggregate is 3-4 times higher than that of concrete. Even for ordinary concretes incorporating different aggregates, the difference in compressive strength between them may be very small. However, when preparing high-performance or ultra-high-performance concrete, the difference in aggregate has a great influence on compressive strength of concrete. As shown in Figure 5.1, when compressive strength of concrete is below 50 MPa, that is, the water/cement ratio is over 0.4, compressive strength of concrete with crushed stone aggregate K or river pebble aggregate R is roughly the same. However, when the water-cement ratio is less than 0.35, when different types of coarse aggregates are used to prepare concrete at the same water/cement ratio, the difference in compressive strength is more obvious. Both the strength and the interfacial structure of the crushed stone aggregate are more favorable than those of the pebble aggregate, so the concrete strength with the former is higher.

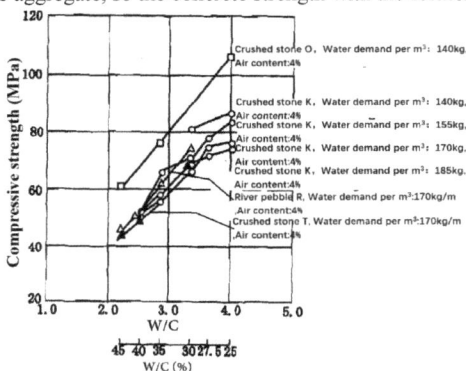

Figure 5.1 Compressive strength and W/C of concrete with different types of aggregates.

So far, concrete has been generally regarded as a two-phase composite material of cement mortar and coarse aggregate, to analyze the stress and strain under external loading to find out the influence of the quantity and quality of concrete constituent materials on strength. However, in fact concrete is composed of three-phases: namely, aggregate, hardened cement paste and interface transition zone, as shown in Figure 5.2. In high-performance or ultra-high-performance concrete, the interface transition zone is a relatively weak link. How to improve the performance of the interface transition layer is the key to improving the strength, durability and impermeability of high-performance or ultra-high-performance concrete. Therefore, it is necessary to study the interaction between aggregate and hardened cement paste, and to study the influence of the type, quantity and quality of aggregate on the interface transition zone.

Figure 5.2 Microstructure of interface between aggregate and cement paste.

5.2   Bonding strength of aggregate and hardened cement slurry

The bonding between aggregate and hardened cement paste is the basic issue to improve the strength of concrete. The bonding theory is one of the relevant theories to be studied.

5.2.1   The bonding theory

When two different kinds of substance are in contact, an adhesion is generated between each other, which is the bonding force. This is caused by the gravitational force of the interaction between the molecules and atoms that make up the substances. The gravitational force between molecules is formed by the van der Waals gravitational force and the bonding force formed by the bond between two hydrogen atoms. When a solid is to be bonded, the adhesive is applied on its surface and solidified, and the solids are bonded together. This phenomenon is called bonding. This bonding phenomenon starts when the liquid adhesive wets the bonded solids. The bonding force is related to the wetting angle, as shown in Figure 5.3. The smaller the wetting angle, the easier the wetting, and the easier the wetting, the greater the bonding force.

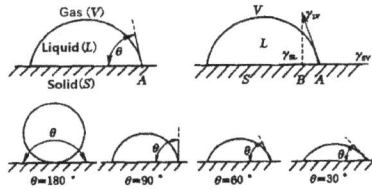

Figure 5.3 Wetting angle and surface tension.

Note: Point A is the intersection point of gas, liquid and solid phases.

γlv—The force between the gas phase and the liquid phase (the interfacial tension of the liquid phase).

γsv—The force between the solid phase s and the gas phase v.

$\gamma s\ l$—The force between the solid phase s and the liquid phase l. When each force is in equilibrium at point A:

$$\gamma sv = \gamma sv + \gamma lv \cos\theta \tag{5.1}$$

adhesion work = solid surface tension + liquid surface tension-interfacial tension between solid and liquid:

$$W_A = \gamma s + \gamma lv - \gamma s\ l \tag{5.2}$$

Substitute (5.1) into formula (5.2), if $\gamma s = \gamma sv$, then $W_A = \gamma lv.\ (1+\cos\theta)$

The smaller the wetting angle θ, the greater the W_A, and the better the bonding performance. In addition, there is a critical surface tension γc in the solid. In order to wet the surface of the object, the interfacial tension between solid and liquid, $\gamma s\ l$, must be greater than γc.

In addition, the Solubility Parameter (*SP*), which is related to the bond strength, has the following relationship with Cohesive Energy Densities (*CED*):

$$SP = (CED)_{1/2} \tag{5.3}$$

$$CED = \Delta E/V = \Delta H - RT/V = d/M.(\Delta H-RT) \tag{5.4}$$

where: ΔE-evaporation energy;

V-molecular volume;

ΔH-latent heat of evaporation;

R—gas constant;

d-density;

M-molecular weight;

T-absolute temperature.

According to the above formula, the cohesive energy CED can be calculated, and then according to the formula (5.3), the solubility factor SP can be calculated. The closer the SPs between the adjacent phases, the easier the wetting and the higher the bonding strength.

The above theory is applicable to the interface between cement paste and aggregate. Okishi et al. in Japan studied the influence of the interface energy of several construction materials on the strength and wetting behavior. Prof. Chen Zhiyuan et al. in China studied the adhesion between cement paste and marble. These are all related to the wetting angle. In order to improve the bonding performance of cement hydrates and aggregate, R. Zimbelmann used an additive to reduce the surface tension of water and increased the bonding strength by 2 to 2.8 times.

5.2.2 Bonding mechanism between cement paste and aggregate

1) Bond strength and failure of concrete

The cement paste in concrete not only binds aggregates together, but also maintains the strength of the matrix between the aggregate particles. This is different from the use of epoxy resin to bond two objects together.

In addition, the bonding of hydrates in cement paste is different from the bonding of organic materials. Because the binder is cement paste, its structure and strength may change with the increase of age and is also affected by the water-cement ratio, curing conditions, and the physical and chemical properties of the aggregate.

In other words, the strength of the interfacial transition zone is related to the following factors:

  i. strength of cement paste,
  ii. strength of the aggregate itself,
  iii. bonding force between the aggregate and the hydrates,
  iv. the cohesion of the hydrates,
  v. Bonding between the hydrates and hardened cement paste.

Therefore, there are various factors influencing failure of concrete, such as hardened cement paste, the interface, the aggregate, or the composite state of all these factors.

Kawamura et al. used a microhardness tester to continuously measure the hardness of both sides of the cement paste and aggregates at the interface, and further explored the physical and chemical characteristics of the interface zone. They also studied the initiation and development of cracks at the interface zone through the microscopic analysis. Tanigawa et al. summarized the cracks related to coarse aggregates in concrete as follows:

  i. due to bleeding, primary cracks formed under the aggregate,
  ii. at the interface between mortar and coarse aggregate, cracks were caused by the difference in temperature deformation;
  iii. cracking occurred in the mortar;
  iv. cracks in the mortar and cracks in aggregate were connected together to form a growing crack;
  v. cracking occurred in the coarse aggregate, etc.

The failure of ordinary concrete originates from the interface between coarse aggregate and hardened cement paste, or from hardened cement paste. However, the failure of high-strength concrete with strength from 80 MPa to 100 MPa is mostly caused by aggregate damage.

2) Interface status

A transition zone is formed at the interface between coarse aggregate and cement mortar, which is characterized by a porosity higher than that of the bulk cement paste (see Figure 5.2).

In the transition zone, especially near the surface of the aggregate, there are almost vertical plate-shaped or layered $Ca(OH)_2$ (represented by CH) crystals. Coarse crystals of CH, ettringite and a small amount of CSH gel are distributed in the middle layer of the transition zone, which is of relatively low strength. In Portland cement concrete, a large amount of CH crystals forms a loose structure on the surface of the aggregate with low strength, low impermeability and bad durability.

5.2.3 Aggregate type and bond strength

1) Aggregate surface polishing and bonding strength of cement paste to polished aggregate

Coarse aggregates of different rocks were polished to form smooth surfaces of them, which were further used to prepare concrete specimens with low water-cement ratio. The bonding strength was determined, of which the results are shown in Figure 5.4.

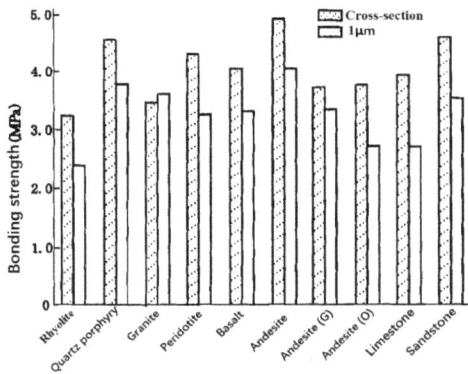

Figure 5.4 Bonding strength between rock and cement paste.

It can be seen from Figure 5.4 that sandstone, andesite and quartz porphyry performed better in terms of bond strength, followed by limestone and basalt.

Unevenness of surface and bonding of different mineral rock specimens

In this case, when the cement paste contained and did not contain silica fume, the bond strength was very different. When silica fume was added, the interface bond strength was very high. At 28 d age, the tensile strength of pure tension reached 11 MPa, but with very high deviation. Attention must be paid to this situation in order to obtain a stable strength of HPC.

The influence of chemical composition and activity on bond strength

Under the conditions of elevated temperature and high pressure, the surface of the aggregate becomes active and aggregate will react with cement paste to produce tobermorite, which has high strength. Therefore, quartz aggregate is often used. Even at room temperature, the strength of concrete with quartz aggregate can reach 10 MPa at 1d age. The bonding strength of limestone aggregate at room temperature is also very good. UHPC prepared with limestone aggregate can obtain compressive strength of 120 MPa.

It was reported for the highest strength of concrete, steel sand was used to replace normal aggregate for preparing mortar. Through a process of the heating to elevated temperature and high pressure, the compressive strength could reach 300 MPa, while the strength of the silica sand mortar specimen under the same conditions was only 220 MPa, i.e. the former was about 38% higher. This shows that aggregate strength is very important when choosing aggregate.

Regarding the strength of the aggregate, the crushing value test method is generally used. The strength of HPC and UHPC has a good correlation with the strength of the coarse aggregate. It is reported that the comprehensive evaluation of the strength and bonding performance of the coarse aggregate is an important factor for producing HPC and UHPC. It can be done during the concrete compressive strength test by changing the amount of coarse aggregate, observing the variation of compressive strength of concrete, so as to comprehensively evaluate the influence of both the quantity and quality of aggregate on the strength of concrete. The test results of HPC with W/B from 0.25 to 0.40 showed that this effect is very obvious. In addition, the shape and particle size of the coarse aggregate also have a great influence on fluidity of fresh concrete and strength of hardened concrete.

5.3 The influence of the apparent density and water absorption of aggregate on the compressive strength of HPC and UHPC

5.3.1 The relationship between the apparent density of aggregate and the compressive strength of HPC and UHPC

Figure 5.5 illustrates the relationship between the apparent density of aggregate and the compressive strength of HPC and UHPC. With the same water-cement ratio, for example, when W/C was 0.25 and the apparent density of the aggregate was lower than 2.50 g/cm^3, the concrete compressive strength was lower, e.g. 50-70 MPa. However, when the apparent density of the aggregate is higher than 2.65 g/cm^3, the strength of concrete with the same W/C of 0.25 could reach 110 MPa. When the apparent density of the aggregate is 2.65-3.00 g/cm^3, with W/C of 0.25-0.35, the strength of the prepared concrete was higher. In other words, for the preparation of HPC and UHPC, the apparent density of the aggregate should be selected to be over 2.65 g/cm^3.

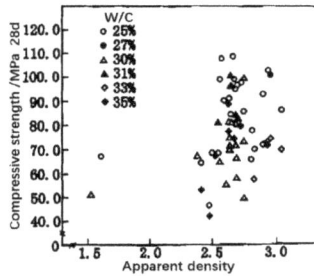

Figure 5.5 The relationship between apparent density of aggregate and strength of HPC, UHPC.

5.3.2 Influence of water absorption of aggregate on the compressive strength of HPC and UHPC

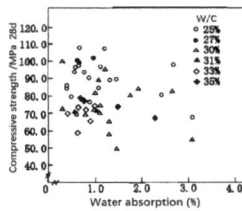

Figure 5.6 The relationship between water absorption of aggregate and compressive strength of HPC, UHPC.

For concrete with the same water-cement ratio, if the aggregate has high water absorption, the concrete will have low strength, as shown in Figure 5.6.

For concrete with a water/cement ratio of 0.25-0.35, when the water absorption of the aggregate is less than 1.0 %, the strength of concrete is high. When the water absorption of the aggregate is greater, the strength of the concrete is relatively low. Thus, for preparing HPC and UHPC, coarse aggregate with water absorption around 1.0% should be selected.

5.3.3 Influence of different types of fine aggregates on compressive strength of HPC and UHPC

Different types of fine aggregates including river sand, mountain sand, washed sea sand, crushed rock and land sand may be used to prepare HPC and UHPC. The compressive strength of mortar prepared with different types of sand is shown in Figure 5.7.

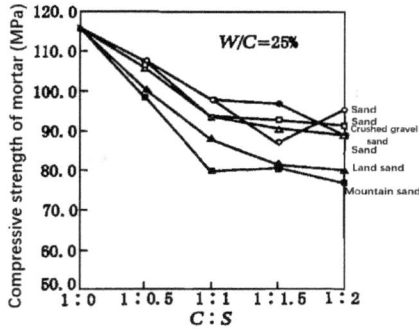

Figure 5.7 Types of fine aggregate and compressive strength of mortar.

The mortar prepared with river sand, gravel sand, or washed sea sand has higher strength than that of the mortar with land sand or mountain sand. Therefore, both HPC and UHPC should use river sand, gravel sand, or washed sea sand.

5.4 Influence of aggregate on strength and deformation of concrete

5.4.1 Compressive strength

In order to determine the strength of the coarse aggregate, the specimens of $5\times5\times5$ cm or $\phi5\times5$ cm were cut from a parent rock. The specimens were soaked in water for 48 h before testing compressive strength. The ratio of the rock strength to concrete strength should be not less than 1.5, or the aggregate crushing value Q_A should be smaller than 10%. Generally, crushed stone is better than pebbles. The compressive strength of the parent rock of the crushed stone is stronger, and the dense hard sandstone and andesite also have high strength. With water/cement ratio of 0.25, the difference of strength of concrete prepared with various coarse aggregates may be about 40 MPa, while the difference of concrete strength with different fine aggregate is about 20 MPa. The influence of different types of aggregate on the strength of concrete is shown in Figures 5.8 and 5.9.

It can be seen from Figures 5.8 and 5.9 that the strength of concrete prepared with hard sandstone sand and hard sandstone gravel was the highest, reaching 120 MPa. The strength of concrete prepared with coarse aggregate of quartz schist and fine aggregate of bauxite was 150 MPa at the age of 4 weeks and 160 MPa at 13 weeks.

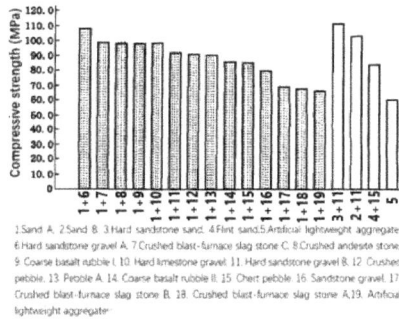

1.Sand A. 2.Sand B. 3.Hard sandstone sand. 4.Flint sand.5.Artificial lightweight aggregate.
6.Hard sandstone gravel A. 7.Crushed blast-furnace slag stone C. 8.Crushed andesite stone.
9. Coarse basalt rubble I. 10. Hard limestone gravel. 11. Hard sandstone gravel B. 12. Crushed
pebble. 13. Pebble A. 14. Coarse basalt rubble II. 15. Chert pebble. 16. Sandstone gravel. 17.
Crushed blast-furnace slag stone B. 18. Crushed blast-furnace slag stone A.19. Artificial
lightweight aggregate.

Figure 5.8 Relationship of different types of aggregate and concrete strength.

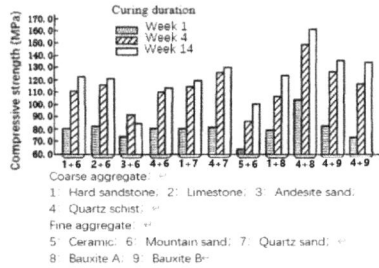

Coarse aggregate:
1: Hard sandstone. 2: Limestone. 3: Andesite sand.
4: Quartz schist.
Fine aggregate:
5: Ceramic. 6: Mountain sand. 7: Quartz sand.
8: Bauxite A. 9: Bauxite B.

Figure 5.9 Relationship of different coarse and fine aggregate and concrete strength.

5.4.2   The relationship between aggregate volume and the compressive strength

What is the appropriate amount of coarse aggregate in 1 m^3 concrete? The test results are shown in Figure 5.10. It can be seen from Figure 5.10 that for gravel concrete, when the amount of coarse aggregate is 300 L/m^3, the difference between concretes with various types of coarse aggregate is not significant. However, when the amount of coarse aggregate is 400 L/m^3, the compressive strength of concrete is significantly different. Compared to the concrete with hard sandstone aggregate, the compressive strength of concrete with crushed pebble is 10 MPa lower. For UHPC with compressive strength over 100 MPa, the hard sandstone aggregate should be used and the volume of coarse aggregate should be about 400 L/m^3.

5.4.3   The relationship between the aggregate volume and the elastic modulus of concrete

Concretes prepared with different coarse aggregates have different modulus of elasticity, as shown in Figure 5.11.

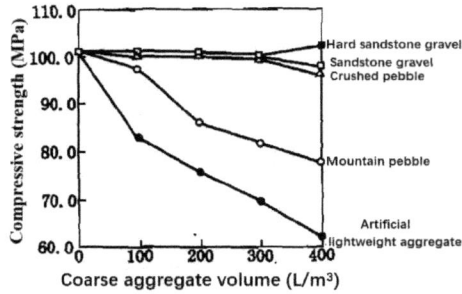

Figure 5.10 Relationship between aggregate volume and compressive strength.

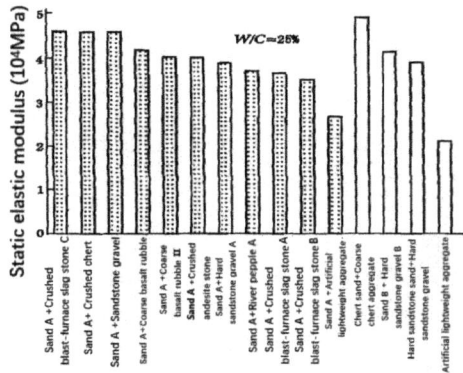

Figure 5.11 Elastic modulus of concrete with different coarse aggregate.

In general, the higher the density and compressive strength of concrete, the higher the static elastic modulus. It is generally believed that the static elastic modulus of concrete increases by 0.5 times of compressive strength. The static elastic modulus of HPC and UHPC need to be tested in case of using different aggregates. It can be seen from Figure 5.11, for W/C of 0.25, the static modulus of elasticity of concrete prepared with river sand and hard slag rock (or sandstone rock, flint rock) can be as high as 4.5×10^4 MPa.

5.4.4   Influence of different aggregates on Poisson's ratio

The concrete prepared with different aggregates has little effect on Poisson's ratio. Within the strength range of 50-100 MPa, the Poisson's ratio of concrete prepared with different aggregates is 0.16-0.26. Even if the types of cement and aggregates are changed, the Poisson's ratio is still within the above range. However, if the concrete with the compressive strength of 90-100 MPa is prepared with cement clinker, the Poisson's ratio will be 0.19-0.25.

5.4.5   Influence of the shape of coarse aggregate on the workability and compressive strength of concrete

When preparing prestressed pipe concrete at strength grade of C80, three types of coarse aggregates were used with the same mix proportion. The workability and compressive strength were compared. The three types of coarse aggregates were: 1#: Sanya crushed rock aggregate with high percentage of needle and flake particles, and crushing value of 6.5%; 2#: Sanya crushed rock with good particle shape, and the crushing value of 8.1%; 3#: Haikou crushed stone with a good particle shape, and the crushing value of 3.6%. The concrete mix proportion is shown in Table 5.1. The workability and strength test results are shown in Table 5.2.

Table 5.1 Mix design of C80 prestressed pile concrete.

NO.	C	MB	BFS	S	G	W	AG
1#	360	60	80	700	1250	107	2.1%
2#	360	60	80	700	1250	107	2.1%
3#	360	60	80	700	1250	107	2.1%

Table 5.2 Workability and compressive strength of concrete.

No	Strength by sun shed curing (MPa)			Strength by outdoor wet curing (MPa)			slump
	1d	3d	7d	1d	3d	28d	
1#	52.3 MPa	72.1	77.7	42.3	70.1	86.5	3.5cm
2#	77 MPa	86.8	92.2	60.7	89.3	107	6.7cm
3#	78.2 MPa	87.1	87	64.2	94.7	107	6-7cm

It can be seen that the concretes of 2# and 3# had better workability and higher strength due to the good shape of coarse aggregates with fewer needle and flake particles. Therefore, when preparing HPC and UHPC, crushed rock aggregate with a better particle shape should be selected.

5.5   Effect of volume and particle size of coarse aggregate on compressive strength of HPC and UHPC

For ordinary concrete with strength grade of C30, the coarse aggregate is intact and undamaged when concrete failed under compression. However, in HPC and UHPC, when the concrete is damaged under compression, the coarse aggregates at the fracture surfaces are almost completely broken and destroyed. Compressive strength tests were conducted with different types, different volumes or different maximum particle size of coarse aggregate to establish a mathematical model for evaluation of the performance of different aggregates.

5.5.1 Test program and raw materials

Raw materials of testing

Cement: ordinary Portland cement of grade 52.5;

naphthalene series superplasticizer;

ultrafine mineral powder, 8000 cm^2/g;

river sand, medium sand;

coarse aggregate: A, hard sandstone gravel; B, quartz schist gravel; C, artificial light aggregate.

Test program: factors and levels of experimentation are shown in Table 5.3.

Table 5.3 Factors and levels of concrete testing

Factor	level			
	1	2	3	4
W/C %	20	25	35	65
Type	A	B	C	/
Amount	0	200	400	/
Max size(mm)	10	15	20	/

Test results

The relationship between the volume content of coarse aggregate V_g, the maximum particle size D_{max}, the mortar strength F_m and the corresponding concrete strength F_c

The influence of V_g on concrete strength F_c is shown in Figure 5.12 and the influence of D_{max} on concrete strength Fc is shown in Figure 5.13.

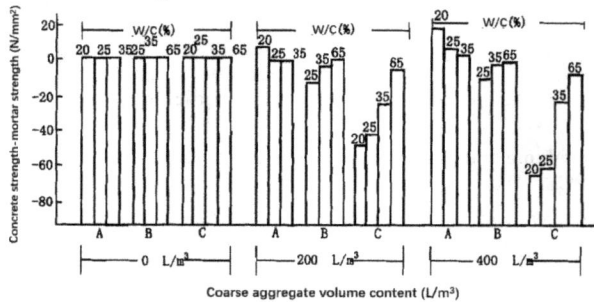

Figure 5.12 Effect of aggregate volume on the compressive strength.

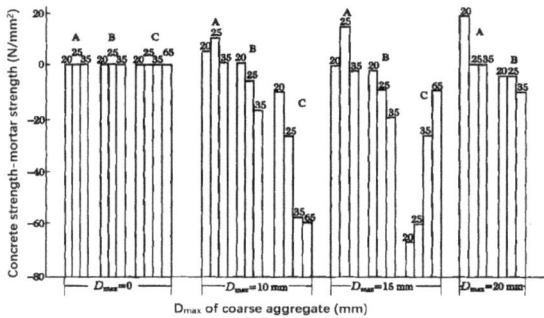

Figure 5.13 Effect of Max aggregate size on the compressive strength.

It can be seen from Figure 5.12 and Figure 5.13 that, in ordinary concrete with W/C=0.65, when the volume content and maximum particle size increase, the strength of the concrete decreases; but as the strength of the concrete increases, for coarse aggregate A, the strength of concrete increases with the increase of Vg and D_{max}, while for coarse aggregate B and C, the situation of ordinary concrete with W/C=0.65 is similar to that of coarse aggregate A. However, when W/C is from 0.25 to 0.35, with the increase of V_g and D_{max}, the compressive strength decreases.

5.5.2 Mathematic models for the influence of aggregate on concrete strength F_c

In ordinary concrete (W/C=0.65), regardless of the type of coarse aggregate, with the increase of V_g and D_{max}, the concrete strength decreases. But with the decrease of W/C: for type A, the strength of concrete with coarse aggregate A changes from low to high; B, the strength of concrete with coarse aggregate B and C drops sharply. The mathematic models based on these results are given in Figure 5.14 and Figure 5.15 respectively.

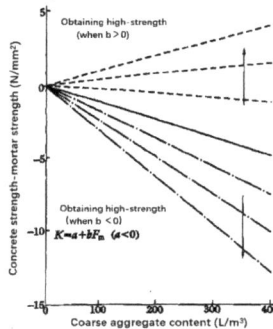

Figure 5.14 Effect of V_g on the concrete (mortar) strength.

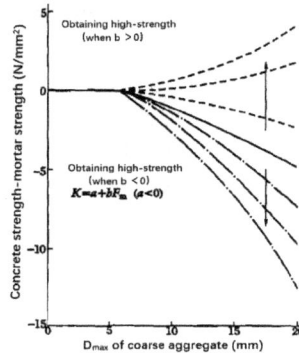

Figure 5.15 Effect of D_{max} on the concrete (mortar) strength.

Figure 5.14 shows the modelling of the effect of V_g on concrete (mortar) strength F_c. When D_{max} is a fixed value, the concrete strength Fc, mortar strength Fm will increase or decrease with Vg. The proportional constant k is the function of the mortar strength F_m. The influence of Vg on compressive strength can be expressed by equation 5.5.

$$Fc1\text{-}Fm = K. \ V_g = (a + b. \ Fm). \ V_g \qquad\qquad 5.5$$

where: Constant $K = a + b. \ F_m$,

a and b are the parameters related to the physical properties of mortar by coarse aggregate, but a<0.

Figure 5.15 shows the model of the influence of D_{max} on the strength of concrete F_c. When D_{max} is 5 mm, there is no effect of particle size, since the mixture with D_{max} of 5 mm is mortar. When D_{max} exceeds 5 mm, its influence on strength becomes an exponential relationship. The effect of D_{max} on strength is the same as that of Vg. There are enhancement effects and negative effects, as shown in the following equation.

$$Fc2 - Fm = c. \ K. \ (D_{max}\text{-}5) \ d = c. \ (a + b. \ Fm) \ (Dmax - 5) \ d \qquad\qquad 5.6$$

where c is the ratio of the influence of coarse aggregate on the strength to the influence of D_{max} on the strength.

Therefore, the influence of Vg and D_{max} on the strength, when comprehensively considered, is as follows:

$$Fc\text{- } Fm = (a + b . \ Fm). \ [Vg + c.(D \ max - 5) \ d \ ]$$

$$= K . \ [Vg + c \ (D \ max - 5) \ d \ ] \qquad\qquad 5.7$$

where: a is a negative value, and mortar and aggregate have the same elastic modulus, a is 0.

b represents the internal flaw sensitivity parameter, which is affected by W/C. In HPC and UHPC, the weak link is the coarse aggregate and the interface between the coarse aggregate and the mortar. When both are strong, b is a positive value; when one of them is a weak link, b is a negative value. For example: for HPC with mineral containing ultrafine powder, the interface layer is strengthened, and the interface strength between the coarse aggregate and the mortar is increased. If the strength of the aggregate is higher than that of the mortar, the strength of the mortar is increased by adding coarse aggregate, and b is a positive value.

c is the proportional parameter of the influence of D_{max} on the strength of concrete.

K is the correlation coefficient of increase in compressive strength (b>0); or the correlation coefficient of decrease in compressive strength (b<0). The relationship between mortar strength and concrete strength is given in Figure 5.16. It can be seen that the strength of aggregate has a considerable influence. Under a certain water/binder ratio, the strength of concrete>the strength of mortar.

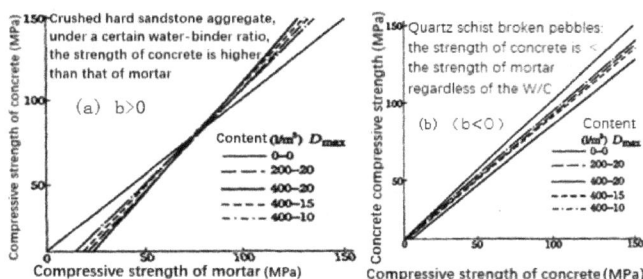

Figure 5.16 Relationship of mortar strength and concrete strength.

5.6   Influence of coarse aggregate on the durability of concrete

Durability is an important performance of HPC and UHPC, and is also an important basis for the design of HPC and UHPC. In this chapter, only the influence of coarse aggregate on the frost resistance and drying shrinkage of concrete is described. The durability of other aspects of HPC and UHPC will be discussed in other chapters.

5.6.1   Frost resistance

Generally, the water/cement ratio of HPC and UHPC is relatively low (W/C<35%), even if no air-entraining agent is added. Its frost resistance is also very good. However, if aggregates with high water absorption are used, even if the water/cement ratio of concrete is relatively low, the frost resistance may be not so good, as shown in Figure 5.17. The water absorption of different aggregates is as shown in Table 5.4.

Table 5.4 Water absorption of different aggregates.

No	Type	Water absorption (%)	No	Type	Water absorption (%)
A	Hard sand stone	0.78	a	River sand	1.69
B C	Andesite	1.49(B) 2.44(C)	b	River sand	3.80
D	Andesite	2.30	c	Crushed rock	4.90

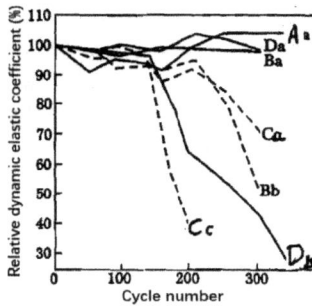

Figure 5.17 Effect of coarse aggregates combined with various fine aggregates on concrete durability.

It can be seen that the concrete (W/C=0.30) with the combination of coarse and fine aggregates Aa, Da, and Ba, although after 300 freeze-thaw cycles, the relative dynamic elastic modulus was still fluctuating around 100%. However, the combination of coarse and fine aggregates of Db, Ca, Bb, Cc concrete (W/C=0.30), after 150-200 freeze-thaw cycles, the relative dynamic elastic modulus dropped rapidly. This reveals the significant influence of the water absorption of the aggregate listed in Table 5.4. Therefore, when preparing HPC and UHPC, attention should be paid to choosing aggregates with low water absorption.

5.6.2   Properties of drying shrinkage

Aggregate has a remarkable influence on the shrinkage of HPC and UHPC. Concrete with coarse or fine aggregate of high-water absorption tends to have high shrinkage value, and the shrinkage of concrete with limestone aggregate is low.

HPC and UHPC have high early shrinkage; but due to the low W/C, the long-term shrinkage value may be the same as or even lower than that of ordinary concrete. According to the concrete code, the total shrinkage of concrete is 5/10,000-7/10,000. If the shrinkage is too large, the concrete may encounter cracking on the surface and this will cause neutralization or accelerated penetration of aggressive ions, which will further cause the corrosion of steel bars. As to HPC and UHPC, attention should be paid to early shrinkage and autogenous shrinkage.

5.7   Selection of maximum particle size of coarse aggregate

When preparing ordinary concrete, the size of aggregates should be used as large as possible to reduce the water consumption of concrete, or to improve the workability of concrete under the same water consumption. However, in HPC and UHPC, how to choose the maximum particle size D of coarse aggregates?

According to H.M. Jennings's recommendation, high-strength aggregates should be selected for the preparation of HPC and UHPC, the maximum particle size D max smaller than 10 mm, and the particle size distribution should be beneficial to form a densely packed state of the aggregate particles. Using coarse aggregate with a small maximum particle size D max, the interface between aggregate and cement paste becomes thinner, and it may be difficult to induce major flaws at the interface. The aggregates in a densely packed state have a low voidage, and the amount of cement paste can be reduced, which is beneficial to strength and durability.

The strength of concrete basically belongs to the failure strength of solid materials, which can be analysed by Griffith theory. For a penetrating elliptical crack in an infinite elastic plate under uniform tension, the following equation is obtained:

$$\partial 0 = \sqrt{2E\gamma /\pi a} \qquad\qquad\qquad 5.8$$

where: E—modulus of elasticity.

γ—surface energy.

$\partial 0$—fracture stress.

2a—the long diameter of an elliptical hole which represents a latent flaw in material.

Equation 5.8 is about the two-dimensional elastic plate model, when it is extended to the three-dimensional elastic plate model, it can be expressed by the following equation:

$$\partial 0 = \sqrt{\pi\, E\gamma /2(1-v)a} \qquad\qquad\qquad 5.9$$

where, v—Poisson's ratio of the material; other symbols have the same meaning as in equation 5.8.

In equation 5.5, if $E=4.5 \times 10^4$ MPa, $\gamma=10 \times 10$ erg/cm^2, $v=0.20$, and the potential flaw size (radius of the elliptical hole.) a can almost be regarded as the D_{max} of the aggregate; set $D_{max} =10$ mm, the corresponding fracture stress in this case is $\partial 10$, and the fracture stress at $D_{max}=20$ mm is $\partial 20$.

Then $\partial 20 = 0.7\, \partial 10$. That is, with the increase of the aggregate particle size, the fracture stress decreases. Therefore, the D_{max} of the coarse aggregate in HPC and UHPC should be reduced as much as possible. it is generally considered that the maximum size should be not more than 20 mm, preferably not more than 10 mm. Our research work has proved that, when D_{max} exceeds 20 mm, K_{1c} decreases as the particle size increases.

5.8   Fineness modulus and quality factor of coarse aggregate

5.8.1   Fineness modulus of coarse aggregate Mz

The fineness modulus M_z of coarse aggregate is also an important indicator for evaluating the quality of coarse aggregate. The calculation method is similar to that of fine aggregate by using the sieves of size 40, 20, 10, 5, 2.5, 1.25, 0.63, 0.315 and 0.16 mm. The coarse aggregate is sieved and the cumulative sieve residue for each sieve is calculated, as shown in Table 5.5, and the M_z is calculated according to the following equation:

$$M_z = [(A2+A3+A4+A5+A6+A7+A8+A9)-6A1] / (100-A1) \qquad (5.6)$$

Table 5.5 Sieve analysis of aggregate.

Sieve (mm)	Cumulated sieve residue (%)	Sieve (mm)	Cumulated sieve residue (%)
40	$A_1 = a_1$	1.25	$A_6 = a_1 + a_2 + a_3 + a_4$
20	$A_2 = a_1 + a_2$	0.63	$A_7 = a_1 + a_2 + a_3 + a_4$
10	$A_3 = a_1 + a_2 + a_3$	0.315	$A_8 = a_1 + a_2 + a_3 + a_4$
5	$A_4 = a_1 + a_2 + a_3 + a_4$	0.16	$A_9 = a_1 + a_2 + a_3 + a_4$
2.5	$A_5 = a_1 + a_2 + a_3 + a_4$		

Because the particle size of the coarse aggregate is larger than 5 mm, the cumulative sieve residue of the sieve with d smaller than 5 mm is the same.

The fineness modulus M_z of coarse aggregate is generally from 6 to 8.

5.8.2 Quality factor of coarse aggregate K

The quality coefficient K is an index for comprehensively evaluating the quality of concrete aggregates. It can be used either for fine aggregates or for coarse aggregates.

$$K = M z (50-p) \qquad (5-7)$$

where: K-quality factor of aggregate.

M z-fineness modulus of aggregate.

p-Percentage of aggregate voidage.

The quality factor K of the aggregate is related to the fineness modulus M z of the aggregate and the voidage P. The large K value means the high quality of the aggregate and good gradation.

5.9 The selection of aggregate for HPC and UHPC

According to the current widely-recognized requirements for HPC, its strength grade should be above C60, and its service life should be more than a hundred years. According to this goal, the selection of aggregates must consider the following issues.

5.9.1 Grading of aggregate

The well-graded aggregate has low voidage, low cement paste consumption, low shrinkage and deformation of concrete, low hydration heat and good volume stability, which are good for both strength and durability of concrete. Therefore, the quality of the aggregates used for HPC and UHPC should be comprehensively evaluated, and the quality factor K of the aggregate should be used.

5.9.2   Physical properties of aggregate

Aggregate should have high apparent density (>2.65) and loose bulk density (>1450 kg/m^3). The water adsorption should be low (about 1.0%). This kind of aggregate has low voidage and high compactness. It also requires particles of good shape, and fewer needle and flake particles. It can reduce the amount of cement paste and improve workability and strength of concrete. Crushed limestone or crushed hard sandstone is commonly used with particle size $\leq$ 20 mm. For UHPC, crushed andesite or diabase rock should be selected, with particle size $\leq$ 10 mm.

5.9.3   Mechanical properties of aggregate

Aggregate should not contain weak particles or weathered particles. It is stipulated in China building materials industry standard JGJ153 that the compressive strength of aggregate rock should be 1.5 times of the compressive strength of concrete. The rock strength should be tested with 50 mm cube specimens or ϕ50x50 mm cylinders in a water saturated state. And the strength should be not less than 80 MPa. The aggregate crushing value should be less than 10%. The elastic modulus of concrete has the following relationship with the elastic modulus of aggregate:

$$Y = 2.50 + 0.20 \ X \qquad\qquad\qquad (5.8)$$

where: Y – elastic modulus of concrete

X – elastic modulus of aggregate

It can be seen that the greater the elastic modulus of the aggregate, the greater the elastic modulus of concrete. Therefore, it is necessary to choose an aggregate with a higher elastic modulus.

5.9.4   Chemical properties of aggregate

Inactive aggregate should be first selected. Aggregate should not contain any clay lump and clay content should be less than 1.0%. It should not contain impurities such as organic matter, sulphide and sulphate.

5.10   Current status and problems of aggregate

At present, China's annual commercial concrete output is about 2 billion m^3, which consumes a huge amount of coarse and fine aggregate. The aggregates used to prepare HPC and UHPC are of even greater concern, i.e., on the one hand, the resources of aggregate are depleted, on the other hand, normal aggregate is of relatively low quality for preparing HPC or UHPC.

5.10.1   Resource of aggregate

Taking the Pearl River Delta area in China as an example, since the reform and opening up of China, the main sources of sand for construction have been from the so-called Sanjiang area, i.e. Dongjiang, Xijiang and Beijiang. Sometimes a certain amount of sea sand is mixed with the river sand. But the resources of the river sand of the Sanjiang area have been basically exhausted after more than 30 years of excavation, and the employment of sand for construction has been transferred to large-scale development of washed sea sand. However, the sea sand resources have also been almost used up after excavation for 10 years. Coarse aggregates have also almost entered the scenery of no further supply. Since most of the coarse aggregates are made of limestone, some green mountains have been severely damaged, and the excavation of natural aggregates has also brought serious pollution to the environment. The government in the Pearl River Delta area ordered the closure of the mountain to stop further excavation. Therefore, the price of aggregates has skyrocketed. It was originally 80-85 yuan (RMB)/m^3, but now the price has been risen to 125 yuan/m^3, and the supply is rather limited. In this case, the solution of the problem of aggregate resource is of the top priority for concrete materials.

5.10.2   Quality problem of aggregate

At present, the quality of aggregates sold in the aggregate market in China is not so satisfactory, with too many needle or flake particles, high voidage, and size generally larger than 25 mm, or even more than 30 mm. It is difficult to use such type of aggregate to prepare HPC and UHPC. Although there are also some good quality aggregates with good shape in the market produced with impact crusher, the price is quite high.

5.10.3   Alkali-aggregate reactivity

Some limestone aggregates in northern China, such as the limestone in the south of Beijing, the limestone quarry in Weifang, Shandong province, and the aggregate of a limestone quarry in Tianjin, contain active SiO_2, or clay limestone, which can react with the K_2O and Na_2O in cement to result in alkali-silica reaction or alkali-carbonate reaction, causing structural damage. For example, alkali-aggregate reaction has been found in Beijing's Siyuan overpass, the original Xizhimen bridge and Tianjin Balitai overpass, etc., where damages to the piers and beams were found. The high cement content in HPC and UHPC may cause a high alkali content in the concrete such as over 3 kg/m^3. If the alkali-reactive aggregate is used, there may be a risk of alkali-aggregate reaction. Therefore, the alkali-reactivity of the aggregate must be strictly tested.

5.10.4   Chloride content in sand

Due to the shortage of river sand in some cities in China, sea sand is often used in concrete. If the chloride ion brought by sea sand into the concrete exceeds 0.3 kg/m^3, the steel bars in the concrete may be corroded, which may further cause cracking of concrete structures, as shown in Figure 5.18.

(a) Cracking caused by steel corrosion

(b) Spalling caused by steel corrosion

Figure 5.18 Steel corrosion and cracking of concrete structures.

Therefore, sea sand should be fully washed in water and used in combination with river sand or manufactured sand to ensure safety. For example, the C50 concrete used in the prestressed reinforced concrete box girder of a sea-crossing bridge in Okinawa, Japan, was made of 50% washed sea sand and 50% manufactured sand, as shown in Figure 5.19.

Figure 5.19 Prestressed reinforced concrete box girder with washed sea sand and manufactured sand.

The chloride ion content of the sand should not exceed 0.1% as specified in the specification. The total chloride ion content in the concrete should not exceed 0.3 kg/m^3, which can ensure the safety of reinforced concrete structures.

Chapter 6

Development and Application of New Types of Superplasticizers

6.1   Introduction

The development and application of new type of superplasticizer is the material basis of UHPC. When superplasticizer is used in concrete, it can greatly reduce the water consumption of concrete when the slump is maintained at a certain value; or the slump of concrete can be greatly increased when the water consumption of concrete is unchanged. In 1962, a naphthalene-based superplasticizer was invented in Japan. This greatly promoted the Japanese concrete technology. With the features of dispersibility, retardation and low air-entraining properties of naphthalene-based superplasticizer, the production of 100 MPa prestressed pipe piles and railway bridge trusses, etc. became possible. This made the span of a bridge larger and noise-free. The research in China on the naphthalene-based high-range water-reducing agent started only in 1974, e.g., the work by Zhang Lu of Tsinghua University, and by Dayu Xiong and Dezhen Fang of the China Institute of Metallurgical Building Science. Around 1980, China's naphthalene-based high-range water-reducing agent was put into production and application.

On the other hand, around 1971, melamine-based superplasticizer was developed in Germany, with which a new flowable concrete was developed. In 1975, Japan introduced the flowable concrete technology and put it into practice in Japan. In 1976, the British Concrete Society summarized these high-range water reducing agents developed in Japan and Germany into the state-of- art report with a term of superplasticizer. Since then, the term of superplasticizer has spread all over the world.

6.2   The classification of superplasticizers

According to the chemical structure, superplasticizers can be divided into five categories: naphthalene series, melamine series, sulfamate series, and polycarboxylic acid series including acrylic acid series and maleic acid series, as shown Figure 6.1. There are 3 types of polycarboxylic acid superplasticizer: (a) acrylic, (b) polyether, (c) maleic acid. The water reduction rate of these superplasticizers is higher than that of other superplasticizers. Thus these types of superplasticizers are usually used in the preparation of UHPC.

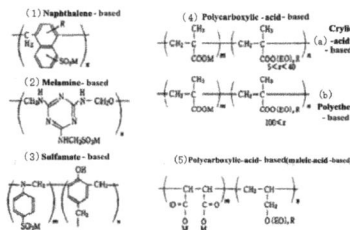

Figure 6.1 Types and structures of superplasticizers

6.3   The Water reducing mechanism of superplasticizer

6.3.1   Dispersion mechanism

When the superplasticizer is mixed into the cement paste, it can disperse the cement particles and release the free water in it to make the cement paste flow easily.

Figure 6.2 shows the relationship between the content of naphthalene-, melamine- and polycarboxylic acid-based superplasticizers and the Zeta potential. With the increase of the content of naphthalene- or melamine-based superplasticizer, the Zeta potential increases, indicating the effect on the dispersibility of cement particles increases with the increase of the dosage. However, when the dosage reaches a certain range, the Zeta potential tends to be stable and no longer increases. The Zeta potential of the polycarboxylic acid superplasticizer is relatively low, equal to only about 50% of the other types, but its dispersibility to cement particles is much higher than that of the former two. This needs to be explained from the adsorption form between the molecules of the superplasticizer and cement particles.

After the superplasticizer was added into the cement paste, the various adsorption forms of the polymer at the interface between the solid phase and the liquid phase are shown in Figure 6.3.

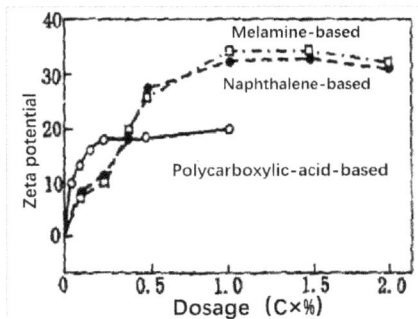

Figure 6.2 Relationship between the content of superplasticizer and Zeta potential

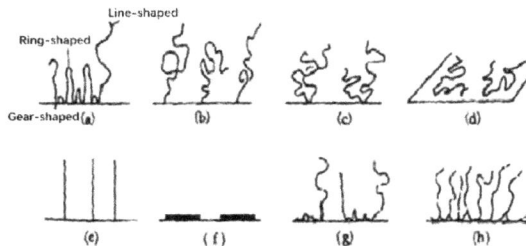

(b) End adsorption, lead type; (c) Point adsorption, 2 leads type; (d) Plane adsorption; (e) Vertical adsorption; (f) Rigid chain horizontal adsorption; (g) Gear type, lead type; (h) Copolymer gear type adsorption

Figure 6.3 Form of superplasticizer molecules adsorbed on the surface of cement particles

This difference of adsorption form on cement particles may be responsible for the influence of superplasticizer on performance and the control of slump loss. The adsorption form of naphthalene series and melamine series on cement particles is shown in Figure 6.3 (f) and the adsorption form of polycarboxylic acid superplasticizer on cement particles is in Figure 6.3 (h). The adsorption form of sulfamic acid-based superplasticizer on cement particles is shown in Figure 6.3 (a) or (g), which is a ring type or line type. The adsorption layer formed on the surface of cement particles is shown in Figure 6-4 (a), and forms a polyline-shaped density distribution, as shown in Figure 6-4 (b). From the straight line to the ring layer at the junction of the line segments, the density changes greatly. The repulsive force between the electrostatic force fields is three-dimensional and has a more considerable dispersion effect.

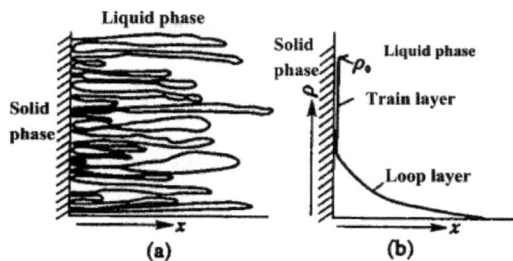

Figure 6.4 Adsorption layer of superplasticizer at the solid-liquid interface.

With the absorption of superplasticizer molecules by cement particles, the full energy curve of the interaction between the particles is shown in Figure 6.5.

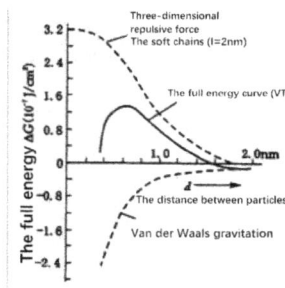

Figure 6.5 Three-dimensional full energy curve of cement particles after adsorbing superplasticizer molecules.

The sum of the three-dimensional repulsive force and van der Waals gravitation is the full energy curve, which is the moving force between particles. When the sum is positive, the particles are in a dispersed state, whereas, when the sum is negative, the particles are in a cohesive state.

Therefore, it is understandable that such a mechanism has an effect similar to that if the DLVO theory, but due to its three-dimensional dispersion effect, the dispersion performance is more remarkable.

6.3.2 Maintaining dispersion mechanism

The maintaining dispersion mechanism means the plasticity retention of mortar or concrete mixed with superplasticizer. The relationship between the time-dependent change of cement paste and Zeta potential after mixing is shown in Figure 6.6. The time-dependent change of concrete slump is shown in Figure 6.7.

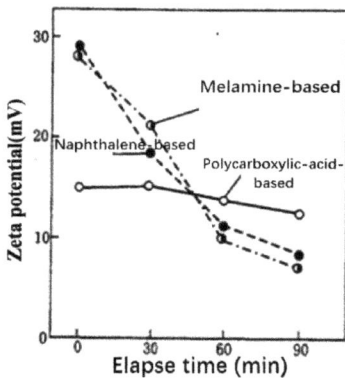

Figure 6.6 Change of cement slurry and change of Zeta potential.

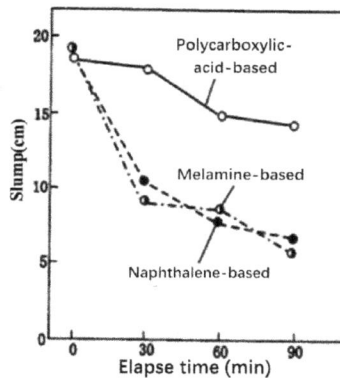

Figure 6.7 Change of concrete slump with time.

Figure 6.6 has an obvious correlation with Figure 6.7. The Zeta potential of naphthalene and melamine superplasticizers decrease rapidly with time, while the slump of concrete prepared with these two superplasticizers is lost with time very quickly, but the Zeta potential of the polycarboxylic acid-based superplasticizer decreases rather little over time, and thus the slump loss is very small over time. This is mainly because the molecules of the superplasticizers and the adsorption type of cement particles are different, and the force of the adsorption layer between the cement particles is also different. The force of the polycarboxylic acid superplasticizer and the adsorption layer of the cement particles is a three-dimensional electrostatic repulsion, and the zeta potential changes little. The function of plasticity maintaining is good.

If components that can maintain dispersion are added to this polycarboxylic acid-based superplasticizer, for example, a polymer compound that is soluble in alkali but insoluble in water, the loss of slump may be more effectively controlled.

6.4 Air-entraining superplasticizer (high-performance AE superplasticizer)

After the superplasticizer is added into the concrete, in addition to the effect of high-efficiency water reduction, it can also bring in certain bubbles, reduce the viscosity of fresh concrete, and have a higher water reduction rate than that of ordinary air-entraining water-reducing agents. It can have good plastic retention function and improve durability as well. The relationship between various water reducing agents and water reduction rate is shown in Figure 6.8.

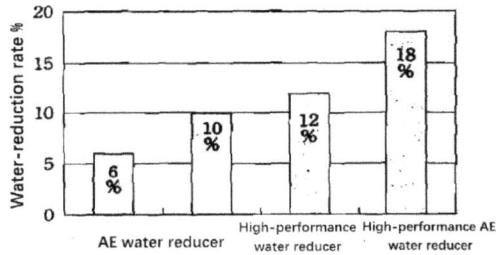

Figure 6.8 Various water reducing agents and its water reduction rate (JIS 6204:2006).

The high-performance AE water reducing agent has a high water reduction rate and good plasticity retention (water reduction rate is more than 18%, and the slump remains unchanged for more than 60 minutes), so it has quickly been promoted and applied and popularized in the market. It is widely used in ordinary concrete, high-strength concrete, high-flow concrete, high-durability concrete, mass concrete, porous concrete, shotcrete and underwater concrete, etc.

6.4.1 Types of high-performance AE (entrained air) superplasticizer

The four representative types of high-performance AE (air-entraining) superplasticizers are naphthalenesulfonate formaldehyde condensate, melamine sulfonate formaldehyde condensate, sulfamate formaldehyde condensate, and polycarboxylic acid series. The basic structural formula is shown in Figure 6.1. The development and application of various high-performance AE superplasticizers in Japan are shown in Figure 6.9. Polycarboxylic acid-based air-entraining superplasticizer has about 80% of market share.

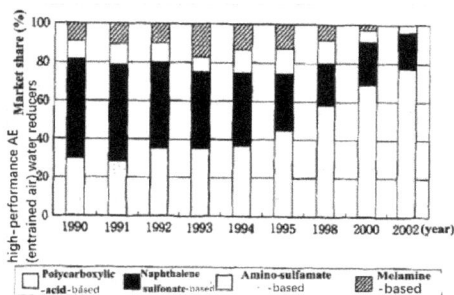

Figure 6.9 Development of different high performance AE superplasticizers in Japan.

6.4.2   Function and effect of high-performance AE (air-entraining) superplasticizer

The structure of the high-performance AE (air-entraining) superplasticizer is shown in Figure 6.1. The main components in its molecule are a long main chain and functional groups, as well as a side-chain structure. The polycarboxylic acid air-entraining superplasticizer is shown as Figure 6.1 (4) and (5), where the main chain plays a role of adsorption, and the side chain forms a three-dimensionally dispersed protective film in water.

The water-reducing effect of the air-entraining superplasticizer is to disperse the cement particles, so that the water constrained between the cement particles becomes free water, and the fluidity increases. The stability of the cement particles after being dispersed is due to both the electrostatic repulsion between the particles and the repulsive force of the three-dimensional barrier. The dispersion and stability of the cement particles by the electrostatic repulsion can be explained according to the DLVO theory. The dispersion of cement particles by Naphthalene-based and melamine-based superplasticizer can be explained by the theory of electrostatic repulsion of DLVO.

The dispersion stability of stereoscopic barriers can be explained based on the theory of Mackor's information superposition effect. The repulsive force of stereoscopic barriers can be calculated based on the accumulated information of the structure of the surfactant, the adsorption form and the thickness of the adsorption layer. The polycarboxylic superplasticizer is adsorbed on the surface of cement particles, and the carboxylic acid-based particles are negatively charged, which hinders the contact of the cement particles and hence the stereoscopic repulsive force is formed, so that the cement particles are further dispersed. It can be seen from Figure 6.10 and Figure 6.11 that the adsorption amount of polycarboxylate acid-based superplasticizer is less than that of naphthalene-based and melamine-based superplasticizer, and the electrostatic repulsion is also smaller. Figure 6.12 illustrates the relationship between the dosage of different superplasticizer and the water reducing rate. The dosage of polycarboxylate acid superplasticizer is very low, but the water reducing rate is very high.

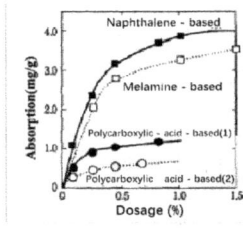

Figure 6.10 Types and absorption of air-entraining superplasticizer.

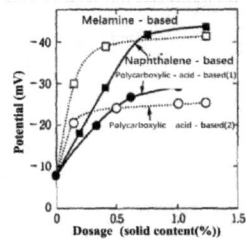

Figure 6.11 Type and Zeta potential of air-entraining superplasticizer.

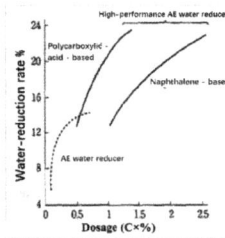

Figure 6.12 Dosage and water reduction rate of air-entraining superplasticizer.

6.4.3   Performance of concrete with high-performance AE superplasticizer

Function to slump loss control

The biggest difference between superplasticizer and other water-reducing agents is that it has an excellent function of slump loss control, as shown in Figure 6.13.

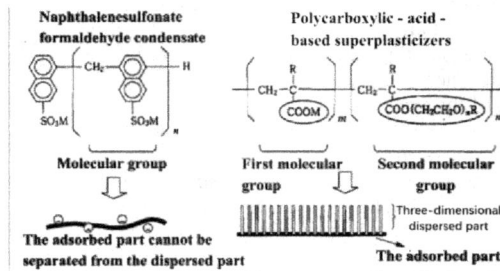

Figure 6.13 Molecular structure of naphthalene and polycarboxylic acid superplasticizers and their adsorption state with cement.

Naphthalene-based and melamine-based superplasticizers are structurally unable to reduce the adsorption rate of cement particles, while polycarboxylic acid-based superplasticizer can adjust their molecular structure, so that the ratio of the first molecular group (m) to the two molecular groups (n) of polycarboxylic acid-based superplasticizer can be changed. The adsorption speed can be changed freely, and the slump loss can be easily controlled, as shown in Figure 6.14 and Figure 6.15 respectively.

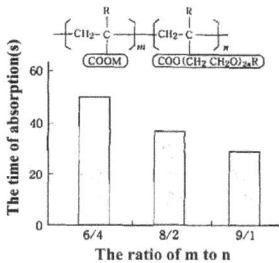

Figure 6.14 Influence of m/n ratio on the adsorption rate.

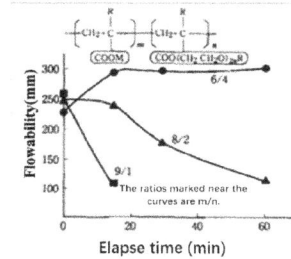

Figure 6.15 Influence of m/n ratio on flowability.

As shown in Figure 6.15, when the m/n ratio = 4/6, the effect of flowability control is very obvious, and there is no change within the first 60 minutes. The air-entraining superplasticizer can achieve a very good slump control of fresh concrete at a low dosage.

Setting time

The concrete with polycarboxylate-based superplasticizer has no retardation effect. As shown in Figure 6.16. When the dosage of carboxylate-based superplasticizer is 0.5 %-1.0 % of the cement content, the initial setting time is 7-9 hours, and the final setting time is 10-12 hours, which is the same as that of a mixture with ordinary air-entrained water reducing agents.

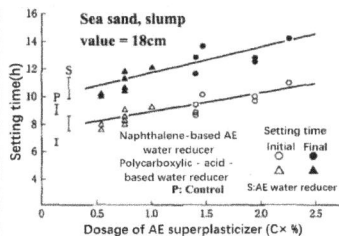

Figure 6.16 Dosage and setting time of high-efficiency air-entraining superplasticizer.

Compressive strength

The basic law of compressive strength obeys the law of water/cement ratio. Under the same water-cement ratio, the compressive strength of concrete mixed with air-entraining superplasticizer is the same as that of concrete with ordinary air-entraining water-reducing agent, as shown in Figure 6.17.

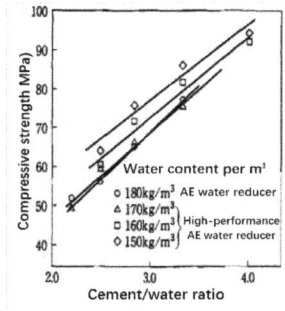

Figure 6.17 Relationship between concrete compressive strength and cement-water ratio of high-efficiency air-entraining water-reducing agent.

However, when the water content is reduced from 180 kg/m^3 to 150 kg/m^3, even if the water/cement ratio is the same, the compressive strength of concrete mixed with air-entraining superplasticizer is higher than that of concrete with other water-reducing agents.

Frost resistance

Generally, the air content of concrete that meets the requirements of frost resistance is 4-6%, and the air content of the concrete with air-entraining superplasticizer is the same. Figure 6.18 shows the frost resistance of concrete with air-entraining superplasticizer. All the concretes with different W/C ratios have excellent frost resistance.

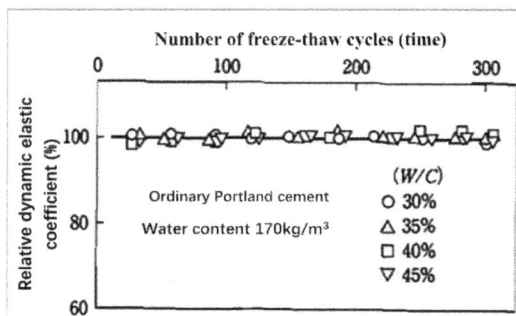

Figure 6.18 Freezing resistance (at 4% air content).

6.5　Sulphamate-based superplasticizer

The sulfamate-based superplasticizer has a high water reduction rate, good slump retention effect, and it can help concrete have good durability. The production process is simpler than that of naphthalene and polycarboxylate superplasticizers. The author once blended sulfamate-based superplasticizer with 35% solid content with naphthalene superplasticizer also with 35% solid content) at a ratio of 1:1 in liquid form, with further addition of 3-5% ultrafine powder, to produce a new type of superplasticizer. This superplasticizer could keep slump retention for 3 hours for C120 ultra-high-performance concrete. The concrete was applied to Shenzhen Jingji Building project, and pumped to a height of 510 m. Thus, it is a promising superplasticizer. The author has produced and applied this superplasticizer in Weifang of Shandong province and Shenzhen.

6.5.1　Production principle

It is formed by heating and condensing aromatic sulfamate and formaldehyde. The molecular formula of its main product is shown in Figure 6.19:

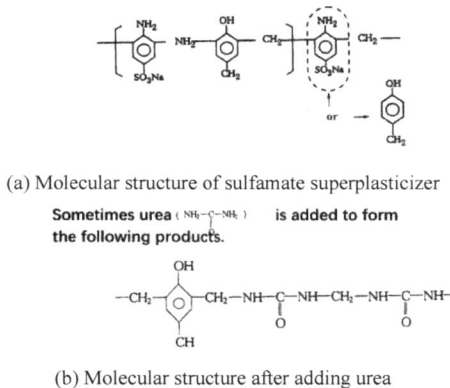

(a) Molecular structure of sulfamate superplasticizer

(b) Molecular structure after adding urea

Figure 6.19 Molecular structure of sulfamate superplasticizer.

Its molecular structure is characterized by more branches, shorter hydrophobe molecular segments, and strong polarity.

6.5.2　Tests on cement paste and concrete

Raw materials

Sulfamic acid series water reducer (AS); and its combination with naphthalene-based superplasticizer AN1 and AN2;

Cement: Onoda 52.5 OPC (1); Dongfanglong 52.5 OPC (2); Daewoo 52.5 OPC (3); and Shaofeng 52.5 OPC (4);

Sand: medium to coarse, density 2.65, apparent density 1.45; grading and organic content meet the requirements of relevant standards;

Aggregate: crushed limestone, Particle size 5-25mm, density 2.65, apparent density 1.46; qualified grading.

Other admixture: preservative (produced in Tianjin); fatty acid water reducer (produced in Nanning), with a solid content of 40 %.

Flowability of cement paste

Under a condition of cement 500 g, W/C of 0.29, and admixture content of 0.7 %, the flowability of various cement pastes with different raw materials and the change of the flowability over time are shown in Tables 6.1, 6.2, and 6.3 respectively.

Table 6.1 Flowability of cement paste and its changes over time (AS)
(Mix proportion: cement 500g, water 145ml, AS 3.5g).

Cement type	Flowability (cm)				
	Initial	30 min	60 min	90 min	120 min
1	25 × 25	25 × 25	25 × 25	26 × 26	25 × 25
2	24 × 24	25 × 25	25 × 25	25 × 25	24 × 24
3	24 × 25	25 × 25	24 × 25	24 × 24	24 × 24
4	26 × 25	25 × 25	25 × 25	25 × 25	24 × 25

Table 6.2 Flowability of cement paste and its changes over time (AN2)
(Mix proportion: cement 500g, water 145ml, AN7.5g)

Cement type	Flowability			
	initial	60 min	120 min	180 min
1	26 × 26	26 × 26	26 × 26	/
2	27 × 26	26 × 26	26 × 26	/
3	25 × 25	25 × 25	25 × 25	24 × 24
4	26 × 26	26 × 26	26 × 26	25 × 25

Table 6.3 Flowability of cement paste and its changes over time (AN1)
(Mix proportion: the same as in Table 6.2)

Cement type	Flowability (cm)			
	Initial	60 min	120 min	180 min
1	26 × 26	26 × 26	26 × 26	/
2	28 × 28	27 × 28	26 × 26	/
3	25 × 25	26 × 26	24 × 24	25 × 26
4	28 × 28	27 × 27	25 × 25	/

It can be seen from the cement paste test: the three sulfamic acid series water reducers, AS, AN1, AN2, all have good compatibility to four types of cement; the initial flowability was high, i.e., after 2 hours, the flowability of the net paste was basically maintained. The water reduction rate is high, and the function of controlling the loss of flowability is good. This is one of the characteristics of sulfamic acid series water reducers. The dosage of AN1 and AN2 is too high, which is more than 2 times of that of AS. This is because these two water reducers are composed of amino series and naphthalene series.

Concrete test

The mix proportion of concrete is shown in Table 6.4, where the solid content of AN1 was 37%, and the solid content of AS was 42%. In the concrete test, the water content in the water reducing agent should be deducted from concrete mix. The mineral admixtures are slag powder and fly ash in the mixture (1), grade II fly ash in (2), and slag powder and silica fume in (3).

Table 6.4 Mix proportion of concrete.

Type	Cement	W/B	Raw material of concrete (kg/m3)						Admixture
			Cement	Mineral powder	Sand	Chipp ing	Crushed rock	Water	
1	Onoda	0.40	300	140(1)	760	/	1000	180	AS0.7%
2	Daewoo	0.43	340	75(2)	800	150	850	180	AN12.0%
3	Zhujiang	0.30	385	165(3)	750	150	850	165	AS 3.0%

The change of slump and flow of concrete with time is as shown in Table 6.5, and the strength, electric flux and chloride ion diffusion coefficient are shown in Table 6.6. It can be seen from Table 6.5 that the slump and flow had no change within the first 60 min. After 120 min, the loss of slump and the loss of flow were also very small. In Table 6.6, the concrete with Onoda Cement 300 kg+140 kg composite mineral admixtures, and W/B=0.40, had 28d strength up to 60 MPa and excellent flowability. There is basically no slump loss in the first 2 hr. The electric flux of the three types of concrete at 56 d is lower than 750 coulomb/6h, and the chloride ion diffusion coefficient at 56 d is lower than 6.2676 x 10^{-9} cm²/s. This is a high-durability concrete.

Table 6.5 Change of slump and flow of concrete.

Cement	Initial	60 min	120 min
Onoda	24.5/680	24.5/630	23.0/480
Daewoo	24.0/630	21.0/620	20.0/560
Zhujiang	23.0/580	22.0/560	22.0/540

Table 6.6 Strength, electric flux and chloride ion diffusion.

Cement	Compressive Strength (MPa)			Electric flux (C)		CL-diffusion($\times 10^{-9}$cm²/s)	
	3 d	7 d	28 d	28 d	56 d	28 d	56 d
Onoda	35.6	46.6	60.7	2605	700	15.3943	6.0216
Daewoo	32.5	42.7	58.5	2760	750	16.1568	6.2676
Zhujiang	56.4	64.8	81.2	1605	495	10.4743	5.0131

The slump of concrete with different water-reducing agents (three water-reducing agents: amino-based, naphthalene-based, melamine-based) is shown in Figure 6.20. After 60 minutes, the concrete slump of naphthalene-based or melamine-based was reduced from the original 20 cm to about 5 cm, while the concrete slump of the amino-based water-reducing agent can remain basically unchanged for 1.5 h.

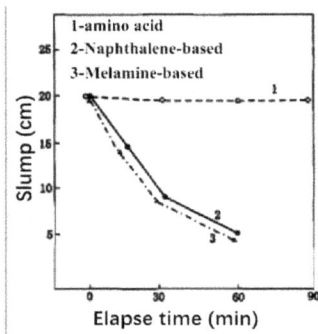

Figure 6.20 Change of slump of concrete with different superplasticizers.

It can be seen from Figure 6.20 that the sulfamate superplasticizer has a good effect in controlling slump loss. This is related to the adsorption form of this water-reducing agent molecule on cement particles, as shown in Figure 6.21.

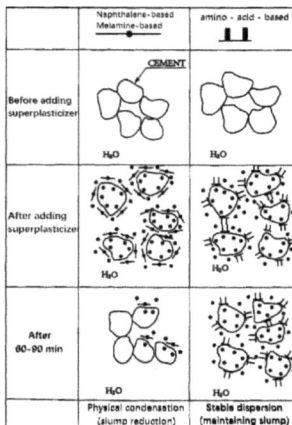

Figure 6.21 Absorption of cement particles on superplasticizer molecular.

6.5.3   Mechanism of sulfamate superplasticizer to control slump loss

As shown in Figure 6.21, the sulfamic acid superplasticizer is absorbed by cement particles by rigid vertical bond adsorption, which has an effect of three-dimensional dispersion, so that the cement particles are stably dispersed, and the slump loss is small over time. However, the naphthalene-based and the melamine-based superplasticizers are of a plane repulsive force, cement particles are prone to agglomeration, and the slump loss is fast.

6.6   Mechanism of slump loss and control of concrete with naphthalene-based and melamine-based superplasticizers

From the dispersion and agglomeration of cement particles, the relationship between the concentration of water reducing agent in the liquid phase and the change of slump and the change of yield value, the mechanism of slump loss and its control can be further clarified.

6.6.1   Cohesion and dispersion of cement particles

After mixing the superplasticizer into the cement paste, the superplasticizer molecules are adsorbed by the cement particles, and an electric double layer potential (Zeta potential) is formed on the surface of the cement particles. The electrostatic repulsion is increased, and the network structure of the cement paste is destroyed. The water constrained by the network structure is released, and the fluidity of the paste increases, as shown in Figure 6.22(a). However, due to physical and chemical dispersion, the number of particles in the cement paste increases. In order to reduce the surface energy of the particles, the adjacent particles come into contact closely and absorb to each other, causing cement particles to agglomerate and the slump is reduced. Hattori Kenichi believed that if force by the Brownian motion of the cement particles, gravity, and mechanical action exceed the potential barrier Vmax caused by the electric double layer potential, the agglomeration of cement particles will occur as shown in Figure 6.22 (b) and (c).

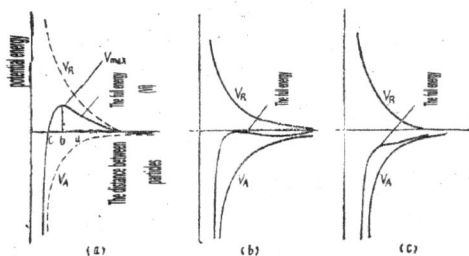

Figure 6.22 Potential energy curve of cement particle interaction.

6.6.2   The relationship between yield value of cement paste and slump loss and residual amount of superplasticizer in the liquid phase

Tests have proved that, after adding superplasticizer to cement paste, the residual amount of superplasticizer (SP) in the liquid phase decreases with time, and yield value of the corresponding cement paste increases, and the slump loss increases, as shown in Figure 6.23 and Figure 6.24.

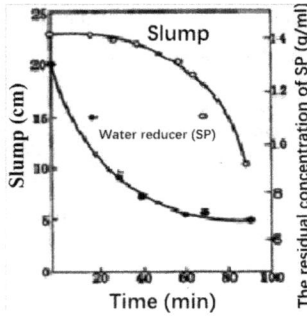

Figure 6.23 Change of slump and residue of superplasticizer with time.

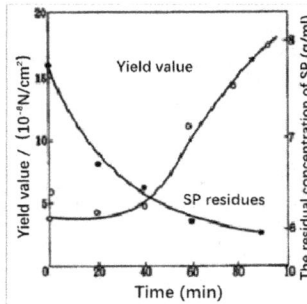

Figure 6.24 Change of yield value of cement paste and residue of superplasticizer with time.

It can be seen from Figure 6.23 and Figure 6.24 that, if the residual amount of superplasticizer in the concrete can be supplemented in time, the yield value of cement paste or the slump loss of concrete can be suppressed, as shown in Figure 6.25.

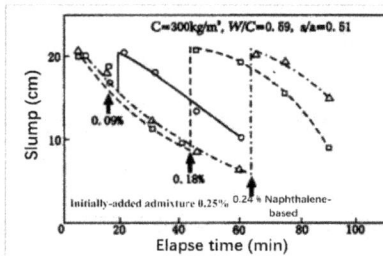

Figure 6.25 Relationship between the recovery of slump and the replenishment of superplasticizer.

The initial slump of the concrete was 20 cm. After 20 min, the slump loss was about 15 cm. After adding 0.09% water reducing agent, the slump was restored to the original state. After 40 min from the initial state, the slump was lost to less than 10 cm. with adding 0.18% of water reducing agent, the slump returned to its original state. By analogy, after 60 minutes from the initial state, 0.24% x C of water reducing agent should be added to make the slump return to its original state.

Figure 6.26 shows the effect of the same superplasticizer on slump recovery when added at different times. Adding superplasticizer within 4-87 min after the concrete was mixed, the extent of slump recovery was different. After 90 min from the initial state, the slump was the same.

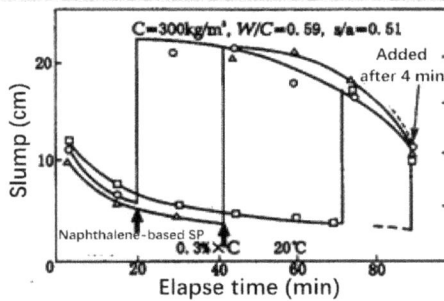

Figure 6.26 Effect of adding time of superplasticizer on slump recovery.

The test of Hattori Kenichi proved that the use of multiple additions of naphthalene-based superplasticizers can maintain the slump unchanged for more than 2 hr and, has no adverse effects on the performance of concrete.

6.6.3   Model of slump loss and recovery

Through the above experimental analysis, a model can be considered according to the results in Figure 6.27.

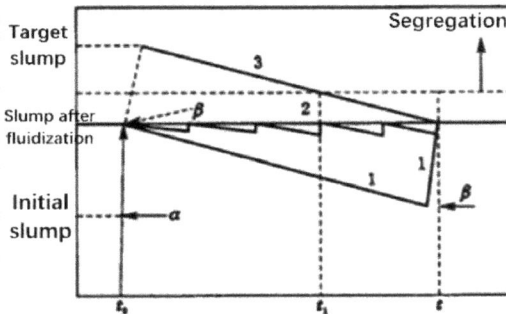

Figure 6.27 Model of slump loss and recovery when superplasticizer is added.

Multifunctional Concrete Technology   Materials Research Forum LLC
Materials Research Foundations **127** (2022)   https://doi.org/10.21741/9781644901991

where:

Curve 1—The slump loss was restored by adding the superplasticizer at the end.

Curve 2—Through multiple additions of superplasticizers, the slump loss was restored;

Curve 3—By adding superplasticizer ß at t0 time, the concrete segregates.

The superplasticizer α is mixed in the concrete, and the concrete reaches the slump after fluidization. After t time, the slump reduced. If the same slump as the initial slump is to restore, it must be mixed with the superplasticizer ß, as curve 1 shown in the figure. The method of adding the superplasticizer ß several times to restore the slump, such as curve 2, is called repeated addition. However, if the mixed superplasticizer of α+ß is added into the concrete at one time, the concrete will segregate, and the quality cannot be guaranteed.

To keep the slump unchanged within T time, superplasticizer must be added into concrete. The superplasticizer can be added according to equation (6.1).

$$A = a (T - T_0) b \qquad\qquad (6.1)$$

where: A—cumulative dosage;

T—slump maintenance time;

T0—starting time of superplasticizer addition;

a—constant;

b—constant determined by the superplasticizer addition scheme.

The superplasticizer addition scheme and slump are shown in Figure 6.28.

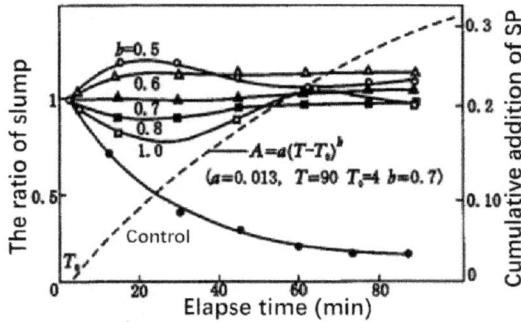

Figure 6.28 Superplasticizer addition scheme and slump change.

If A = the amount of cementitious material X 0.3% (solid content of superplasticizer), T = 90 min; b is changed in the range of 0.5, 0.6, 0.7, 0.8 and 1.0 to adjust the slump change. In fact, the difference in b indicates that the amount of superplasticizer added within 4-30 min is different. The result is shown in Figure 6.28. When b=0.7, after 90 min, the slump changes smoothly and the loss is very small, while for b=0.5 to 0.6, during 20-30 min, the slump changes too greatly. For b=0.8, 1.0, the slump is too small. Therefore, according to the equation A = a (T- T0) b, b=0.7 is the most appropriate to maintain the slump by adding superplasticizer. At this time, the curve of the time-dependent addition of superplasticizer is shown by the dotted line in Figure 6.28. Without addition of superplasticizer continuously, the slump changes over time, as shown in the blank curve in Figure 6.28, where the slump loss cannot be controlled.

6.7 Control of the slump loss of concrete by carrier fluidizing agent

After the natural zeolite powder adsorbs the superplasticizer, the natural zeolite powder becomes the carrier of the water reducing agent. When mixed into mortar or concrete, it can keep the mortar or concrete plastic for 3-4 hours, which is convenient for construction, including pouring and moulding. Natural zeolite powder adsorbed superplasticizer to become carrier fluidizer. It has been used in concrete construction in many projects.

6.7.1 Experiment of the adsorption and desorption of natural zeolite powder to superplasticizer

Raw materials

Natural zeolite powder: clinoptilolite produced in Weifang, Shandong, cation exchange capacity 120 meg/100g, fineness 600 m^2/kg;

Superplasticizer: (1) Naphthalene series, Type A, powder, water reduction rate of about 20%; (2) Sulfamate series, B, solution, solid content about 45%, water reduction rate $\geq$ 25%; (3) Polycarboxylic acid series, C, solution, solid content about 40%, water reduction rate $\geq$30%;

Cement: Mitsubishi brand P.O 42.5; Weifang Cement (vertical kiln) P.O 32.5; Yuexiu brand P.O 52.5;

Aggregate: river sand, medium sand, well graded, apparent density 2.60, loose bulk density 1450 kg/m^3; Limestone crushed stone, particle size 5-20 mm, well graded, apparent density 2.60, loose bulk density 1450 kg/m^3.

Water: drinking water.

Experiment of the adsorption and desorption of zeolite powder to superplasticizer

Prepared were four samples and each had 100 g of natural zeolite powder, and then added the superplasticizer solution with 45% solid content respectively. After every 30 minutes the solution containing zeolite powder is filtered, washed, dried, and weighed to obtain the weight of zeolite powder after soaking for 30 min, 60 min, 90 min and 120 min respectively. Calculated adsorption of the zeolite powder to superplasticizer is shown in Table 6.7.

Table 6.7 Isothermal adsorption capacity of zeolite powder to superplasticizer (g/10g).

Time (min) Type of superplasticizer	30	60	90	120
A	2.5	3.5	5.6	5.7
B	2.6	3.6	5.5	5.8

Table 6.8 Isothermal discharge amount of zeolite superplasticizer in aqueous solution (g/10g).

Time (min) \\ Type of superplasticizer	30	60	90	120
A	2.3	3.3	4.1	4.5
B	2.4	3.4	4.0	4.6

After the zeolite powder was adsorbed and saturated with the superplasticizer, 100 g of the obtained zeolite superplasticizers A and B were sampled, put into the aqueous solution, and the discharge amount of the superplasticizer was measured at 30 min, 60 min, 90 min and 120 min, as shown in Table 6.8.

According to Table 6.7 and Table 6.8, plot the adsorption and discharge curves of natural zeolite powder to water reducers A and B, as shown in Figure 6.29 and Figure 6.30. It can be seen from these figures that the adsorption capacity of natural zeolite powder for water reducer A and B increased almost linearly before 90 minutes, and after 90 minutes, it basically reached a saturated state, and the saturated adsorption value for water reducer was about 5.6-5.8 g/10 g. With further time passing, the adsorption amount increased slightly. The discharge amount of zeolite water reducing agent in water increased with the increase of time, but the extreme value of its discharge amount was below the saturated adsorption value.

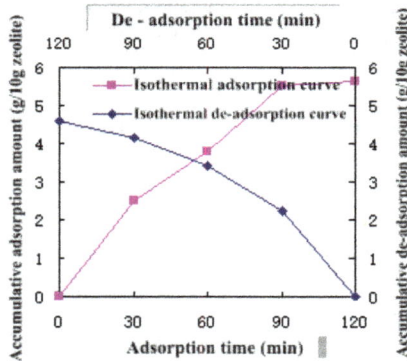

Figure 6.29 Adsorption and discharge curve of zeolite powder to water reducer A.

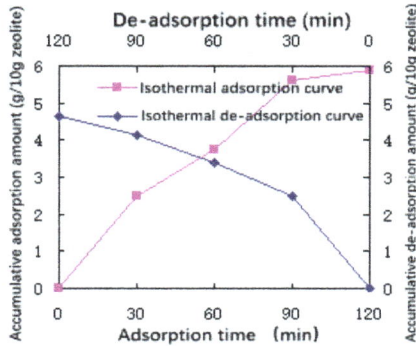

Figure 6.30 Adsorption and discharge curve of zeolite powder to water reducer B.

Effect of zeolite water reducer on fluidity of cement paste

Cement paste was prepared with 500 g of cement, 150 ml of water, and 10 g of zeolite water reducer. The fluidity test of the paste was carried out according to China standard, GB/T-8077-2000, and the changes over time were measured. The results are shown in Table 6.9.

It can be seen that the zeolite water reducing agent can effectively control the fluidity loss of cement paste. At 120 min, the fluidity of Mitsubishi cement paste was not reduced, but increased by 9.2%, while for Weifang shaft kiln cement, the fluidity decreased by about 20%, which might be caused by the excessively high content of f-CaO or C_3A in the cement.

Table 6.9 Fluidity change of cement paste over time (mm).

Cement	Change of fluidity				
	initial	30 min	60 min	90 min	120 min
Mitsubishi P.O.42.5	238	240	240	270	260
Weifang P.O.32.5	220	200	200	180	175

The slump change with time and concrete strength

The mix proportion of concrete is shown in Table 6.10. The measured concrete slump change with time within 2 hours and the strength at various ages are shown in Table 6.11. It can be seen from the test results that, after 2 hr, the concrete slump dropped from 20 cm to 16.5 cm, a decrease of 17.5%, but the slump was still more than 16 cm, which could still meet the requirements of pumping construction.

Table 6.10 Mix design of concrete.

W/C	Amount of raw materials per m³ (kg/m³)						Cement
	Cement	Fly ash	Sand	Aggregate	Water	Zeolite admixture	Mitsubishi
0.40	350	100	750	1100	180	11.25 (2.5%)	

Table 6.11 The change of slump and compressive strength.

Change of slump (cm)			Compressive strength (MPa)		
Initial time (11:12)	70 min (12:30)	120 min (1:20)	3d	7d	28d
20	19	16.5	16.4	25.6	42

Mechanism for the effect of zeolite water reducing agent on control of concrete slump loss

Both the slump loss of concrete and the increase in the apparent yield value of cement paste are related to the decrease in the residual amount of superplasticizer in the liquid phase (actually the amount of the superplasticizer adsorbed by cement particles) as shown in the test results in Figure 6.31 and Figure 6.32 respectively. The changes of Zeta potential with and without zeolite water reducing agent is compared for the cement paste at W/C of 0.4, and the result is shown in Figure 6.34. The absolute value of the Zeta potential of the cement paste without zeolite water reducing agent decreased rapidly, from 5.5 mv to 4.7 mv at 30 min, and to 4.2 mv at 1 hr. However, the cement paste containing zeolite water-reducing agent had an initial 5.95 mv and basically maintained at a level of 6 mv within 140 min, and the cement paste maintained a good dispersion state and fluidity. Zeolite water reducing agent also has a great influence on the shear strength and structural viscosity of cement paste.

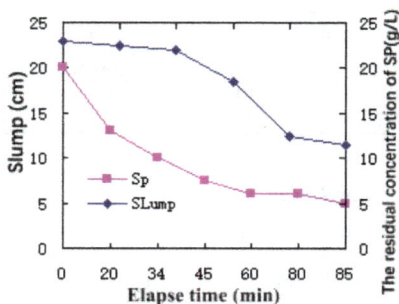

Figure 6.31 Change of slump with time and residual concentration of superplasticizer.

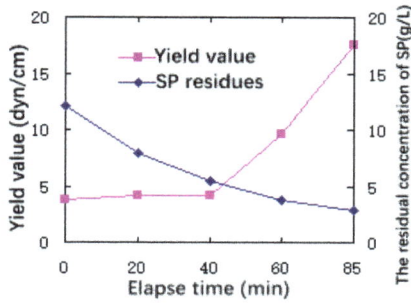

Figure 6.32 Changes of cement slurry yield value with time and residual concentration of water reducer.

The cement paste was prepared at W/C of 0.30, adding 0.6% naphthalene superplasticizer or 2.0% zeolite superplasticizer. A Warming mixer was used to agitate the cement paste, and a Fanning viscometer was used to measure the shear strength and structural viscosity of the cement paste within 40 min. The results are shown in Figure 6.33.

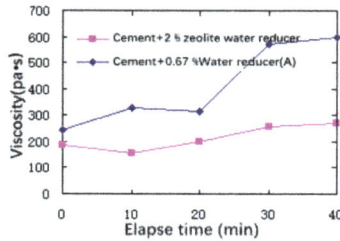

(a) Change of structural viscosity

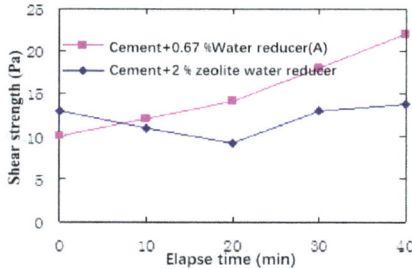

(b) Change of shear strength

Figure 6.33 Change in structural viscosity and shear strength with time of cement slurry with or without zeolite admixture.

Materials Research Forum LLC

https://doi.org/10.21741/9781644901991

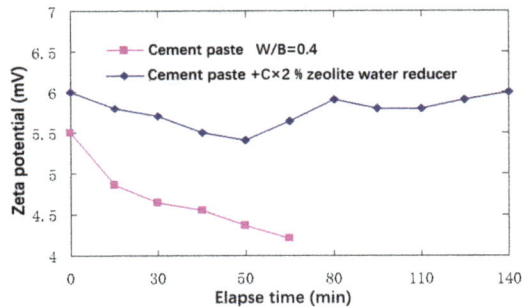

Figure 6.34 Changes of zeta potential on the surface of cement particles with time.

It can be seen from the results in Figure 6.34 that the zeolite water reducing agent can slowly discharge the adsorbed superplasticizer into cement paste, maintaining the adsorption amount of superplasticizer on the surface of cement particles, thereby maintaining the zeta potential on the surfaces of cement particles and, keeping cement particles in a dispersed state. The structural viscosity and shear strength of the cement slurry are also in a stable state, and the slump loss of the concrete can be controlled.

6.8   Construction application of carrier fluidizing agent

Case 1: Slump loss control of high-strength and high-fluidity concrete

Raw materials for the test including: cement - Handan 52.5 Portland cement; coarse aggregate - river pebble 5-40 mm; fine aggregate - medium sand, river sand; naphthalene-based water-reducing UNF

The mix proportion of concrete is shown in Table 6.12.

Table 6.12 Mix proportion of high strength/high flow concrete (kg/m^3).

NO	W/C (%)	C	water	sand	aggregate	M-Ca	UNF	CFA
1	35	500	175	675	1100	/	CX0.6%	/
2	35	500	175	675	1100	CX0.25 %	CX0.6%	CX1.0%
3	38.8	450	175	680	1100	/	CX1.0%	/
4	38.8	450	175	680	1100	/	CX1.0%	CX1.5%

In Table 6.12, C represents the amount of cement (kg/m3), CFA—control slump loss additive (zeolite water reducing plasticizer), UNF—superplasticizer, M-Ca— Lignosulfonic water reducer. In No. 2 mix, CFA was 1.0% of the cement amount, and in No. 4 mix, CFA was 1.5% of the cement amount. The change of slump with time is shown in Table 6.13. It can be seen that, if the

CFA dosage was 1.0% or 1.5%, the slump of concrete could be controlled without change within 2 hr.

Table 6.13 Change of slump with time.

NO.	The slump change (cm) with time(min)									
	0	20	30	40	50	60	70	80	9	120
1	18			14						
2	20	20		21		21				19
3	19		12							
4	21		21			20			20	19

Case 2: Application of C35 ordinary concrete in construction of Wuhan Shengshihuating project

Table 6.14 Mix proportion of concrete for construction application (kg/m³)

material concrete	Cement	Fly ash	Slag	River sand	Aggregate	water	FDN	CFA
Reference concrete	230	90	100	710	1080	180	9.0	—
CFA concrete	230	90	100	710	1080	180	—	11.76

Table 6.15 Slump change with time (Slump/flow value)

change concrete	0 hr.	1 hr.	2 hr.
Reference concrete	205 mm / 360 mm	160 mm / —	—
CFA concrete	190 mm / 510 mm	195 mm / 500 mm	190 mm / 480mm

Table 6.16 Setting time and compressive strength of concrete

property concrete	Setting time (hr.)		Compressive strength (MPa)		
	Initial	Final	3d	7d	28d
Reference	13.5	18	21.3	29.8	45.1
CFA concrete	10	14.5	24.8	32.8	51.5

Mix proportion for concrete production, change of concrete slump and flow value with time, and concrete compressive strength are shown in Table 6.14, Table 6.15, and Table 6.16, respectively. Photos of the construction site of the Shengshihuating project and the feeding mode of CFA are shown in Figure 6.35 and Figure 6.36 respectively.

Figure 6.35 Construction site of Shengshihuating project.

Figure 6.36 Application of CFA at the construction site.

It can be seen from Table 6.14, Table 6.15, and Table 6.16 that the slump and flow value of the concrete mixed with CFA could remain basically unchanged for more than 2 hr, ensuring the requirements of concrete construction. The initial setting and final setting time were also shorter than those of the reference concrete. The strength at 3 d, 7 d and 28 d were all 10% higher than that of the reference concrete. The reason is that the carrier zeolite powder participates in the hydration reaction of cement.

Case 3: Application of CFA in ultra-high-performance concrete

In October 2008, in the Guangzhou Zhujiang New Town West Tower project, the research and development of the ultra-high-performance concrete with strength grade C100 and the ultra-high pumping were carried out. In order to reduce the viscosity of concrete and control the change of concrete workability with time, 2% CFA (polycarboxylate water-reducing agent with 23% solid content) was added to the concrete, which can make the workability of fresh concrete unchanged for 3 hr, as shown in Table 6.17 and Figure 6.37.

Table 6.17 Workability retention of ultra-high-performance concrete with CFA

Time	Slump (cm)	Flow value (cm)	Flow time（s）
Initial	24	64/62	6.0
60 min	25.5	61/67	4.4
120 min	23	56/57	5.3
180 min	24	52/54	9.0

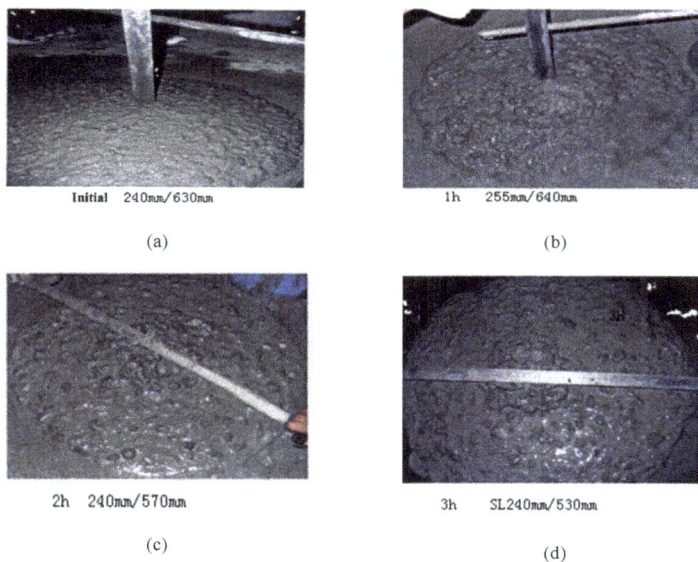

Initial 240mm/630mm

(a)

1h 255mm/640mm

(b)

2h 240mm/570mm

(c)

3h SL240mm/530mm

(d)

Figure 6.37 Change of slump and slump flow.

The ultra-high-performance concrete with the use of zeolite water-reducing plasticizer CFA could also retain plastic for more than 3 hr, ensuring the construction of ultra-high pumping concrete.

In another research and development of C120 ultra-high performance concrete and its ultra-high pumping test of the Kingkey Building project, which was a few years prior to the Guangzhou Zhujiang New Town West Tower project, zeolite powder was added into the superplasticizer solution, so that the C120 ultra-high performance concrete kept plastic for 3 hr, and pumped from the ground to a height of 510 m.

6.9   Method and effect of adding superplasticizer into concrete

When mixing concrete, there are two methods of adding superplasticizer: simultaneous mixing and post-mixing. In the simultaneous mixing method, when mixing concrete, the superplasticizer is first dissolved in the mixing water, and the water is poured into the mixer. The post-mixing method is to mix sand, crushed stone, cementitious materials and water in the mixer to form fresh concrete, and then superplasticizer is added separately. Due to the different mixing methods of the superplasticizer, the effect is also different. In the case of the same amount of superplasticizer, the post-mixing method has a higher slump and better fluidity than the simultaneous mixing, or it can save superplasticizer than the simultaneous -mixing method. The reason is related to the initial hydration of cement.

6.9.1 Adsorption capacity of superplasticizer during cement hydration

Different types of minerals in cement have different compositions, and hydration products have different adsorption capacities for superplasticizers. The adsorption of melamine-based superplasticizer (SMF) by cement minerals is shown in Figure 6.38.

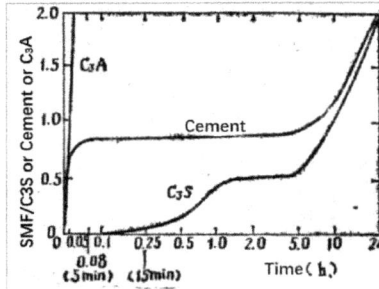

Figure 6.38 Adsorption curves of different cement minerals to SMF during cement hydration.

V.S. Ramachandran measured the amount of adsorption of SMF in the solution for cement samples with water/solid = 2 at different times. It was found that, in a water medium, C_3A adsorbed SMF very strongly. It could adsorb a considerable amount of SMF within a few seconds. C_3S has a very low adsorption capacity for SMF in the initial stage of hydration.

The experiment of Hattori Kenichi proved that, when the apparent equilibrium concentration of adsorption in the solution was 0.2%, temperature was 20 °C, and the adsorption time was 10 min, without the addition of gypsum, the hydrates of C_4AF and C_3A were effective for naphthalene-based superplasticizer with a absorption of 100 mg/g. However, when gypsum was added, C_4AF and C_3A reacted with gypsum to form hydrates on the surfaces of cement particles, so that it was not easy to adsorb superplasticizers on these surfaces, and the adsorption capacity was only 10 mg/g, or even lower, as shown in Figure 6.39.

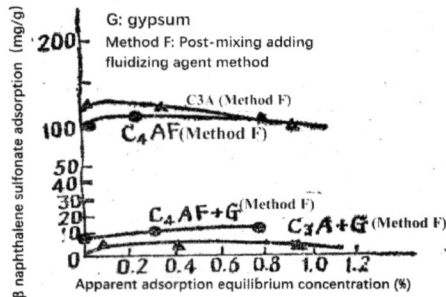

Figure 6.39 Adsorption capacity and curve with and without gypsum.

105

Due to the fast hydration speed of C_4AF and C_3A, the adsorption capacity of hydrates on the superplasticizer is large, so the amount of superplasticizer remaining in the liquid phase is small. Therefore, the superplasticizer that can be used during C_3S hydration is small. It is not beneficial to increasing the zeta potential on the surface of cement particles. However, due to the addition of gypsum, the initial hydrates of C_4AF and C_3A are changed, and the adsorption amount of superplasticizer is reduced, and there are more efficient residues of superplasticizer in the liquid phase, so that the amount of superplasticizer absorbed by the hydrates of other two main mineral components of cement C_3S and C_2S is high, which means that the zeta potential on the surface of the cement particles increases and the barrier value Vmax increases as well, so that the cement particles are more effectively dispersed. That is why post-mixing method can result in higher slump and better flowability for concrete compared with the simultaneous method.

6.9.2   Choosing appropriate types of cement for improving slump loss

The above test also shows that the use of medium and low heat cement with high C_2S content is effective in controlling the slump loss of concrete.

Meyer and Perenchio proposed that the content of C_3A in cement minerals, the content of SO_3 in the solution, and the type and number of chemical admixtures are related to the formation of ettringite and slump loss. Khalil and Ward suggested that, in concrete with superplasticizers, the increase in the content of SO_3 with the increase in gypsum will delay the hydration of C_3A, which can improve the slump loss. The slump loss is caused by the initial hydration and the consumption of superplasticizers. Therefore, if the content of C_3A in cement is high, the amount of adsorbed superplasticizer will be high, and hence the slump loss will be high.

6.10 Summary

Superplasticizer is an indispensable component of high-performance concrete and ultra-high-performance concrete. Superplasticizer has been developed from naphthalene-based, melamine-based, sulfamate-based to polycarboxylate-based, and the water reduction rate of it in concrete has been continuously improved. The W/B ratio of concrete has also been continuously reduced, so that the strength of concrete can reach more than 200 MPa, and the concrete can also be constructed via pumping and self-compacting. The continuous improvement of the water reduction rate and the continuous improvement of control of slump loss are the material basis for the improvement of concrete technology.

If the naphthalene-based water-reducing agent is used to prepare high-performance concrete, when W/B is lower than 0.30, the concrete mixture may be very viscous, and C60 may be an extreme strength grade of HPC. However, with the polycarboxylate-based superplasticizer, UHPC can be prepared with compressive strength of 200 MPa. Moreover, it has multiple functions such as vibration-free and self-compacting. This is caused by the different molecular structure of the water reducing agent and the different adsorption form of the cement particles.

In concrete, whether the cement particles can be dispersed and whether the fresh concrete can retain plasticity, the key is the zeta potential formed on the surface of the cement particles after adsorbing the water reducing agent molecules and its maintaining duration. The graft copolymer polycarboxylate-based superplasticizer not only has a high zeta potential, but also can maintain the zeta potential value for a relatively long duration. Therefore, it has a high water reduction rate and a long-term plasticity maintaining function.

Maintaining the zeta potential value between the cement particles can maintain the fluidity of the concrete. CFA can also be used to continuously supplement the water-reducing agent adsorbed on the surface of the cement particles, which can also keep the concrete plastic for a long duration.

There are two methods for adding superplasticizer to concrete: 1) simultaneous mixing method: adding superplasticizer to water, and then mixing with the cement, sand, and gravel of the concrete; 2) post-mixing method: mixing water with cement, sand, and crushed stone first and then after 1-1.5 min, adding superplasticizer. For the same slump, the post mixing method can save 30 % superplasticizer compared to simultaneous mixing method.

Chapter 7

Concrete Shrinkage Cracks and Application of Shrinkage-Reducing Admixtures

7.1 Analyses of causes for concrete cracks due to shrinkage

Concrete is a multi-phase composite formed through the bonding produced via cement pastes with sands and coarse aggregates. Cracking of concrete due to shrinkage is mainly caused by the shrinking of the cement paste.

7.1.1 Cracks caused by the shrinkage of cement pastes

Cements paste mixed using 315 g cement and 200 cm³ water (w/c=0.64) has an initial volume of 100+200=300 cm³ if taking into the consideration the relative density of cement as 3.15 (relative to water) and the initial volume of cement: 315/3.15=100 cm³.

Assuming that, after complete hydration, 1 cm³ cement produces a volume of 2 cm³ hydrates, cement may have different capillary pore volume at different degree of hydration, even though cement has hydrated completely, there is still capillary porosity of 33%, as shown in Figure 7-1, which may cause the shrinkage cracks of the cement paste.

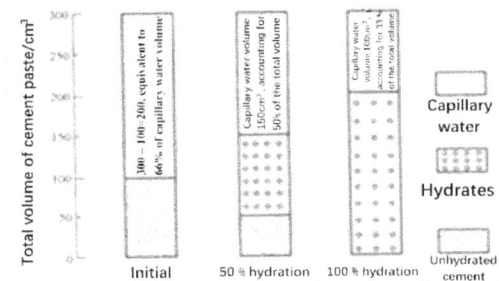

Figure 7.1 Cement 100 cm³, W/C=0.64, change of capillary pore volume at different degree of hydration.

As to cement pastes at different water/cement ratios, after achieving the same degree of hydration, although the solid hydrates of cements are the same, the capillary pore volumes are different. For example, four types of cement pastes mixed with W/C=0.70, 0.60, 0.50 and 0.40 respectively, assuming that the degree of hydration is 100%, the solid hydrate volumes after complete hydration are all 200 cm³. However, the contents of capillary pore volumes in hardened cement pastes at different W/C ratios are different, which are 120 cm³ (37 % of the total volume), 88 cm³ (30% of the total volume), 57 cm³ (22% of the total volume) and 26c m³ (11% of the total volume), respectively, as shown in Figure 7.2.

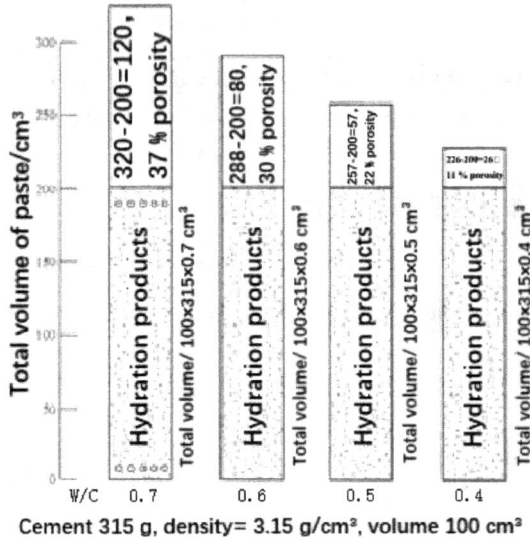

Figure 7.2 For full hydration of cement, the change of W/C causes the change of capillary pore volume in cement paste.

When exposed to an environment with relative humidity lower than 100%, cement paste will start to lose water and shrink afterwards. Shrinking stress will be generated if cement paste is restrained during the process of shrinking, which will cause cracking if the shrinking stress is higher than tensile strength of cement paste, as shown in Figure 7.3.

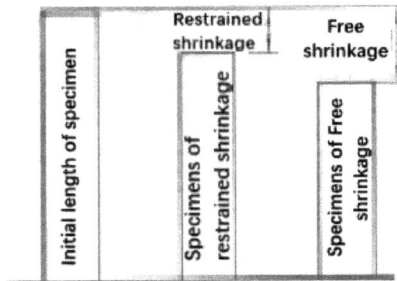

Figure 7.3 Comparison of free shrinking and restrained shrinking of testing samples.

In concrete, sand and coarse aggregate are bonded together by cement hydrates. When hardened cement paste has lost, both the water which exists in capillary pores and is kept by pore hydrostatic tension, and the water that are physically adsorbed by calcium silicate hydrates, as the products of hydration, shrinkage will be caused. Concretes will then encounter cracking when concretes are restrained during the shrinking, as shown in Figure 7.4.

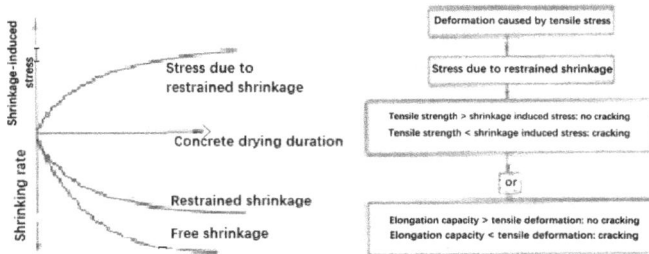

(a) Shrinkage and shrinking-induced stress of concrete (b) Shrinking and cracking of concrete

Figure 7.4 Shrinkage and cracks of concrete testing samples under the conditions with and without restraints.

It is estimated that shrinkage rate of completely dried of hardened cement paste may be as high as 10.000×10^{-6}. Experimentally measured shrinkage rate is 4.000×10^{-6}. This is because aggregates in concrete can be assumed to have no shrinkage, hence the actual measured shrinkage rate is roughly in the range of $(200 - 1000) \times 10^{-6}$.

Some researchers believed that more or less cracks originally exist in concrete, which means concrete is bound to have original cracks, and it is unavoidable that cracks exist in concrete. This is an intrinsic feature of cement-based materials.

7.1.2   Causes and conditions of concrete drying shrinkage

It should be the responsibilities of concrete professionals to study the causes of concrete cracking, make use of corresponding technical measures to prevent concrete from cracking, reduce the formation of harmful cracks and take remedy measures to repair harmful cracks.

The causes and conditions of concrete cracking due to dry shrinkage can be shown in Figure 7.5. From the figure, the countermeasures in control concrete cracking can be found. Concrete cracks due to dry shrinkage can be controlled and determined by cracking conditions, namely, the free shrinkage rate, constraint rate, constrained shrinkage rate, constrained shrinkage stress, tensile strength and elongation capability. The conditions of concrete cracking are due to the causes both internally and externally, including structural factors (structural member size and the spacing between reinforcing bars), and so on. In other words, it is necessary to control the shrinkage and cracking of concrete in terms of the quantity and quality of the constituent materials of the concrete, structural design, construction environment and conditions.

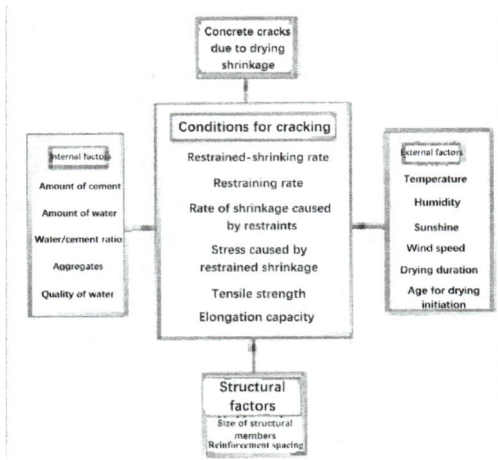

Figure 7.5 The main reasons and conditions for drying shrinkage and cracking of concrete.

7.2 Concrete constituent materials and dry shrinkage

7.2.1 Types of cement and drying shrinkage

Drying shrinkage of different types of cement is shown in Figure 7.6.

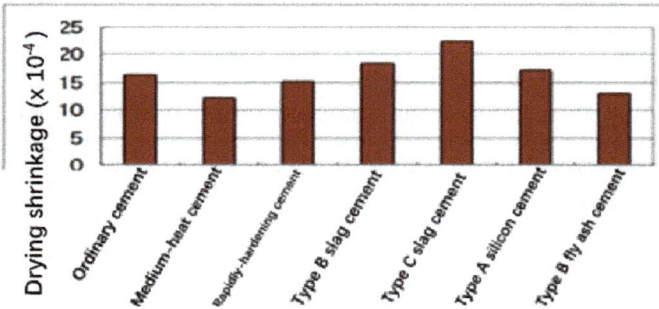

Figure 7.6 Dry shrinkage of different types of cement

Note:

Type B blast cement: slag content 30-60%.

Type C blast cement: slag content 60-70%.

Type B fly ash cement: fly ash content: 10-20%.

Type A siliceous material cement: siliceous material content 10-20%.

From Figure 7.6, it can be found that Slag cement has the biggest dry shrinkage, while Fly ash cement with Fly ash content of 10-20% has the least dry shrinkage.

7.2.2   Fly ash content and concrete shrinkage

As to concrete made by using cement blended with fly ash, as the content of fly ash increases, the drying shrinkage of the concrete decreases, as shown in Figure 7.7.

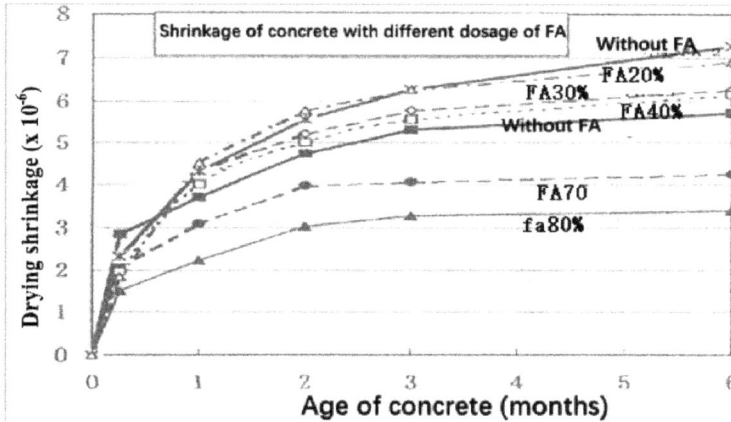

Figure 7.7   *Shrinkage of concrete made by using cement blended with different contents of fly ash.*

7.2.3   Slag content and concrete shrinkage

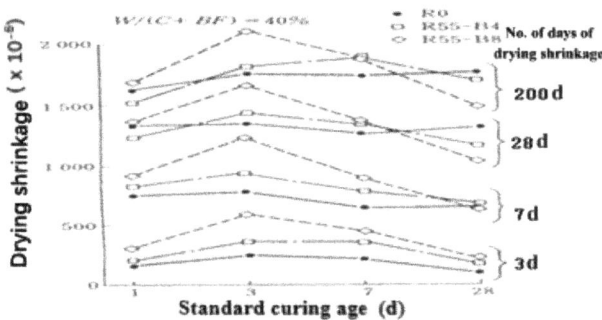

Figure 7.8 *Drying shrinkage of slag concrete under standard curing.*

It is shown in Figure 7.8 that, after standard curing for various age, concrete specimens prepared with different contents of slags, were exposed to different duration of drying, i.e. 3 d,7 d, 28 d and 200 d and tested to obtain drying shrinkage of them. The specimens cured for 28 days under standard conditions have a 3-day drying shrinkage value greater than that of the reference concrete. The 7-day drying shrinkage value is the same as that of the reference concrete. The 28-day and 200-day drying shrinkage values were lower than that the reference concrete.

7.2.4   The drying shrinkage value of different rocks

The aggregates of concrete are processed from different rocks. Knowing the shrinkage of the parent rocks is helpful to know the shrinkage of aggregate. The drying shrinkage values of different rocks are shown in Figure 7.9.

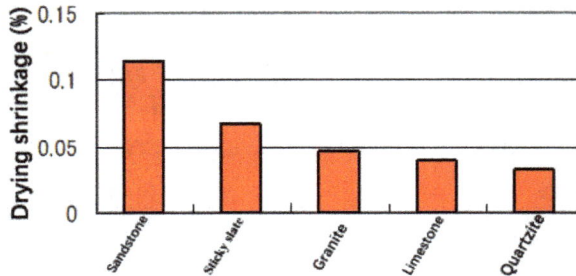

Figure 7.9 Drying shrinkage values of different rocks.

Different types of rock aggregates have different amount of drying shrinkage. Ordinary sandstone and slate series have the largest shrinkage coefficients, followed by igneous rock series, and finally limestone dichotomy series. When the water absorption of aggregate is large, the mass reduction rate after drying is also large. Hence the drying shrinkage value of the concrete increases. As long as the water absorption rate of the aggregate is within the range of a specified value, it has little effect on the drying shrinkage of the concrete. The larger the static elastic coefficient of the aggregate, the smaller the impact on the drying shrinkage of concrete. The drying shrinkage becomes smaller after the amount of aggregate in concrete increases.

7.3   Examination of drying shrinkage and crack patterns of drying shrinkage

7.3.1   Examination and allowed value of concrete drying shrinkage

Cracks of concrete members caused by drying shrinkage, whether harmful or not to structural members, can be examined by the following equation.

$$V_p = (S_p / SK) \leq 1.0 \qquad\qquad\qquad 7.1$$

where: Sk—Ultimate deformation caused by drying shrinkage in concrete (500-700 μm);

Sp—Measured deformation in concrete caused by drying shrinkage;

V p—Safety factors of Sp precision, which are generally taken as 1.0-1.3.

If the measured deformation of concrete caused by drying shrinkage meets the requirement in equation 7.1, the structural member may be assumed in a safe state.

7.3.2 Patterns of cracks caused by the shrinkage of concrete

Cracks caused by shrinkage appear at the locations of windows and walls, shown in Figure 7.10.

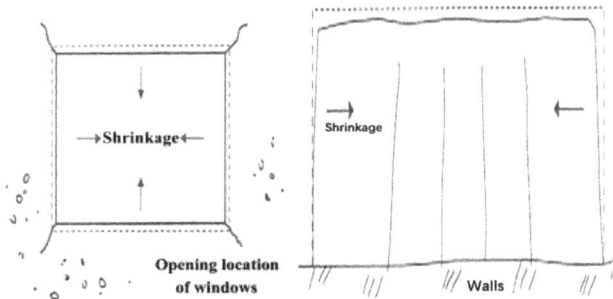

Figure 7.10 Cracks caused by shrinkage at locations of windows and walls.

Cracks caused by the restraints of steel reinforcements, shown in Figure 7.11.

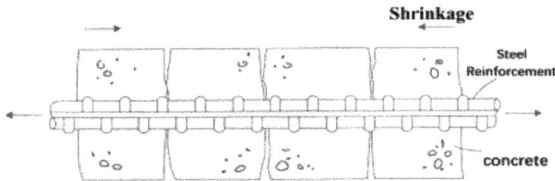

Figure 7.11 Cracks caused by the restraints of steel reinforcements.

Cracks occur when the internal shrinkage is restrained as shown in Figure 7.12. As the surface layer shrinks due to the drying, tensile stress will be developed, when the shrinkage-induced tensile stress is greater than the tensile strength of concrete, cracking occurs in concrete.

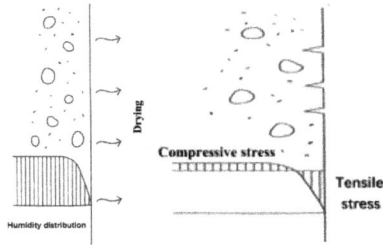

(a) Evaporation of water on concrete surface (b) Tensile stress is developed at the surface layer of concrete

Figure 7.12 Cracks caused by the evaporation of water on concrete surface.

Cracks due to drying shrinkage coupling with external restraints

A cantilever beam can have free shrinkage. Its deformation caused by drying shrinkage=ΔL / L is shown in Figure 7.13. Such a beam will not encounter cracking when drying shrinkage occurs.

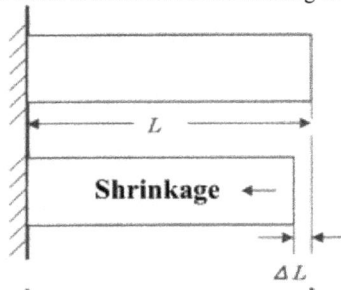

Figure 7.13 Shrinkage in a cantilever beam.

Conceptual diagram of cracking due to drying shrinkage coupling with external restraints is given in Figure 7.14.

A beam fixed at two ends, its deformation due to drying shrinkage $\varepsilon s= \Delta L / L$

Ifσs= E. εs < σt, No cracking will be caused.

Ifσs= E. εs > σt, cracking will be caused.

where,σs: Drying shrinkage induced stress;

σt: Tensile strength of concrete;

E: Young's modulus of concrete.

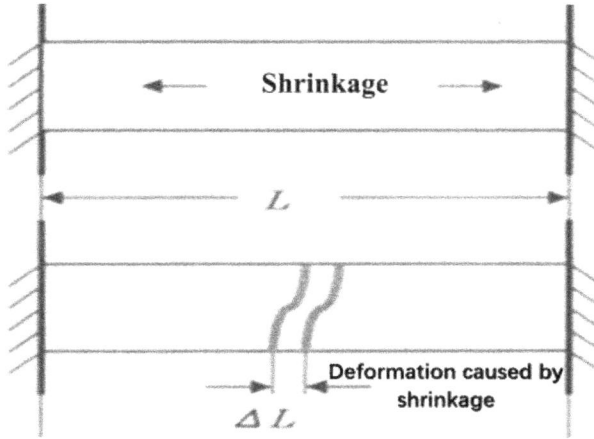

Figure 7.14 A beam fixed at two ends will encounter cracking when the shrinkage-induced stress is higher than tensile strength of concrete

7.4   Cracking caused by autogenous shrinkage of concrete

A super-high-rise building is located in the CBD of Zhujiang New Town, Guangzhou. The main building is 530 m high and has a total building area of 506988 m^2. It was constructed by China State Construction Engineering Corporation. The main tower structure, a framed structural system with the core tube, is formed through 8 giant columns connected by a spatial ring truss. The 8 giant columns are made of steel pipe filled with high-strength concrete. The ring truss, a space ring truss, is connected to the 8 giant columns for forming a closed structural system. In addition to the use of C80 high strength of concrete for construction of the outer wall of the core tube at the lower part of this tower building, a double layered steel plate shear wall (shown in Figure 7.15) is embedded inside the concrete. When the shear walls of concrete were demolded, it was found that the walls were full of cracks. In the worst case, there were more than 100 cracks on a wall as shown in Figure 7.16.

In Figure 7.16(a), the wall has about 27 cracks. The average width of all the cracks is less than 0.2 mm, and the depth of all the cracks is less than 4 cm. Vertical, horizontal and oblique cracks were found on the wall, and there were craze cracks in some areas. The longest diagonal crack was 3.1 m, and the longest vertical crack was 2.3 m.

Figure 7.15 Outer walls of the core tube cast with C80 concrete and embed internally with double-layered steel plate shear walls.

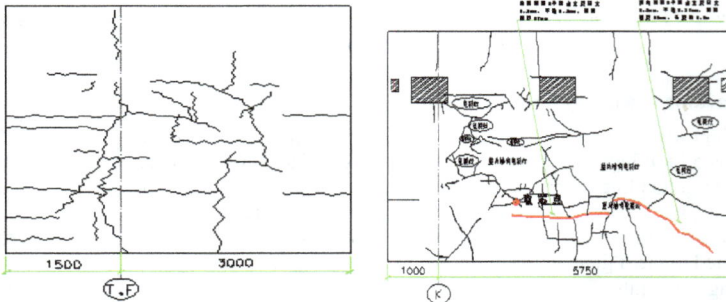

(a) Cracks on No. 1 west wall (external wall) (b) Cracks on external walls of western direction

Figure 7.16 When the concrete of shear walls was demolded, it was found that the walls were full of cracks.

In Figure 7.16(b), the wall had about 120 cracks. The width of cracks was about 0.1 mm to 0.2 mm. Some fine cracks could only be observed after spraying water on the wall at initial stage after demoulding, and they are now healed. These cracks were mainly craze cracks and located at central part of the wall. There were 6 vertical cracks with a length of about 0.5 m to 1.5 m located at the top of the wall. Two of the deepest and longer cracks had a width of about 0.2 mm and a depth of less than 5 cm. The red lines represent the two longer and deeper cracks. This type of cracks is under a condition that there is no mass exchange between the concrete and the surrounding environment. After the concrete has set and hardened, the cement in the concrete continues to hydrate, which absorbs the capillary water in the concrete and makes the capillary enter a self-vacuum state. When the capillary tensile stress generated exceeds the tensile strength of the concrete, the concrete encounters cracking. For high-strength concrete or ultra-high-strength concrete, the amount of cement used in concrete is high and the water/binder ratio is relatively low, which makes concrete more prone to autogenous shrinkage and cracking.

7.4.1 Theoretical explanation of concrete cracking caused by autogenous shrinkage - capillary tension mechanism

As shown in Figure 7.17, the deformation caused by capillary tension starts at about 40% relative humidity.

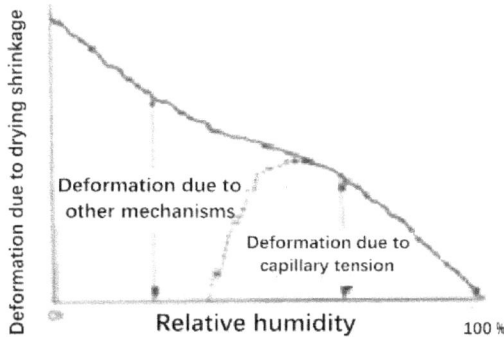

(a) Relationship between deformation caused by drying shrinkage and relative moistures

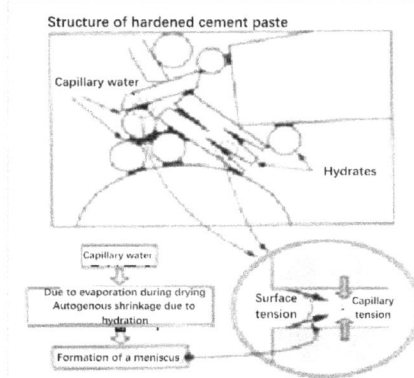

(b) The shrinkage mechanism of capillary tension

Figure 7.17 Formation range and mechanism of concrete autogenous drying shrinkage.

According to the relationship between the deformation caused by drying shrinkage and relative humidity shown in Figure 7.17(a), if the relative humidity is within the range of about 100-40%, deformation may be caused. It can be seen from Figure 7.17(b) that the loss of the capillary water is due to the evaporation of water due to drying or the absorption of water in the capillary pores due to hydration of cement, which forms a self-vacuum effect that generates drying shrinkage and also capillary tension. If the sum of the two exceeds the tensile strength of concrete, the concrete will encounter cracking. In the absence of external mass exchange conditions (such as the cracking of concrete when the mold is yet to be removed), the shrinkage cracks are called cracks caused by autogenous shrinkage.

7.4.2   The effect of shrinkage-reducing admixture on shrinkage reduction of concrete

The relationship between the drying shrinkage deformation and surface tension of concrete with shrinkage-reducing admixture is shown in Figure 7.18. The drying shrinkage is highly correlated with the surface tension, i.e. the drying shrinkage decreases linearly with the decrease of the surface tension. Figure 7.19 shows the autogenous shrinkage of cement paste and the effect of adding shrinkage-reducing admixture. From Figure 7.19, it can be clearly seen that the autogenous shrinkage of the concrete mixed with two different shrinkage-reducing admixtures is significantly lower than that of reference concrete. Autogenous shrinkage of concrete is due to the hydration of cement, which consumes the interlayer water in the hydrates, which generates a drying state inside the concrete and hence causes the shrinkage of concrete.

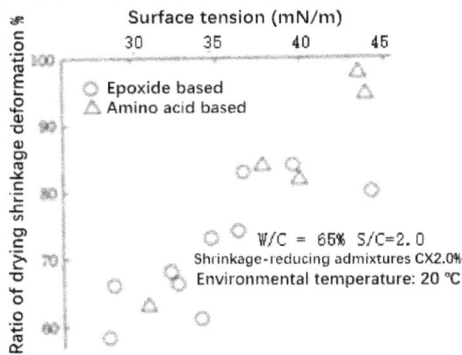

Figure 7.18 Relationship between the ratio of shrinkage deformation and surface tension of cement paste mixed with shrinkage-reducing admixture.

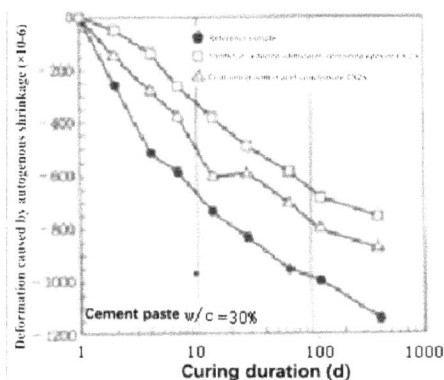

Figure 7.19 Deformation of concrete caused by autogenous shrinkage and the effect after adding shrinkage-reducing admixtures.

7.5   Performance of concrete mixed with shrinkage-reducing admixture

7.5.1   Performance of fresh concrete and the setting condition of concrete

The slump and air content of concrete may be changed by shrinkage-reducing admixture. This can be adjusted by adding water-reducing and air-entraining admixtures. The setting time may be slightly longer.

7.5.2   Concrete hardening condition

1. Strength

The compressive strength of concrete mixed with shrinkage-reducing admixture has slightly slower strength development, but it has no adverse effect. The strength of the concrete is slightly lower if comparing with the reference concrete. The other strength properties are still related to compressive strength.

2. Change of length

The shrinkage of concrete mixed with shrinkage-reducing admixture is shown as Figure 7.20. There is a very little difference in the effect of reducing shrinkage between various sorts of shrinkage-reducing agent at the same dosage. It is estimated that the final deformation due to drying shrinkage is about 60-80% of that of reference concrete. In addition, it can be found from Figure 7.21 that the higher the dosage of shrinkage-reducing admixtures added, the more significant the effect on shrinkage reduction. It has even a significant effect on the autogenous shrinkage of high-strength concrete.

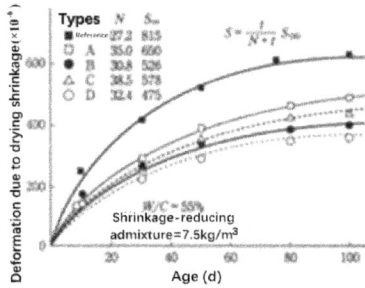

Figure 7.20 Effect of the various brands of shrinkage reducing admixtures on the reduction of drying shrinkage.

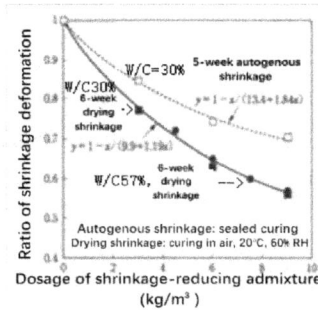

Figure 7.21 Relationship between shrinkage induced deformation and mixing dosage of shrinkage-reducing admixtures.

3. Resistance to cracking

The restraint cracking tests were carried out in accordance with a Japanese test method for restraint cracking due to drying shrinkage, and the effect of shrinkage-reducing admixture on cracking resistance is shown in Figure 7.22. With the amount of shrinkage-reducing admixture increasing, the time at which the cracking of concrete began became longer. According to the conditions of the constraints, it is possible to inhibit the occurrence of cracking. No. 3 and No. 6 concretes had no cracking, since shrinkage-reducing admixture at a dosage of 4.6 kg/m^3 and 5.6 kg/m^3, respectively, were added in the two concretes.

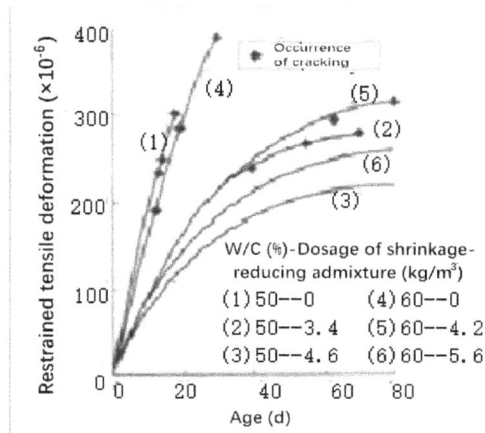

Figure 7.22 Test for cracks of concrete due to the shrinkage with and without shrinkage-reducing admixtures.

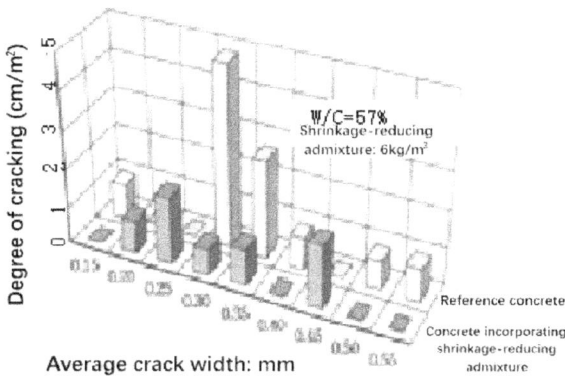

Figure 7.23 Investigation results of concrete cracking.

Figure 7.23 shows the results of a case study on actual structures. A concrete slab of 250 mm thick incorporating shrinkage-reducing admixture was built on a loose soil ground. The effect of the shrinkage-reducing admixture on cracking of concrete was measured. It can be seen from Figure 7.23 that the crack length of the reference concrete slab was 11 cm/m^2, but the concrete slab containing shrinkage reducing admixture was only 4 cm/m^2 approximately. Moreover, in the latter slab, the average crack width was much smaller, and the number of cracks of larger width was reduced, which confirms the significant effect of shrinkage-reducing admixture on shrinkage of concrete.

7.5.3   Impact on concrete durability

There is no difference in durability between concrete containing shrinkage-reducing admixture and reference concrete. However, the durability of the concrete structure containing shrinkage-reducing admixture is better than that of reference concrete structure, this is because both the number and the width of cracks in concrete are reduced, hence the durability is improved.

7.5.4   Shrinkage-reducing admixture sold on the market

The representative shrinkage-reducing admixtures available on the market are shown in Table 7.1. Tianjin Construction Research Institute in China has developed a shrinkage-reducing admixture named "SRS". It is effective, with a shrinkage-reduction rate of 30-70% at early age and that of 20-40% at long term age of concrete.

Table 7.1 Representative admixtures for reducing concrete shrinkage on the market (Mukimoto Market)

Type	Main Ingredient	Appearance	Density	Standard Dosage	Application
A	Low alcohol epoxide	Colourless transparent liquid	1.0	6kg/m3	Concrete
B	Polyether	Colourless transparent liquid	1.0	CX 2-6%	Concrete
C	Glycol ether	Light yellow liquid	0.97	CX1-4%	Concrete
D	Propylene-Ethanol-Ethers	Light-coloured liquid	0.93	CX2%	Concrete

7.6   Inhibition of cracking caused by concrete shrinkage by using SRA series of shrinkage-reducing admixture

7.6.1   Mix proportion and performance of concrete tested

Tianjin 42.5 MPa ordinary Portland cement, medium size sand, crushed stone (particle size 5-10 mm), and SRA-E type of shrinkage-reducing admixture at a dosage of 2% were used in the test. The mix proportion of the concrete is given in Table 7.2. The test results are given in Table 7.3.

Table 7.2 Mix proportion of tested concrete (kg/m³).

NO.	W/B	Material Composition					
		Cement	Sand	Coarse Aggregate	SP-5	SRA-E	Water
1	0.46	380	796	1099	0.6%	0	175
2	0.32	550	724	1001	1.1%	0	175
3	0.32	550	724	1001	0.7%	2.0%	175

Table 7.3. Performance of fresh concrete and strength of hardened concrete

NO.	Slump change over time (mm)			Compressive strength (MPa)			
	Initial	1 h	2 h	3 d	7 d	28 d	90 d
1	210	165	120	26.4	39.6	46.9	47.3
2	240	215	195	42.6	58.3	66.8	69.5
3	255	240	240	37.9	57.7	68.6	72.5

7.6.2 Summary of shrinkage-caused cracking in the test on concrete slab

Crack area and the maximum crack width are used as the evaluation criteria for cracking behaviour in the slab test. The test results are shown in Table 7.4.

Table 7.4 Results of slab testing.

NO	Description of cracks						
	Occurrence h:m	Average width mm	Maximum mm	Length cm	No of cracks	Crack area mm^2	Control %
1	2:31	0.51	1.5	49.7	1	393.5	-
2	1:50	0.74	1.45	24	1	244.4	Benchmark
3	2:35	0.12	0.2	43.5	1	63.75	73.9

It can be seen from Table 7.4 that the crack area of concrete No. 3 containing shrinkage-reducing admixture was only about ¼ of that of reference concrete No. 2. The control of cracks on concrete No. 3 was about 73.9%. Adding SRA-E shrinkage-reducing admixture has no adverse effect on other properties of concrete.

7.7 Self-curing agent made with ultrafine natural zeolite powder for inhibiting cracking of UHPC caused by autogenous shrinkage

Natural zeolite with zeolite content higher than 60% is selected, and then crushed and ground to a specific surface area greater than 6000 cm^2/g, which can be used as self-curing agent to inhibit cracking of UHPC caused by autogenous shrinkage.

7.7.1 Formation of UHPC structure

After UHPC is poured and compacted via vibration, due to the difference in density between constituent materials, settlement and floating may occur inside the concrete. For example, sands and gravels tend to settle, while water tends to rise. When the water flows upward, a capillary passage may be formed, and floating paste may also appear on the upper surface, which is a main cause for inhomogeneity in concrete structure.

7.7.2 Cracks of UHPC caused by autogenous shrinkage

After initial setting of paste in UHPC, it begins to harden, but the cement in the concrete continues to hydrate, and the cement particles need to absorb the surrounding water. The water in capillary pores is the source for the continued hydration of the cement, so that the capillary pores lose gradually the adsorbed water and enter a vacuum state, generating capillary tension, which may cause cracking of concrete once it exceeds tensile strength of concrete. Due to the large cement amount in UHPC, such cracking caused by autogenous shrinkage may be rather severe.

7.7.3 How to inhibit cracking caused by autogenous shrinkage by using ultrafine natural zeolite powder

Ultrafine natural zeolite powder may be used as one of constituent materials of UHPC, generally at a dosage of 20-30 kg/m^3. During mixing of concrete, ultrafine natural zeolite powder is added and it can absorb water, which is evenly distributed in the concrete. After concrete sets and hardens, the cement particles continue to hydrate, which absorbs the surrounding water, which may be supplied by the zeolite powder evenly distributed in the concrete, so as to avoid excessive surface tension caused by loss in capillary water and hence inhibit cracking caused by autogenous shrinkage. In addition, the ultrafine natural zeolite powder in UHPC has another effect, i.e. it can participate in hydration and hardening of the cementitious materials, and improves both strength and durability of UHPC.

Multifunctional Concrete Technology   Materials Research Forum LLC
Materials Research Foundations **127** (2022)   https://doi.org/10.21741/9781644901991

Chapter 8

Corrosion Inhibitors Applied to Reinforced Concrete

In order to inhibit the carbonation in the reinforced concrete, a sort of admixture i.e. corrosion inhibitor has been used. Concrete is a strong alkaline substance, with pH of 12.5-13. The surface of rebar in reinforced concrete may be covered with a blunt film (r-Fe_2O_3) with a thickness of 2-6 nm. However, if the chloride ion content on the surface of the rebar is greater than or equal to 0.3 g/m^3 to 0.6 kg/m^3, the blunt film on the surface of rebar will begin to be damaged, forming electrochemical corrosion. Therefore, it is necessary to mix a corrosion inhibitor in reinforced concrete.

8.1   Types of corrosion inhibitor

Corrosion inhibitor is a kind of admixture to inhibit corrosion. It is adsorbed on the surface of the rebar through physical or chemical action, and inhibits the anode or cathode reaction on the rebar, thus it can inhibit the corrosion of the rebar. The corrosion inhibitors for anodic corrosion resistance include chromate, nitrite, orthophosphate, and silicate benzoate etc. The corrosion inhibitors for cathodic corrosion resistance include all kind of alkali, such as NaOH, Na_2CO_3, and NH_4OH etc, which can increase the pH value of the medium and reduce the solubility of iron ions. Aniline and its chlorine group, alkyl group and nitro group substituents, as well as p-hydrothiophene benzene can also be used as corrosion inhibitors, of which the test results are also very good. In addition, there are a mixed type of corrosion inhibitor, the electron density distribution in the molecule therein makes this corrosion inhibitor be absorbed both at the cathode and at the anode, so that it can inhibit either the corrosion of the cathode or the corrosion of the anode.

8.2   Effect of corrosion inhibitors

8.2.1   Corrosion of rebar

In general, concrete is highly alkaline, with pH of about 12-13. Under such a condition, the blunt film (r-Fe_2O_3) on the surface of rebar is stable. However, if the corrosion reaction occurs on the surface of the rebar, more flaws will be produced at this film. Under the condition of available water, the corrosion reaction of rebar is an electrochemical reaction. Microcell corrosion occurred on the steel surface, which further promoted the corrosion of rebars. The damaged part of the rebar becomes the anode and the intact part becomes the cathode. Therefore, corrosion begins when the following reactions occur under neutral or alkaline conditions, as shown in the Figure 8.1.

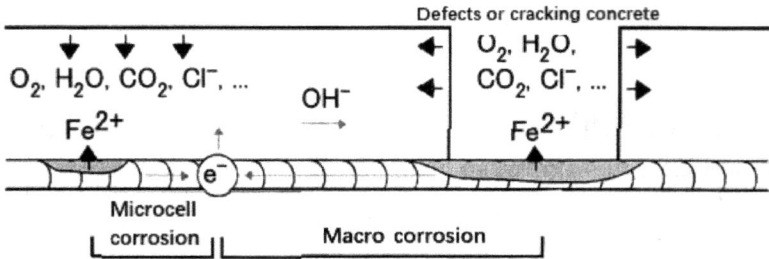

Figure 8.1 Rebar corrosion reaction mechanism.

1) The reaction process of microcell corrosion in the anode part (damaged part) is as follow:
Reaction at the micro anode (ferrite):

$$Fe - 2e^- = Fe^{2+}$$

$$Fe^{2+} + 2OH^- = Fe(OH)_2$$

$$Fe(OH)_2 + 1/2\, H_2O + 1/4\, O_2 \longrightarrow Fe(OH)_3$$

2) Reaction at the cathode (impurity):

$$2H+ + 2e^- \longrightarrow H_2$$

$$1/2\, O_2 + 2e^- + H_2O \longrightarrow 2OH^-$$

The feature of microcell corrosion is that the two corrosion processes, both at the cathode and at the anode, occur simultaneously in different areas of the same metal, and the corrosion current flows directly in the metal. In an anode cell, the anodic process forms the corrosion. The electrochemical inhomogeneity of metal surface can easily cause microcell corrosion.

In the presence of chloride ion, in order to keep the balance of anionic cation and cation at a flaw on the surface of a rebar, chloride ions coming either from the constituent raw materials of concrete or from an aqueous solution outside via diffusion through the concrete accumulates at the flaw. At the same time, due to the decomposition of Fe^{2+} in water, pH value decreases, the corrosion at the flaw is further developed, so that microcell corrosion gradually develops into macrocell corrosion.

8.2.2   Corrosion of rebar in the presence of chloride ions

In the presence of chloride ions, rebar in normal concrete may be stable due to passive film protection. However, even in this alkaline environment, the passive film is vulnerable to damage and the rebar is vulnerable to corrosion in the presence of chloride. In case of neutralization of concrete, the pH value decreases, which will accelerate the corrosion process. After the rebar in concrete is corroded, rust is generated, and the volume expands 2.5 times, resulting in expansion pressure and cracking of concrete. Once cracking occurs in a concrete member, both moisture and the oxygen can penetrate into the concrete, which will accelerate the corrosion of the rebar and worsen the properties of a concrete structure significantly.

The relationship between the Cl⁻ content brought by sand into and the corrosion degree of rebar in the concrete is shown in Figure 8.2. The more Cl⁻ content in the sand, the more Cl⁻ content will promote the corrosion of rebar. However, if there is also Cl⁻ infiltration from outside into the concrete interior, the pH value is reduced more greatly, and corrosion is accelerated.

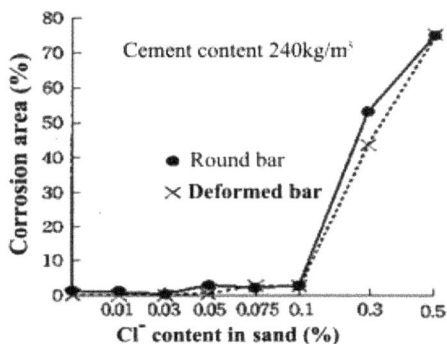

Figure 8.2 Relationship between content of Cl⁻ induced by sand in concrete and rebar corrosion.

8.2.3   Mechanism for effect of corrosion inhibitors in concrete

Corrosion reaction is an electrochemical reaction, i.e. the anode reaction and the cathode reaction take place at the same time. Corrosion of rebar can be prevented if one or both reactions can be inhibited. The role of corrosion inhibitors is to inhibit the electrochemical reaction.

Most of the corrosion inhibitors sold on the market are nitrite, which can inhibit electrochemical anodic reaction.

On the surface of rebar in concrete, oxidation happens between nitrite (NO_2^-) and Fe^{2+} to form Fe_2O_3, so that Fe^{2+} originally moving outward from the anode is blocked, whereas Fe_2O_3 precipitates on the surface of rebar and forms a passive film. The reaction is as follows:

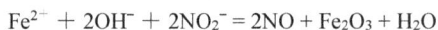

$$Fe^{2+} + 2OH^- + 2NO_2^- = 2NO + Fe_2O_3 + H_2O$$

If there is a large quantity of $2NO_2^-$, it can enhance the inhibition of the corrosion. Once the passive film seals the anode part, the consumption of $2NO_2^-$ will stop.

8.2.4   Results of application of corrosion inhibitors

Figure 8.3 shows the corrosion of rebar in the concrete after 10 years of test, with or without corrosion inhibitors used in concrete.

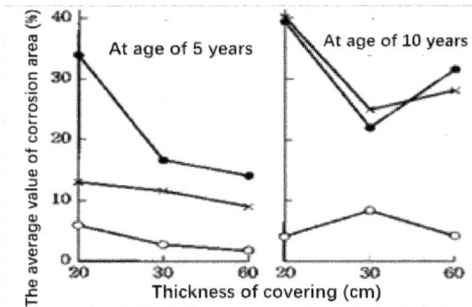

(a) Reinforcement in concrete for outdoor exposure test (Okinawa, Japan)

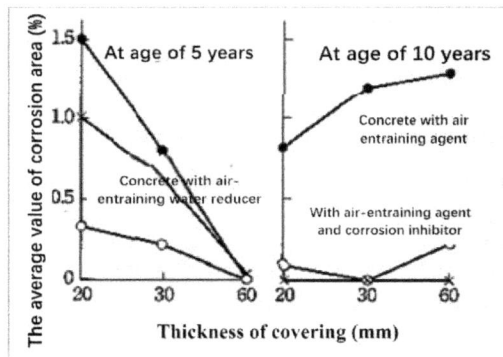

(b) Outdoor exposure test for concrete (roof of material house)

Figure 8.3 Outdoor test of concrete component with different admixtures.

In Figure 8.3 (a), the initial Cl⁻ content was 0.035-0.085% of sand by mass, after 10 years, the Cl⁻ content on the surface increased to 3.3%, and the Cl⁻ content in the interior increased to 2.7%.

In Figure 8.3 (b), after 10 years, the Cl⁻ content increased to 0.24-0.27%.

8.2.5   Concrete incorporating corrosion inhibitors

A corrosion inhibitor with nitrite as the main constituent was mixed into concrete. In Table 8.1, the difference in performance between concretes with and without the corrosion inhibitor is shown.

Table 8.1 Comparison of concrete with and without corrosion inhibitors.

Target Slump		7.5 (cm)			18 (cm)						
CL⁻ content (S %)		0.1	0.1	0.1	nil	nil	0.1	0.1	0.1	0.3	0.3
Corrosion inhibitors (kg/m³)		Nil	3	6	nil	3	nil	3	6	nil	6
W/C %		58.6	58.6	57.9	61.3	61.3	61.3	61.3	60.7	61.3	61.3
C		280	280	280	300	300	300	300	300	300	300
W		164	164	162	184	184	184	184	182	184	184
Slump (cm)		7.0	7.7	7.5	18.6	18.8	18.3	18.0	18.2	18.5	18.5
Air content %		1.5	1.5	1.5	1.6	1.6	1.7	1.5	1.5	1.7	1.6
bleeding(cm³/m²)		0.15	0.14	0.14	0.28	0.26	0.27	0.25	0.25	0.2	0.2
Setting (h: mm)	Initial	4 ; 50	4 : 35	4 : 25	6 : 45	6 : 45	6 : 05	6 : 10	6 : 00	5 : 35	5 : 45
	Final	7 : 15	7 : 00	6 : 50	9 : 00	8 : 55	7 ; 50	7 : 50	7 : 45	7 : 20	7 : 20
Compressive strength(N/mm²)	3 d	15.5	17.6	18.2	10.1	11.8	12.7	14.0	14.1	14.7	16.1
	7 d	22.8	24.2	24.9	17.6	19.4	20.1	21.7	22.4	22.0	22.8
	28 d	33.5	35.0	35.6	31.5	32.8	31.3	32.2	33.1	30.4	31.6
Flexure Strength (N/mm²)	7 d	4.18	4.71	4.55	3.54	3.89	3.71	3.97	3.93	4.09	4.15
	28 d	4.55	4.71	4.78	4.34	4.58	4.35	4.58	4.53	4.41	4.48
Expansion x/10,000	3	6.0	6.1	6.2	6.6	6.6	6.7	6.9	6.9	7.0	7.3
	6	7.0	7.1	7.1	7.4	7.3	7.7	7.8	7.8	7.9	8.0

It can be seen from Table 8.1 that (1) the setting time of concrete containing the corrosion inhibitors was relatively short; (2) compressive strength at 3 d increased slightly; other properties, such as air content, bleeding and expansion, were roughly the same. After years of site exposure test, it has been proved that calcium nitrite is stable in concrete and can be adsorbed on rebar for a long time. Compound multi-functional RI-1 corrosion inhibitor developed by China Metallurgical Building Research Institute, was used in SanShandao gold mine project in Shandong province. After 7 years of accelerated testing of long term corrosion resistance, the results show that the time for beginning of expansion-caused cracking in concrete can be extended 2-5 times.

NS-1, a non-nitrite anodic compound corrosion inhibitor, was developed by Nanjing Academy of Water Sciences. It is superior to nitrite-based corrosion inhibitor in the environment of chloride pollution or carbonation.

Corrosion inhibitors have been used in the concrete projects in the United States, Canada, Italy, Japan and the Middle East etc. During 15 years from 1993, corrosion inhibitors have been used in concrete for more than 20 million m³ in construction projects. Calcium nitrite corrosion inhibitor has been widely used in bridge panels and prefabricated prestressed components.

8.2.6   Precaution for corrosion inhibitor application

1) In general, when adding corrosion inhibitor, we should pay attention to the Cl⁻ content in concrete, which should be lower than 0.6 kg/m³, so that corrosion inhibitors at a general dosage may be effective.

2) When corrosion inhibitor and other admixtures are added at the same time, the pH value of concrete should not be affected. This requires prior testing.

Chapter 9

Research and Experimental Application of Cl⁻ Solidifying Agent in Concrete

On the one hand, the Cl⁻ in the concrete is brought in by the constituent materials of the concrete, and the Cl⁻ allowed to be brought in by the constituent materials in the concrete is shown in Table 9.1. On the other hand, the external Cl⁻ penetrates into the concrete structure through diffusion. The diffusion and penetration of Cl⁻ in the capillary pores may be divided into combined Cl⁻, adsorbed Cl⁻ and free Cl⁻ (see Figure 9.1); the former two are collectively referred to as solidified Cl⁻; only free Cl⁻ can enter into concrete, and endanger the safety of the structure.

Table 9.1 Allowable chloride ions in concrete constituent materials.

Quality Class of concrete	Cl- content in concrete	1	2	3
		Fine aggregate	Water	1+2
High	Limit	0.04%	0.033%	
	Amount	800kgX0.04%=320 g	200kgX0.033 %= 66 g	386 g
Normal	limit	0.1 %	1 g/l (residue)	
	Amount	800kgX0.1 %=800 g	200kgX1%=200 g	100 g

Calculation based on the data given in Table 1: fine aggregate 800 kg/m³, water 200 kg/m³, for high quality concrete and for normal concrete, the allowable chloride content is 386 g/m³ and 1000 g/m³, respectively.

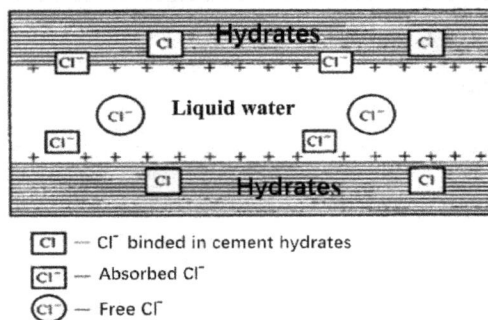

Figure 9.1 The state of Cl⁻ existing in the concrete capillary pores.

Chloride ions penetrate through the capillary pores into the concrete structure. When the chloride ion content on the surface of a steel rebar is ≧ 0.3 kg/m³, the rebar begins to encounter microcell corrosion. It gradually develops into macrocell corrosion, which gradually damages the rebar.

9.1 Solidifying of Cl⁻ by chloride ion solidifying agent

The main components of the chloride ion solidifying agent studied in this experiment are: nitrite-hydro-calcium-aluminate ($3CaO \cdot Al_2O_3 \cdot Ca(NO_2)_2 \cdot nH_2O$). It can interact with the Cl⁻ which either existing in the concrete or diffused and penetrated from external into the concrete, combined to form a new hydrate Friedel salt. This will convert the free Cl⁻ to solidified Cl⁻ and eliminate the risk of corrosion of rebar caused by accumulation of Cl⁻.

9.1.1 Mechanism of Chloride Ion Solidifying Agent

The reaction of chloride ion solidifying agent ($3CaO \cdot Al_2O_3 \cdot Ca(NO_2)_2 \cdot nH_2O$) with Cl⁻ is as below:

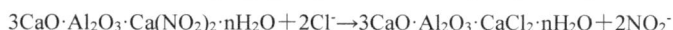

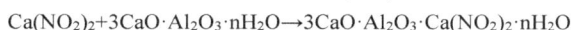

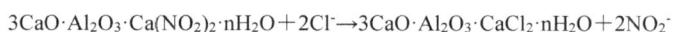

$$3CaO \cdot Al_2O_3 \cdot Ca(NO_2)_2 \cdot nH_2O + 2Cl^- \rightarrow 3CaO \cdot Al_2O_3 \cdot CaCl_2 \cdot nH_2O + 2NO_2^-$$

$$2NO_2^- + Ca^{2+} \rightarrow Ca(NO_2)_2$$

$$Ca(NO_2)_2 + 3CaO \cdot Al_2O_3 \cdot nH_2O \rightarrow 3CaO \cdot Al_2O_3 \cdot Ca(NO_2)_2 \cdot nH_2O$$

$$3CaO \cdot Al_2O_3 \cdot Ca(NO_2)_2 \cdot nH_2O + 2Cl^- \rightarrow 3CaO \cdot Al_2O_3 \cdot CaCl_2 \cdot nH_2O + 2NO_2^-$$

NO_2^- can be recycled.

The solidifying agent reacts with Cl⁻ to produce Fluoride salt ($3CaO \cdot Al_2O_3 \cdot CaCl_2 \cdot nH_2O$), which turns the dissociated Cl⁻ into a compound, which means that Cl is solidified.

9.1.2 Regeneration cycle of NO₂⁻ 's

NO_2^- reacts with Ca^{2+} in concrete to form $Ca(NO_2)_2$, and $Ca(NO_2)_2$ reacts with $3CaO \cdot Al_2O_3 \cdot nH_2O$ in hydrated of cement to produce $3CaO \cdot Al_2O_3 \cdot Ca(NO_2)_2 \cdot nH_2O$ (Hydrocalumite). Hydrocalumite reacts with Cl⁻ in concrete to produce Friedel salt and release NO_2^-.

$$3CaO \cdot Al_2O_3 \cdot Ca(NO_2)_2 \cdot nH_2O + 2Cl^- \longrightarrow 3CaO \cdot Al_2O_3 \cdot CaCl_2 \cdot nH_2O + 2NO_2^-$$

Therefore, it can inhibit the corrosion of rebar caused by chloride ion accumulation from the inside and outside of concrete, and it can be recycled.

9.2 Research for Dongying Yellow River Bridge Project in Shandong

According to investigation, the reinforced concrete bridge in Weifang, Shandong is severely damaged by salt. A study of chloride ion solidifying agent has been carried out for Dongying Yellow River Bridge project. The test plan is shown in Table 9.2.

Table 9.2 Test on Chloride ion solidifying agent

NO.	Size (mm)	W/B	Materials	Steel bar	Cl-solidifying agent	Salt content
1	25×25×285	0.5	C=400, S=900	φ4x250mm	control	/
2	25×25×285	0.5	C=392 S=900	ditto	8 g (CX2%)	/
3	25×25×285	0.5	C=384 S=900	ditto	16 g (CX4%)	/
4	25×25×285	0.5	C=376 S=900	ditto	16 g (CX4%)	8 g (2%)

In Table 9.2: No. 1 was the control specimen, No. 2 was the specimen with Cl⁻ solidifying agent at 2% of the cement content, No. 3 was the specimen with Cl⁻ solidifying agent at 4% of the cement, and No. 4 was the specimen with Cl⁻ solidifying agent accounts at 4% of the cement content together with sea salt at a dosage of 2%.

Rebar: Φ=4 mm, l=250 mm, the surface was derusted, and then embedded in the test specimens, the thickness of the rebar cover was $\geq$ 10 mm. After standard curing for 7 days, the specimens were immersed in a 3% NaCl solution, and the immersion solution was replaced once a month.

The specimen taken out from the immersion solution after 1 year and 2 months are shown in Figure 9.2.

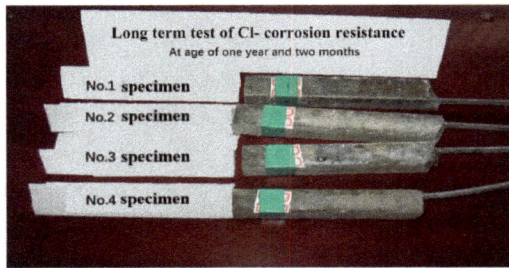

Figure 9.2 Surfaces of rebars of the test specimens taken out from the salt solution.

When the test specimen was just taken out from the salt solution, no corrosion was found on the surface of the rebar. After exposure to air for one day, it was found that the rebar in 1# and 4# specimens were rusted, as shown in Figure 9.3.

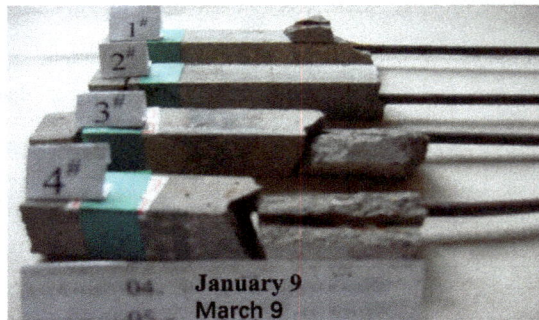

Figure 9.3 Corrosion of 1# and 4# specimens after being stored in the air for 1 day.

After one more year, the test specimens, immersed for 2 years and 2 months in the salt solution, were taken out from the solution, as shown in Figure 9.4. When it was just taken out, no corrosion on the surface of the rebar was found, but after exposure to air for 1 day, the surfaces of 1#, 4# specimens were corroded, and the surface of 2# specimens (solidifying agent content 2%) was slightly corroded. The specimen with Cl⁻ solidifying agent at 4% of the cement content were still intact.

The Cl⁻ solidifying agent was used at 4% in the underwater bridge pier concrete for this project as a trial.

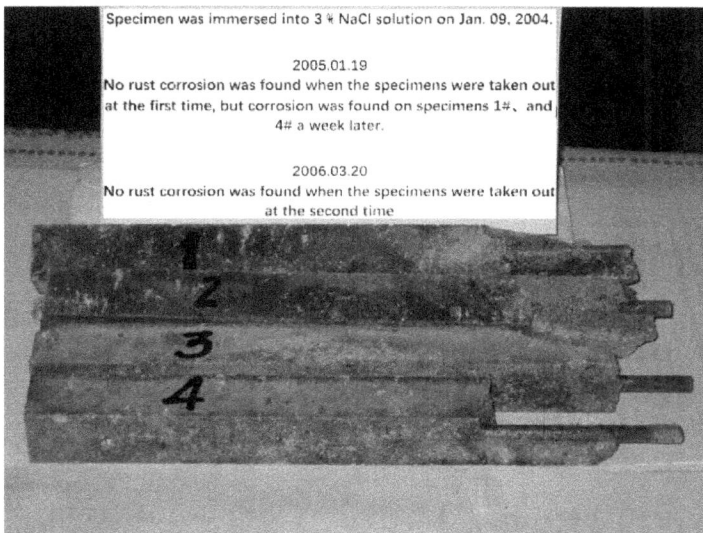

Figure 9.4 Surfaces of rebars after the specimens being taken out at age of 2 years and 2 months.

9.3   Mortar test on Mediterranean sea sand

Mediterranean sea sand was provided by IKRAM Company, Malaysia. At the same time, river sand from Yongding River in Beijing was used for the comparative test. PO 42.5 cement was used. The mortar mix design is given in Table 9.3.

9.3.1 Specimen production, rebar embedment and potential measurement

The experiments are shown in Figure 9.5 and Figure 9.6 respectively.

Table 9.3 Test of mortar prepared from Mediterranean sea sand

No	Mortar mix design	condition
1	River sand mortar cement:sand = 1 : 3 W/C=0.5 specimen size: 40X40X160 mm φ6 mm rebars embedded in the center of specimen after derusting.	After demouling, the specimens were standard cured for 7 days, then immersed in saturated Ca(OH)₂ solution for 3 weeks. The half-cell potential was measured. Then the specimens were soaked in 3% NaCl solution, and the steel corrosion was measured.
2	Sea sand mortar Mortar mix design and specimen size are as above, but river sand was replaced by sea sand.	ditto
3	The mortar mix design is as No2, but 10% Chloride ion solidifying agent was added.	ditto

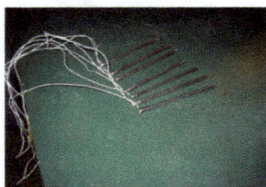

(a) Rebar preparation   (b) Casting   (c) Half-cell measurement

Figure 9.5 Sea sand mortar testing

The test specimens were immersed in 3% NaCl solution, and the wet-dry cycle was performed 10 times. At age of 6 months, the relationship between the potential and time was measured, as shown in Figure 9.6.

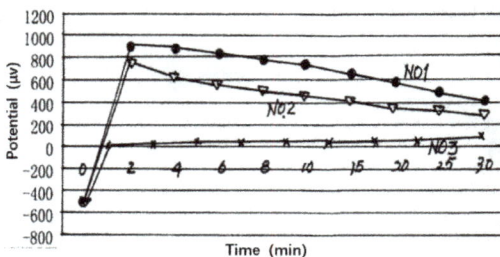

Figure 9.6 Potential vs. time curve of different mortar specimens.

The NO.1 and NO.2 curves indicate that the rebars in the test specimens were corroded, and the potential curve initially rose, and then gradually decreased after reaching the peak. While the NO.3 curve was relatively stable, indicating that the rebars in the test specimens were not corroded. This is due to the addition of Cl⁻ solidifying agent.

9.3.2 Corrosion of rebars embedded in mortar specimens prepared from Mediterranean sea sand

As shown in Figure 9.7, (a) is a sea sand specimen with 10% Cl⁻ solidifying agent, and no rust was found after soaked in the salt water for 6 months during the dry and wet cycle. (b) is sea sand + 3% NaCl+ 10% Cl⁻ solidifying agent, no rust was found as well after 6 months of dry and wet cycles in salt water. (c) Sea sand specimens without solidifying agent, were found to be rusty; (d) Sea sand without solidifying agent + NaCl 3%, the specimens were rusty.

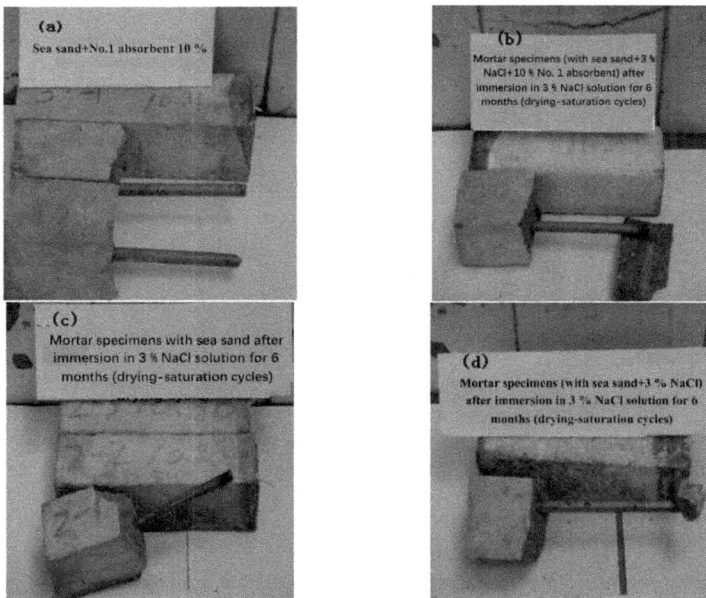

Figure 9.7 Effect of solidifying agent for Mediterranean sea sand

9.4 Field-exposed test of sea sand reinforced concrete specimens

In 2009, using sea sand blended with river sand, a series of trial-produced reinforced concrete columns were placed at a seaside field of Baoan in Shenzhen to observe the corrosion of steel bars in concrete attacked by internal and external chloride ions.

9.4.1   Raw materials for sea sand concrete column test

Cement: PO 42.5

Sea sand: sea sand from Shenzhen sea beach

River sand: from Pearl River Basin

Chloride ion solidifying agent: Own produced

NaCl

9.4.2   Concrete mix design

C40 concrete mix proportion is listed in Table 9.4.

Table 9.4 Mix proportion of concrete column test with chloride ion solidifying agent (kg/m³)

NO.	Ceme nt	Wate r	Sla g	Fly ash	Fine aggregate	Coarse aggregat e	Cl- solidifying agent	Na Cl
1	300	168	50	70	700 (sea sand)	1050	3%	1%
2	300	168	50	70	700(sea sand)	1050	3%	
3	300	168	50	70	700(sea sand)	1050		
4	300	168	50	70	700(river sand)	1050		
5	300	168	50	70	700(50% sea sand and 50% river sand)	1050		

Note: Nos. 1, 2 and 3 were the specimens using sea sand; No. 4 using river sand; No. 5 using half river sand and half sea sand.

9.4.3   Concrete specimens

Two test column specimens for each mix proportion were prepared with the size of 100×40×20 cm with a steel cage inside the specimens, and the covering layer was 3 cm thick. Part of the mortar was screened out from the concrete for each mix proportion, and three specimens of 4×4×16 cm were prepared, and ϕ10 circular rust-removing rebars were placed inside. One set of specimens were stored in open air; the other set of specimens was placed at the beach.

The open-air test specimens in Baoan Sea are shown in Figure 9.8. Among them, No. 1 and No. 2 were mixed with 3% Cl⁻ solidifying agent, and the specimen of No. 1 was also mixed with 1.0% NaCl; No. 3 was also a sea sand specimen, but contained no salt or Cl⁻ solidifying agent. No. 4 was a river sand specimen, and No. 5 was a specimen with half sea sand and half river sand. These specimens had undergone five years of open-air exposure test, and no chloride ion corrosion deterioration had been found. No cracks were found during the inspection at age of 10 year in 2020. It proved that the internal rebars were still not corroded. The Cl⁻ solidifying agent developed by the author was still being developed and tested in Singapore and Malaysia.

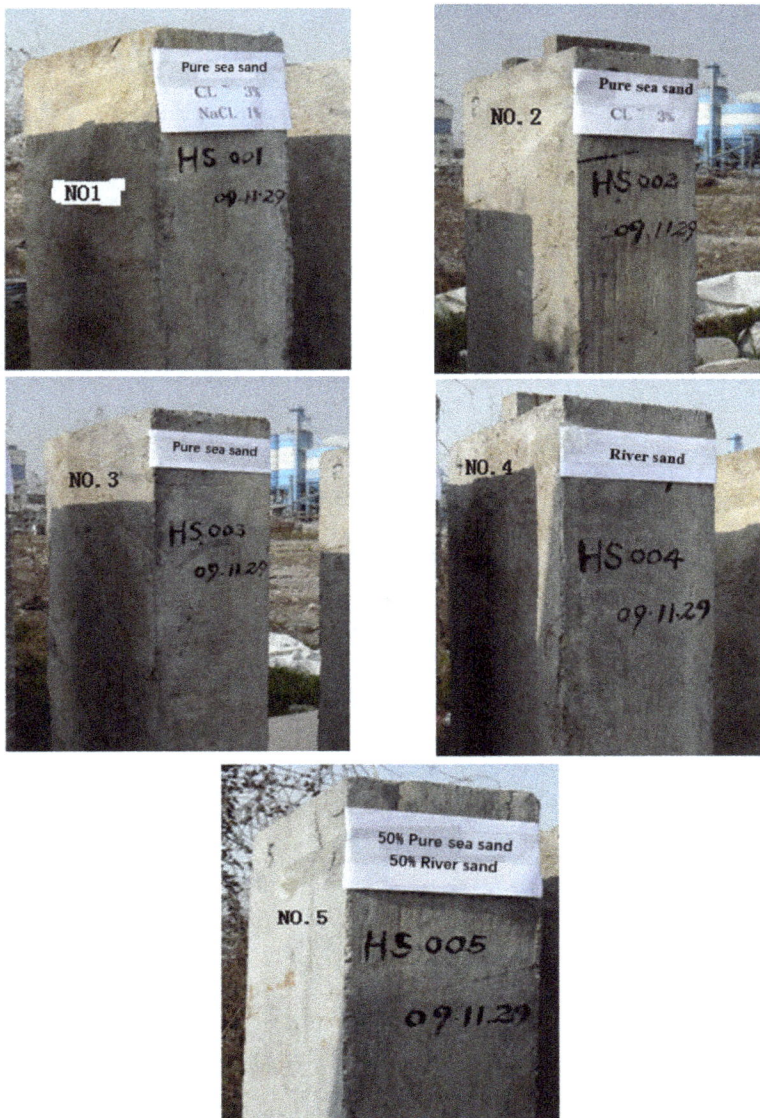

Figure 9.8 Observation of steel corrosion in sea sand reinforced concrete specimens

9.5   Conclusions of the experimental study

According to the results of mortar and concrete tests on Cl⁻ solidifying agent, conclusions are drawn as follows:

1)   With addition of Cl- solidifying agent in mortar or concrete, the internal Cl⁻ or the external Cl⁻ penetrated from the outside via diffusion can be combined into solid product, so that rebars therein are protected from the corrosion of Cl⁻.

2)   The main component of the chloride ion curing agent is nitrite type hydroaluminate ($3CaO \cdot Al_2O_3 \cdot Ca(NO_2)_2 \cdot nH_2O$), which can react with chloride ions to form Friedel salt ($C_3A \cdot CaCl \cdot nH_2O$). At the same time, it also releases nitrite ions (NO_2^-) which react with Ca^{2+} (calcium ions) in the cement to form $Ca(NO_2)_2$;

3) $Ca(NO_2)_2 + 3CaO \cdot Al_2O_3 \cdot nH_2O \rightarrow 3CaO \cdot Al_2O_3 \cdot Ca(NO_2)_2 \cdot nH_2O$. The regeneration cycle has been realized.

Chapter 10

Microbeads and its effect

Microbeads has gained a large number of applications in China: for example, Jinzhu Concrete Products Company used large quantity of microbeads to produce C70 high performance concrete for the construction of Shenzhen Cultural Plaza; Zhongzheng Precast Pile Factory used microbeads to produce C80 high performance concrete for casting and production of PHC piles; Jingji Building used microbeads to produce C120 ultra high performance concrete. It is a new type of powder material with special properties. Internationally, it is known as modified fly ash. Its particle size distribution curve is between that of fly ash and silica fume, as shown in Figure 10.1 (a). It is the remaining ultrafine particles sieved from the original coarse fly ash, after deducing coarse particles.

10.1 The characteristics of microbeads

10.1.1 Particle size distribution of silica fume, microbeads, and fly ash

The particle size distribution of silica fume, microbeads, and fly ash is shown in Figure 10.1 (a). It can be seen that the particle size distribution curve of the microbeads is between the particle size distribution curve of silica fume and that of the original fly ash. The surface area of common fly ash is comparable to that of normal cement. Thus, the three-component composite of cement, microbeads, and silica fume, can effectively reduce the voidage of the binder.

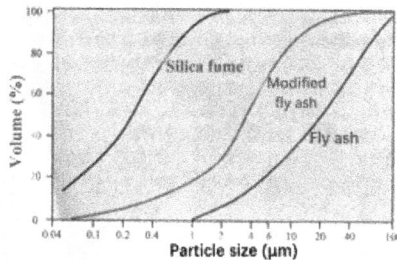

Figure 10.1 (a)The particle size distribution of silica fume, microbeads and fly ash.

10.1.2 Average particle size distribution.

Microbeads is ultra-fine powder with an average particle size ≤1.2 micron. The average particle size distribution curve of microbeads is shown in Figure 10-1 (b). As shown in Figure 10.1 (b): the mean particle size ≤ 0.2 micron accounted for 27.23%, and 0.2-1 micron accounted for 42.43%; that is, the mean particle size ≤ 1.0 micron accounted for a total of 69.66%.

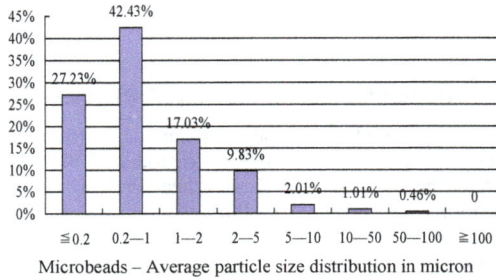

Figure 10.1 (b) Average particle size distribution of microbeads.

Cement has a surface area of about 4000 cm^2/g and an average particle size of about 10-20 micron, while silica fume has a surface area of about 200,000 cm^2/g and an average particle size of about 0.1 micron. Therefore, in a three-component mixture consisting of cement, microbeads, and silica fume, microbeads fill the voids between cement particles, and silica fume fills the voids between microbeads, thus achieving a densely filled powder system.

10.1.3 Particle size analysis

The particle size analysis curve is shown in Figure 10.2.

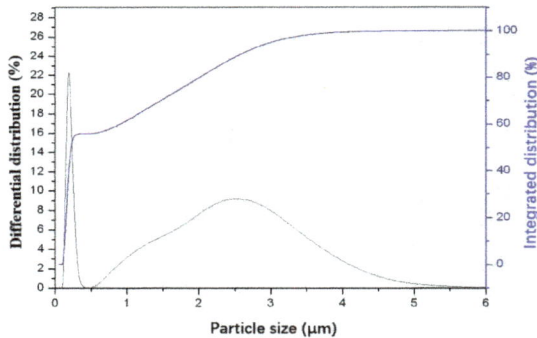

Figure 10.2 Particle size analysis curve.

For the cumulative curve (*blue*): D10 (*10th percentile*) is 0.14 micron; D25 is 0.17 micron; D50 is 0.21 micron; D75 is 1.76 micron; and D90 is 2.51 micron. For the differential curve: between 0-0.5 micron, D50 is 0.18 micron; and between 0.5-5 micron, D50 is 1.91 micron.

10.1.4 Scanning electron microscope

The scanning electron microscope image is shown in Figure 10.3. The particles of the microbeads are spherical, like beads, and flow easily.

(a) (x 10000)   (b) (x 20000)

Figure 10.3 Scanning electron microscope image.

10.1.5 The chemical composition of microbeads

The chemical composition of microbeads is shown in Table 10.1.

Table 10.1 Chemical composition of microbeads (%)

Chemical composition	SiO_2	CaO	MgO	Al_2O_3	Fe_2O_3	Na_2O	K_2O	SO_3	Carbon content
Microbeads	56.5	4.8	1.3	26.5	5.3	1.4	3.28	0.65	<1

As can be seen from Table 10.1, the main chemical components of microbeads are SiO_2 and Al_2O_3, both forming a total content of 83% and contains relatively high soluble SiO_2 and Al_2O_3. Soluble silicate comprises about 8.67% of the total silicate, and soluble aluminate is about 15.42% of the total aluminate. In addition, there is soluble iron content of 3.49%. Therefore, the chemical reactivity of microbeads is relatively high.

10.1.6 Comparison of microbeads with other minerals

According to data from Shenzhen Tongcheng New Materials Technology Co., Ltd., the activity and material characteristics of different ultrafine minerals are shown in Table 10.2.

Microbeads is spherical ultra-fine powder, particle size range between 0.1–5 μm; it has high coefficient of reactivity, good water reduction, and good flowability. Kunming University of Technology carried out an experiment to compare the performance of cement mortar at fixed water/cement ratio using different microbeads replacements of 6%, 12%, 18%, 24% and 30% respectively, with silica fume dosage of 10% and cement grade 42.5. The relative expansion and relative compressive strength of 28 days and 56 days are shown in Figure 10.4.

Table 10.2 Comparison of the properties of several ultra-fine mineral.

Characteristics	Fly ash	Ground granulated blastfurnace slag	Microbeads	Silica fume
Main particle size distribution, μm	5～30	10～40	0.1～5	0.1～0.5
Particle shape	Mostly spherical	Angular	Spherical	Spherical
Compactability	Medium	Poor	Good	Good
Water reduction	Good	Poor	Excellent	Poor
Flowability	Good	Good	Excellent	Poor
Coefficient of reactivity	Poor	Good	Good	Good
Anti-cracking	Good	Medium	Excellent	Poor

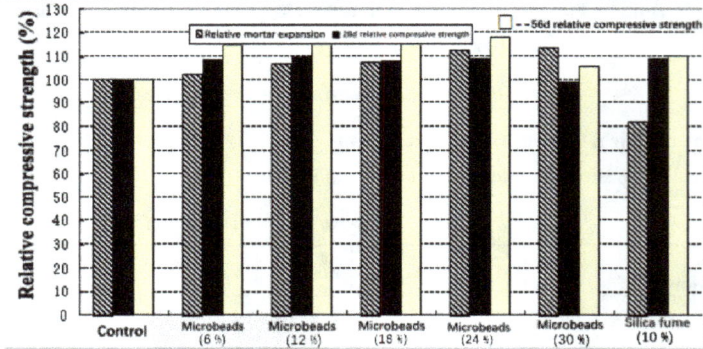

Figure 10.4 Relative mortar expansion and compressive strength of 10% silica fume with different percent addition of microbeads.

10.1.7 Physical characteristics

1. Particle shape: fully spherical; 2. Fineness: average diameter 1.2 μm; 3. Particle density: 2.52 g/cm^3; 4. Bulk density: 0.67 g/cm^3; 5. Moisture content: ≤0.1%; 6. Standard consistence required water ratio: ≤95 per cent; Standard mortar required water ratio: about 77-85% (high water reduction property); Crystalline structure: all amorphous (i.e. glass); Radioactive exposure index: 0.16 (standard ≤1.3).

Thus, microbead is a new type of ultra-micro-powder material. It is an ultra-fine (sub-micron) all-spherical powder. Microbeads has high activity index, is light weight, and has good heat insulation, good electrical insulation, high and low temperature resistant, corrosion resistant, radiation protection, good sound insulation, wear resistant, high compressive strength, good flowability, good thermal stability, unique resistance to thermal effects, waterproof and fireproof, non-toxic and many other excellent properties. It can be used as a new high-quality active mineral blend for high-performance concrete.

10.2 The flowability and strength of microbeads-cement mortar

10.2.1 The flowability of microbeads-cement mortar

The effect of microbeads on the flowability of cement mortar was studied using different brands of admixtures at different dosage, as shown in Table 10.3.

Table 10.3 Study on effect of microbeads on the flowability of cement mortar.

(WZ-microbeads)

Cement	500	475	450	400	300	250
WZ	0	25	50	100	200	250
W/B	0.3	0.3	0.3	0.3	0.3	0.3

The admixtures were: a naphthalene-based admixture at dosage of 0.8%, an ammonia-based water reducing agent (solid content of 38.5%) at dosage of 0.8%, and a polycarboxylate superplasticizer (solid content of 20%), at dosage of 0.6%.

At the same water content, the mortar flowability is shown in Tables 10.4, 10.5, and 10.6.

Table 10.4 Mortar flowability (mm) at the same water content (with naphthalene-based admixture at dosage of 0.8%).

No.	Cement + Microbeads	Admixture 0.8%	W/B=0.3 Water (ml)	Mortar flowability (mm)	
				Initial	After 1 hr
1	500+0	4 g	150	196	155
2	475+25	4 g	150	216	175
3	450+50	4 g	150	230	200
4	400+100	4 g	150	233	200
5	300+200	4 g	150	233	202
6	250+250	4 g	150	230	214

Table 10.5 Mortar flowability (mm) at the same water content (with ammonia-based water reducing agent).

No.	Cement + Microbeads	Admixture 0.8 %	W/B=0.3 Water (ml)	Flowability	
				Initial	2 hr
1	500+0	3.9 g	147.6	220	180
2	475+25	3.9 g	147.6	250	220
3	450+50	3.9 g	147.6	270	240
4	400+100	3.9 g	147.6	240	210
5	300+200	3.9 g	147.6	170	110
6	250+250	3.9 g	147.6	140	100

Table 10.6 Mortar flowability (mm) at the same water content (with polycarboxylate superplasticizer).

No.	Cement + Microbeads	Admixture 0.6%	W/B=0.3 Water (ml)	Flowability	
				Initial	2 hr
1	500+0	3 g	148	195	210
2	475+25	3 g	148	205	220
3	450+50	3 g	148	205	210
4	400+100	3 g	148	185	165
5	300+200	3 g	148	115	100
6	250+250	3 g	148	100	0

As shown in Tables 10.4, 10.5, 10.6: (1) For the Naphthalene-based water-reducing agent, when the microbeads dosage was between 5% to 50%, the flowability was higher than that of the control mix; the flowability of the mortar was the highest when the microbeads dosage was between 20-40%; the flowability was reduced when the microbeads dosage was above 50%; (2) For the ammonia-based water reducing agent, when the amount of microbeads dosage was between 5% to 20%, the flowability of the mortar was higher than the flowability of the control mix; and when the microbeads dosage was more than 40%, the flowability of the mortar decreased noticeably; (3) For the polycarboxylate water reducing agent, when the microbeads dosage was between 5% to 10%, the flowability of the mortar was higher than the flowability of the control mix; when the microbeads dosage exceeded 20%, the flowability of the mortar decreased significantly. Among the three types of high-efficiency water reducing agents, the highest flowability was achieved by the ammonia-based water reducing agent at 270 mm.

Changes to water consumption at similar flowability

When the mortar is achieving the same flowability, the water consumptions were different, as shown in Tables 10.7, 10.8 and 10.9.

Table 10.7 Changes in water consumption when the flowability is the same (naphthalene-based admixture).

No.	Cement + Microbeads	Admixture	Water (ml)	Flowability	
				Initial	1 h
1	500+0	5 g	150	215	180
2	475+25	5 g	144.5	215	190
3	450+50	5 g	140	205	165
4	400+100	5 g	135	210	175
5	300+200	5 g	132	213	185
6	250+250	5 g	130	215	205

Table 10.8 Changes in water consumption when the flowability is the same (ammonia-based water reducing agent)

No.	Cement + Microbeads	Admixture	Water (ml)	Flowability	
				Initial	1 h
1	500+0	4 g	147.5	260	250
2	475+25	4 g	140	260	260
3	450+50	4 g	134	257	245
4	400+100	4 g	128	252	240
5	300+200	4 g	170	200	0

Table 10.9 Changes in water consumption when the flowability is the same (polycarboxylate superplasticizer)

No.	Cement + Microbeads	Admixture	Water (ml)	Flowability	
				Initial	1h
1	500+0	3.5 g	147	200	220
2	475+25	3.5 g	140	200	225
3	450+50	3.5 g	135	195	210
4	400+100	3.5 g	138	195	170
5	300+200	3.5 g	160	145	110
6	250+250	3.5 g	180	145	100

From Tables 10.7, 10.8, 10.9: (1) For naphthalene-based admixture, the initial flow rate was maintained at about 215 mm, and water consumption was decreased with increase in microbeads dosage, from 150 ml to 130 ml; (2) For ammonia-based water reducing agent, the initial flow rate remained at about 260 mm; when microbeads dosage amount was 10%, the flow rate was still maintained at about 260 mm; when microbeads dosage was 20%, the flow rate was slightly reduced to about 252 mm; when microbeads dosage was further increased to 30-40%, although the water consumption increased significantly from 147.5 to 170 ml, the flow rate decreased from 260 mm to 200 mm; and when the amount of microbeads was increased to 50%, the water consumption increased further to 180ml, but the flow rate was only 210 mm; (3) For polycarboxylate superplasticizer, the initial flow rate was maintained at about 200 mm, but when the amount of microbeads dosage was further increased to 40-50%, the flow rate was reduced and the water consumption increased; this is slightly similar to ammonia-based water reduction agents. This may be caused by the difference in absorption mechanism of the ammonia-based water reducer and the polycarboxylate water reducer from that of the naphthalene-based water reducer.

10.2.2　The strength of the mortar

1. The strength of the mortar under the same flowability and different water consumption is shown in Tables 10.10, 10.11, and 10.12.

Table 10.10 Mortar strength under the same flowability and different water consumption (naphthalene-based water reducing agent).

NO.	Cement + Microbeads	Admixture	Water (ml)	Strength （MPa）			
				7 days		28 days	
				Flexural	Compressive	Flexural	Compressive
1	2000+0	6 g	600	11.3	61.6	12.5	81.5
2	1900+100	6 g	578	11.9	67.3	12.5	84.9
3	1800+200	6 g	560	11.9	59.9	12.5	58.8
4	1600+400	6 g	540	12.0	58.0	12.5	60.5
5	1200+800	6 g	528	10.5	55.3	10.6	59.6
6	1000+1000	6 g	520	8.3	46.0	9.0	58.5

Table 10.11 Mortar strength with the same flowability and different water consumption (ammonia-based water reducing agent)

NO.	Cement + Microbeads	Admixture	Water (ml)	Strength （MPa）			
				7 days		28 days	
				Flexural	Compressive	Flexural	Compressive
1	2000+0	6 g	588	12.5	64.1	12.5	85.7
2	1900+100	6 g	560	12.5	66.4	12.5	86.2
3	1800+200	6 g	536	12.5	66.2	12.5	86.5
4	1600+400	6 g	512	12.5	64.0	12.5	81.9
5	1200+800	6 g	680	6.4	45.7	8.2	59.4
6	1000+1000	6 g	720	6.1	34.8	6.1	44.8

Table 10.12 Mortar strength with the same flowability and different water consumption (polycarboxylate water reducing agent)

NO.	Cement + Microbeads	Admixture	Water (ml)	Strength （MPa）			
				7 days		28 days	
				Flexural	Compressive	Flexural	Compressive
1	2000+0	6 g	588	12.5	75.5	11.7	98.2
2	1900+100	6 g	560	12.5	71.1	12.5	99.5
3	1800+200	6 g	540	12.5	66.9	12.5	94.7
4	1600+400	6 g	552	12.5	65.3	12.5	91.8
5	1200+800	6 g	640	8.7	53.1	9.3	54.7
6	1000+1000	6 g	720	6.7	36.8	6.2	54.3

From Table 10.10 (Naphthalene-based water-reducing agent): When microbeads replaced 5% of cement, the 7-day and 28-day mortar flexural and compressive strength were 11.9, 12.5, 67.3, and 84.9 MPa, respectively; all were higher than those of the control samples at 11.3, 12.5, 61.6, and 81.5 MPa respectively.

From Table 10.11 (Ammonia-based water-reducing agent): When microbeads replaced 5-10% of cement, the 7-day and 28-day mortar flexural and compressive strength were 12.5, 12.5, 66.4, and 86.5 MPa, respectively, all were higher than those of the control samples at 12.5, 12.5, 64.1, and

85.7 MPa respectively. This series of test proved that 7-day flexural strength was relatively high, reaching 12.5 MPa.

It can be found in Table 10.12 (Polycarboxylate water-reducing agent) that, when microbeads replaced 5% of cement, the 28 days old mortar flexural strength and compressive strength were equivalent to those of the control samples.

It can be seen that the different water-reducing agents have different strength enhancement effect on the mortar. In this experiment, the ammonia-based water-reducing agent is the best.

2. The strength of the mortar under the same water consumption is shown in Tables 10.13, 10.14, and 10.15.

Table 10.13 Mortar strength under the same water consumption (naphthalene-based water reducing agent).

NO.		Cement + Microbeads	Admixture	Water (ml)	Strength (MPa)			
					7 days		28 days	
					Flexural	Compressive	Flexural	Compressive
1	Control	2000+0	8 g	600	12.3	67.1	12.4	71.2
2	5%	1900+100	8 g	600	12.2	66.9	12.5	64.3
3	10%	1800+200	8 g	600	11.9	63.8	12.5	62.5
4	20%	1600+400	8 g	600	10.7	59.6	12.5	77.1
5	40%	1200+800	8 g	600	7.9	50.4	9.3	60.3
6	50%	1000+1000	8 g	600	7.0	43.0	7.9	58.9

It can be seen from Table 10.13 (Naphthalene-based water reducing agent) that, when the amount of cement replaced by microbeads was 20%, the 28-day flexural and compressive strength of the mortar paste was higher than the control mortar paste strength.

Table 10.14 Mortar strength under the same water consumption (ammonia-based water reducing agent)

NO.		Cement + Microbeads	Admixture	Water (ml)	Strength (MPa)			
					7 days		28 days	
					Flexural	Compressive	Flexural	Compressive
1	Control	2000+0	8 g	595	11.8	59.8	12.5	77.9
2	5%	1900+100	8 g	595	12.0	56.0	12.5	77.5
3	10%	1800+200	8 g	595	11.2	54.6	12.5	70.9
4	20%	1600+400	8 g	595	11.8	50.9	12.5	65
5	40%	1200+800	8 g	595	8.9	48.6	9.8	65.7
6	50%	1000+1000	8 g	595	7.3	36.2	8.7	60.1

It can be seen from Table 10.14 that, when the microbeads replaced 5% cement, the flexural strength and the compressive strength of the mortar paste containing microbeads at 28 days were equivalent to those of the control.

Table 10.15 Mortar strength under the same water consumption (polycarboxylate water reducer).

NO.		Cement + Microbeads	Admixture	Water (ml)	Strength (MPa)			
					7 days		28 days	
					Flexural	Compressive	Flexural	Compressive
1	Control	2000+0	6 g	595	12.5	69.4	12.5	82.5
2	5%	1900+100	6 g	595	11.8	59.5	12.5	82.6
3	10%	1800+200	6 g	595	12.3	56.8	12.5	88.3
4	20%	1600+400	6 g	595	12.5	55.3	12.5	74.8
5	40%	1200+800	6 g	595	8.6	47.5	8.9	64.7
6	50%	1000+1000	6 g	595	6.8	43.4	7.2	57.9

It can be seen from Table 10.15 (polycarboxylate water reducing agent) that, when microbeads replaced 5-10% cement, the flexural strength and compressive strength of the 28-day-old mortar containing microbeads were higher than those of the control mortar.

10.2.3 Mortar strength

Then microbeads replace cement at 17% and 25% to prepare mortar, and when the flowability of the mortar is roughly the same, the mortar containing microbeads can appropriately reduce water consumption. The mortar composition is shown in Table 10.16, and the strength is shown in Table 10.17.

Table 10.16 Proportion of control mortar and microbead mortar

NO	Water consumption (g)	Cement (g)	Microbeads (g)	Standard sand (g)
1	225	450	/	1350
2	199.5	373.5	76.5 (17%)	1350
3	180	337.5	112.5 (25%)	1350

Table 10.17 Strength of control mortar and microbead mortar

NO.	Compressive strength (MPa)			Flexural strength (MPa)		
	3 d	7 d	28 d	3 d	7 d	28 d
1	28.8		49.0	5.8		8.2
2	33.7	42.7	57.6	7.6	8.46	10.0
3	33	42.3	63.6	7.2	8.23	10.4

It can be seen that the strength of the mortar containing 17% and 25% of the microbeads is higher than the strength of the control mortar. In particular, the strength of the mortar containing 25% microbeads is higher.

10.2.4 Strength of concrete

In this section, the effects of microbeads and different powders on the fluidity and strength of high performance concrete are studied.

Raw materials

Composite superplasticizer (combination of naphthalene-base and ammonia-base);

Fly ash: grade-1 (provided by Lijian Commercial Mixing Station);

Ultrafine mineral powder (BFS) p8000, Jinan, Shandong;

Silica fume (SF), Guizhou Elkem;

Cement, Golden Eagle, PII52.5

Composite powder

(1) SF+WZ=70+180=250 Kg/m^3

(2) SF+FA=70+180=250 Kg/m^3

(3) SF+BFS=70+180=250 Kg/m^3

In the mix proportion: SF=silica fume; WZ=beads; FA=fly ash grade-1; BFS=superfine mineral powder.

Mix proportion and performance of concrete

The concrete mix composition is shown in Table 10.18, the properties of fresh concrete are shown in Table 10.19, and the compressive strength of concrete is shown in Table 10.20.

Table 10.18 Concrete mix proportion

NO.	Cement	Compound (250)		Water	Sand	Coarse Aggregates		Composite superplasticizer 4.0%
						5-10	10-20	
1	450	(1)	WZ	135	700	285	665	28
2	450	(2)	FA	135	700	285	665	28
3	450	(3)	BFS	135	700	285	665	28

Table 10.19 Performance of fresh concrete

NO.	Slump	Slump flow	Inverted slump cone flow time	Notes
1	260	600	7 sec	Better effect
2	260	610	6 sec	Slow setting, could not demould on time
3	270	610	13 sec	Sticky, Stick to the bottom

Table 10.20 Concrete strength test

NO.	Strength （MPa）		
	3 days	7 days	28 days
1	60	76.7	95.6
2	58.6	71.3	87.9
3	73.8	83.2	97.9

According to the properties of fresh concrete in Table 10.19, the composite ultrafine powder of slag and silica fume make the fresh concrete too sticky, stick to the bottom, and be difficult to construct; while the composite ultrafine powder of fly ash and silica fume causes retardation. The microbeads and silica fume composite superfine powder make concrete have good flowability and be convenient for construction, and as shown in Table 10.20, the 28-day strength of concrete is relatively high.

10.3 Microbead high performance concrete

10.3.1 Microbead high performance concrete mix

The raw materials used in microbead high-performance concrete are the same as those in the previous section, and the concrete mix proportion is shown in Table 10.21.

Table 10.21 Mix proportion for C70 HPC (kg/m³).

C	WZ	FA	BFS	W	S	G1	G2	AG
380	40	120	80	140	750	760	190	2.6%

10.3.2 The performance of fresh concrete

The fresh concrete is shown in Table 10.22.

Table 10.22 Performance of fresh concrete.

Time	Slump （mm）	Slump flow （mm）	Inverted slump cone flow time （sec）
Initial	270	700×730	8
1h	265	700×690	9
2h	260	690×680	10

The plasticity maintaining of concrete is mainly achieved by using water-reducing agent. The water-reducing agent developed by the author has high-efficiency water-reducing effect, with 2-3 hours of plasticity maintaining, no retardation, and has a strengthening effect. The application of microbeads can reduce the water consumption of concrete, which is conducive to the plasticity maintaining of fresh concrete.

10.3.3 Strength of concrete

The strength of concrete at different ages is shown in Table 10.23. By using PO 42.5 cement, and WZ+FA+BFS composite powder, without silica fume, HPC of C70 to C80 can be prepared.

Table 10.23 Strength of Concrete (MPa).

3d	7 d	28 d
55	70	84

10.4 Microbead ultra-high performance concrete

In China, concrete with compressive strength over 100 MPa is called UHPC. It has been researched and applied in construction of a couple of super-high-rise building structures. The three-component combination of cement, microbeads and silica fume is used as a cementitious material, which has better technical and economic effects to prepare UHPC compared to those of other ultra-fine powders.

10.4.1 Raw materials

Cement—P·II52.5R; Ground granulated blast furnace slag with a specific surface area of 8500 cm^2/g;

Bead—ultra-fine fly ash, with a specific surface area of 12540 cm^2/g, loss on ignition 1.7%, water requirement ratio 92%; Silica fume—specific surface area 2×10^5 cm^2/g, SiO_2 content 90%; polycarboxylate superplasticizer; Coarse aggregates, two gradations 5-9 mm and 9-16 mm mixed at ratio 3:7, crushing value less than 6%; Washed sea sand, with fineness modulus 2.8 and the clay content less than 0.6%; Natural anhydrous anhydrite powder—specific surface area 5200 cm^2/g, composition as $CaSO_4 \cdot 0.022H_2O$; HCSA—sulfa-aluminate expansion agent

10.4.2 Concrete mix proportion

The mix proportion of microbead ultra-high performance concrete is shown in Table 10.24.

Table 10.24 Mix proportion of C120 UHPC.

No.	Water-binder ratio	Cement	Microbeads	Silica fume	Sea sand	Coarse aggregates	Water	Water reducing agent
601	0.20	500	200	50	750	1000	130	3.0%
602	0.187	500	170	80	700	1000	140	3.2%
603	0.187	500	170	80	700	1000	140	4.0%
604	0.187	500	200	50	700	1000	140	3.5%
605	0.187	500	200	50	700	1000	140	2.0%
606	0.187	500	200	50	700	1000	140	Approx. 4.0%

No. 601 uses BASF's naphthalene-based water reducing agent with a solid content of 40%; Nos. 602, 603, 604, and 606 use a naphthalene-ammonia compound water-reducing agent; No. 605 uses Sika polycarboxylate acid admixtures with solid content of 40%.

10.4.3 Performance of fresh concrete

The performance of fresh concrete is shown in Table 10.25.

Table 10.25 Performance of fresh concrete.

No.	Slump (mm)				Slump flow (mm)				Inverted slump cone flow time (sec)			
	Initial	1h	2h	3h	Initial	1h	2h	3h	Initial	1h	2h	3h
601	130											
602	270	260	230	225	680*700	640*700	590*590	540*560	5	7	9	12
603	260	250	245	255	660*690	660*650	650*650	670*620	4	5	7	5
604	265			265	710*740			670*710	4			5

Note: No. 604 has a four-hour slump of 260 mm, a slump flow of 670*730, and an inverted slump cone flow time of 7 seconds.

No. 602 uses concrete admixture at a relatively low dosage of 3.2%, and the inverted slump cone flow time after 3 hours is relatively long, about 12 seconds.

The slump, slump flow and inverted slump cone flow time of concrete Nos. 603 and 604 meet the requirements of this study.

By using self-developed sulfamic acid water reducing agent and BASF's naphthalene water reducing agent (solid content 40%), and mixing a small amount of zeolite powder for adsorbing water reducing agent, C120 concrete with excellent slump retention can be prepared to meet the requirements of long-distance transportation and ultra-high pumping.

10.4.4 Autogenous shrinkage of concrete

In the C120 concrete mix, natural anhydrous calcium sulfate was added with specific surface area 5200 cm^2/g, composition $CaSO_4 \cdot 0.022H_2O$; and sulfa-aluminate expansion agent HCSA are added. The early shrinkage is compared. The proportion of concrete test mixes is shown in Table 10.26. The workability of each type of concrete mix is set at slump of 250 mm or more, slump flow of 680 mm or more, and inverted slump cone flow time within 15 s. The results of autogenous shrinkage measurement are shown in Table 10.27 and Figure 10.5.

Table 10.26 C120 autogenous shrinkage comparison test mix composition.

Concrete category	W/B	Cement	Ground granulated blastfurnace slag	Micro beads	Silica fume	$CaSO_4$	HCSA	Water reducing agent	Sand	Coarse aggregates
Control mix	0.185	550	42	84	24	0	0	3.8%	750	950
Mixed with 10 % anhydrous calcium sulfate	0.185	550	42	84	24	70	0	4.2%	750	950
Mixed with 6 % HCSA	0.185	550	42	84	24	0	42	4.2%	750	950
Mixed with 10 % HCSA	0.185	550	42	84	24	0	70	4.4%	750	950

Table 10.27 C120 concrete autogenous shrinkage value/10^{-6}

Time	Control mix	Mixed with 10% $CaSO_4$	Mixed with 6% HCSA	Mixed with 10% HCSA
0	0	0	0	0
1	13	17	5	30
2	14	30	56	124
3	17	30	92	162
4	21	69	82	142
5	31	136	71	95
6	40	181	69	54
7	48	196	68	23
8	57	204	66	2
9	66	210	69	-14
10	74	216	71	-26
11	81	222	73	-40
12	90	226	72	-47
13	96	230	74	-55
14	103	234	74	-63
15	111	237	75	-70
16	117	241	77	-76
17	124	245	79	-82
18	129	244	80	-86
19	132	248	81	-89
20	139	250	83	-93
21	141	253	85	-95
22	149	256	86	-98
23	150	259	88	-101
24	157	261	90	-103

2 d	231	312	115	-116
3 d	269	339	138	-122
4 d		348	157	-103
5 d		355	166	-76
7 d		369	172	-34
9 d		380	176	-22
11 d		400	179	-55
14 d		425	196	-90
17 d		442	202	
19 d		To be tested	207	

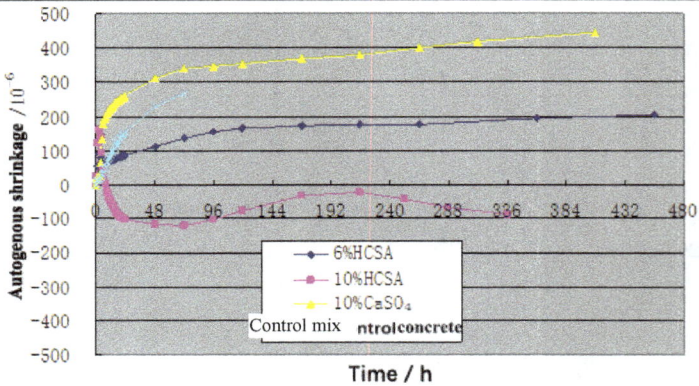

Figure 10.5 Concrete autogenous shrinkage curve.

It can be seen from Table 10.27 and Figure 10.5 that:

1) HCSA has a very good effect in inhibiting the autogenous shrinkage of concrete. After adding 6% of HCSA, the 3 d autogenous shrinkage is reduced from 269×10^{-6} to 138×10^{-6}, which has about 50% reduction, and after the dosage is increased to 10%, the specimen at 14 d was still in a slightly expanded state, about 90×10^{-6}. To inhibit the autogenous shrinkage of UHPC, the dosage of HCSA should be less than 5%.

2) The addition of natural anhydrous $CaSO_4$, however, increases the autogenous shrinkage. After adding 10% natural anhydrous calcium sulfate, the 3 d autogenous shrinkage increased from 269×10^{-6} to 339×10^{-6}, with an increase of 26%.

10.4.5 Compressive strength of hardened concrete

The compressive strength of concrete is shown in Table 10.28.

Table 10.28 Compressive strength of concrete

No.	Compressive strength, MPa			
	3 d	7 d	28 d	56 d
601	85.8	97.5	130	140
602	87.1	102	135	145
603	84.3	102	140	150
604	75.1	96	120	130

10.5 Changes in mechanical properties of microbead ultra-high performance concrete with fibre

The original C120 concrete is mixed with fibre to improve the mechanical properties. The polypropylene fibre added is of a dosage of 1.0 kg/m^3, and 2.0 kg/m^3 and a Shenzhen organic fibre is of a dosage of 2.0 kg/m^3.

10.5.1 Compressive strength and static elastic modulus of prism specimens

The static compression elastic modulus test set-up is shown in Figure 10.6; the test results are shown in Table 10.29.

Figure 10.6 Static compression elastic modulus test set-up.

Table 10.29 Test results of compressive strength and static compressive modulus of concrete prism.

Specimen No.	Age, d	Specimen size (mm)	Compressive strength of prism (MPa)		Static compression elastic modulus (MPa)	
			Individual value	Average value	Individual value	Average value
Control mix	28	100×100×300	115.2	138.0	5.20×10^4	5.20×10^4
			138.0		5.18×10^4	
			138.8		5.22×10^4	
Fibre 1 Kg	28	100×100×300	127.2	128.7	5.10×10^4	5.23×10^4
			126.8		5.21×10^4	
			132.0		5.37×10^4	
Fibre 2 Kg	28	100×100×300	137.2	136.6	5.22×10^4	5.30×10^4
			139.0		5.22×10^4	
			133.6		5.45×10^4	
Fibre2 Kg Shenzhen	28	100×100×300	104.8	116.7	5.18×10^4	5.25×10^4
			116.4		5.23×10^4	
			128.8		5.33×10^4	

The compressive strength of the prisms incorporating the fibre is lower, but the static elastic modulus is higher.

10.5.2 Concrete fracture toughness test

The C120 concrete fracture parameter test is shown set-up in Figure 10.7. Four types of concrete i.e. the control, with fibre of 1 kg, fibre of 2 kg and fibre of 2 kg (Shenzhen) were measured to obtain the load-crack opening displacement (P-CMOD) curves and to calculate concrete softening relationship (σ-w) curve, which are shown in FigureS 10.8-10.11, respectively. The corresponding fracture parameter values were calculated, including cracking strength and tensile strength, bending strength, fracture energy and brittleness parameter index. Three specimens per batch were used for each type of concrete.

(a) Closed-loop hydraulic servo testing machine (b) Deformation sensor placed at the opening of a pre-cut notch the specimen

Figure 10.7 Fracture parameter test of C120 ultra-high performance concrete.

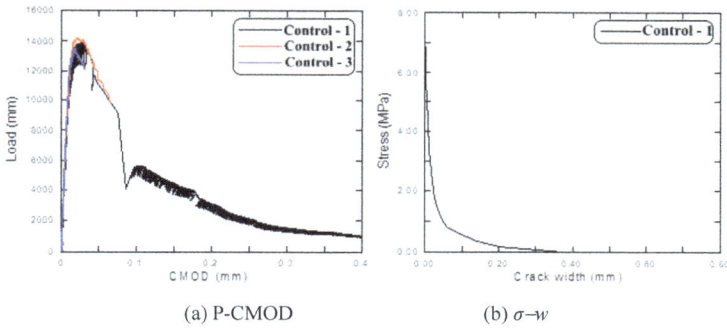

(a) P-CMOD

(b) σ–w

Figure 10.8 P-CMOD curve and σ-w curve of control mix.

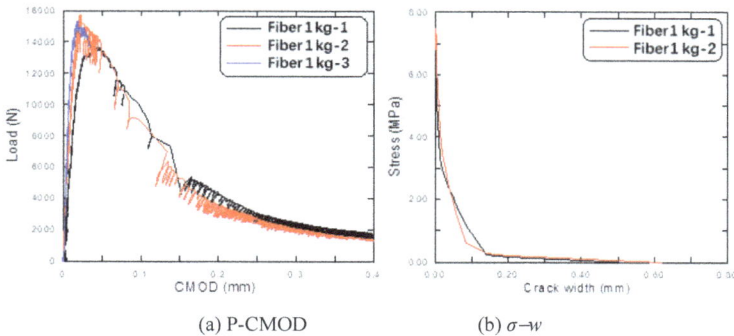

(a) P-CMOD

(b) σ–w

Figure 10.9 P-CMOD curve and σ-w curve of concrete mixed with 1 kg fibre.

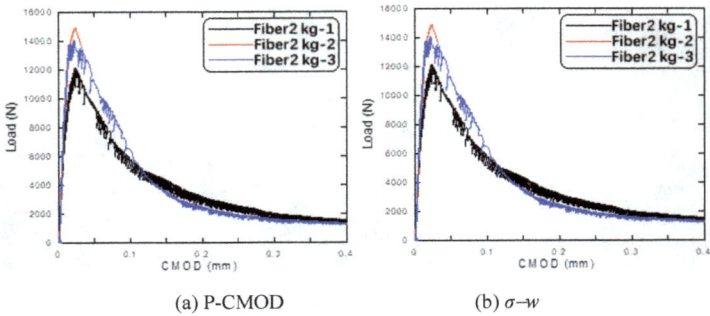

(a) P-CMOD

(b) σ–w

Figure 10.10 P-CMOD curve and σ-w curve of concrete mixed with 2kg fibre.

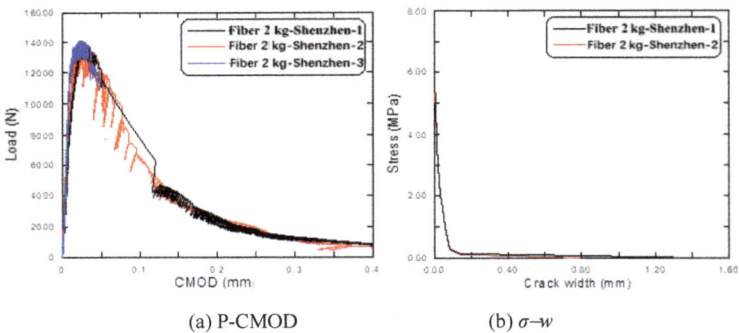

(a) P-CMOD

(b) σ–w

Figure 10.11 P-CMOD curve and σ-w curve of concrete mixed with 2kg Shenzhen fibre.

According to the complete P-CMOD curve measured by the test, the corresponding fracture parameters can be calculated. Table 10.30 shows the fracture parameter values of different types of concrete.

Table 10.30 Summary of concrete fracture parameters.

	Specimen No.	Elastic modulus (GPa)	Cracking Load (N)	Bending resistance Load (N)	Cracking strength (MPa)	Bending resistance strength (MPa)	Tensile strength (MPa)	Fracture energy (J/m²)	Characteristic length (cm)
Control	1 #		5000	14000	6.82	9.07	6.86	213.8	23.5
	2 #		4600	14500	6.27	9.39	-	-	-
	3 #	51.9	4700	13800	6.41	9.59	-	-	-
	Average		4767	14100	6.5	9.35	6.86	213.8	23.5
Fibre 1kg	1 #		4800	13800	6.55	8.94	6.67	291.8	33.8
	2 #		5000	15500	6.82	10.04	7.08	302.9	27.9
	3 #	52.2	5000	15300	6.82	9.91	-	-	-
	Average		4933	14867	6.73	9.63	6.88	297.35	30.9
Fibre 2kg	1 #		4800	12000	6.55	7.78	6.56	322.8	39.2
	2 #		5000	15000	6.82	9.73	-	-	-
	3 #	52.2	5200	14000	7.09	9.08	7.22	322.1	33.4
	Average		5000	13667	6.82	8.86	6.89	322.5	36.3
Fibre 2kg Shenzhen	1 #		5100	14000	6.95	9.07	6.97	300.3	30.0
	2 #		5000	14000	6.82	9.07	6.84	321.8	30.7
	3 #	52.0	5200	14000	7.09	9.07	-	-	-
	Average		5100	14000	6.95	9.07	6.91	311.1	30.4

It can be seen: ① From Figure 5 to Figure 8 that, between different specimens of the same type of concrete, the P-CMOD curves and the σ-w curves are similar in shape and show little difference, indicating that the dispersion between the test specimens is not big. ② The cracking load of different types of concrete is basically the same. This is because the matrix of the material is the same, and the fibre comes into effect only after occurrence of the concrete cracking. ③ With the increase of the fibre content, the tensile strength of the concrete is basically unchanged. ④ With increase of fibre content, the fracture energy of concrete increases. At the same dosage, the different types of fibre result in different fracture energy of the concrete. The order of the fracture energy increased is as follows: control<fibre 1 kg<fibre 2 kg (Shenzhen)<fibre 2 kg; ⑤ The brittle characteristic length is a physical parameter that characterizes the brittleness of concrete. The shorter the characteristic length, the greater the brittleness of the concrete. According to the test results, it can be seen that with the increase of fibre content, the toughness of concrete increases. The order of the brittle characteristic length is as follows: control<fibre 1 kg<fibre 2 kg (Shenzhen)<fibre 2 kg, and the corresponding order of concrete toughness is control<fibre 1 kg<fibre 2 kg (Shenzhen)<fibre 2 kg.

10.6 Fire resistance of microbead ultra-high performance concrete

As the strength of the concrete increases, the density increases. Under the action of high temperature, the water inside the concrete evaporates, causing the steam pressure to increase, but the water vapor is not easy to be discharged from the internal concrete to the outside; as the heating continues and temperature increases, the steam pressure further increases, and when the steam pressure exceeds the tensile strength of the concrete, the concrete will encounter explosive spalling, causing the structure to lose its load-bearing capacity.

The fire resistance test adopts the method of real-time measurement. That is, loading, heating, and test observation are performed at the same time.

10.6.1 Types of fire test specimens

The specimens included reference group, group with 1 kg/m^3 polypropylene fibre, group with 2 kg/m^3 polypropylene fibre, and group of reinforced specimens. Prisms of sizes as 100 mm×100 mm×300 mm, cylinders of sizes as Φ100×150 mm, Φ100×300 mm, were used. The three types of specimens are shown in Figure 10.12. There were two heating jackets for heating the specimens, one was for prism, the other was for cylinder, as shown in Figure 10.13.

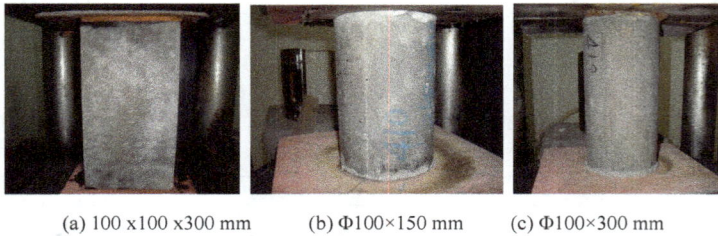

(a) 100 x100 x300 mm   (b) Φ100×150 mm   (c) Φ100×300 mm

Figure 10.12 Shape and size of fire test specimens.

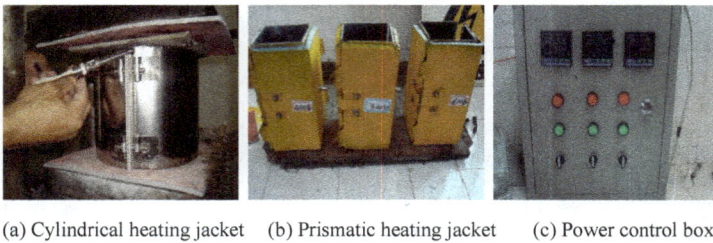

(a) Cylindrical heating jacket   (b) Prismatic heating jacket   (c) Power control box

Figure 10.13 Heating device and control equipment of the test piece.

10.6.2 Real-time measurement

The so-called real-time measurement means that the specimen was heated by an electric heating jacket during the compression process. (1) Loading: constant loading on the specimen at 30% of the ultimate strength of concrete. (2) Heating: through electric heating, observe the changes of the test specimen at three temperatures of 200 °C, 300 °C, and 400 °C. (3) Constant temperature: keep the target temperature at constant for 30-40 min. (4) Observation: At each target temperature for a duration of 30-40 min, observe the change of the specimen. The reference specimens, the 1 kg/m^3 and the 2 kg/m^3 fibre specimens are shown in Figures 10.14-10.16, respectively.

(a) Reference specimen at 200 °C constant temperature and constant load (b) Reference specimen at 300 °C constant temperature and constant load

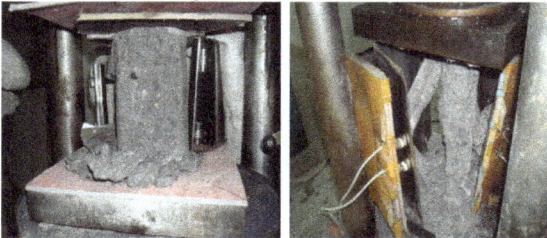

(c) Reference specimen at 300 °C constant temperature and constant load (d) Reference specimen at 400 °C constant temperature and constant load

Figure 10.14 Change in the reference specimen under constant load under different temperature effects.

(1) 200 °C   (2) 300 °C   (3) 400 °C   (4) 400～500 °C

Figure 10.15 Change in the specimens mixed with 1 kg/m^3 polypropylene fibre under constant load at different temperatures.

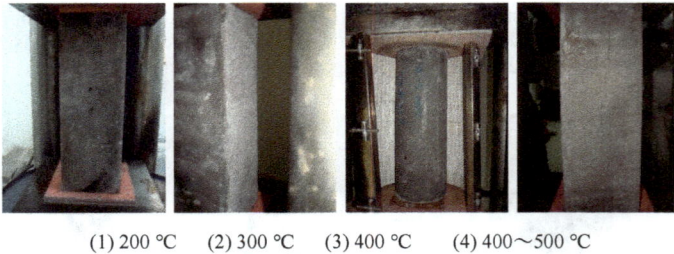

(1) 200 °C   (2) 300 °C   (3) 400 °C   (4) 400～500 °C

Figure 10.16 Change in the specimens mixed with 2 kg/m³ polypropylene fibre under constant load at different temperatures.

10.6.3 Approach to improve the fire resistance of UHPC

High-strength and high-performance concrete is prone to explosive spalling under high temperature. During heating, the width of the cracks on the surface of a specimen increases, and the number of cracks increases, and it may eventually spall, causing the specimen breaking into pieces, so that it cannot continue to withstand load.

The fire resistance of concrete can be significantly improved by incorporating a certain amount of polypropylene fibre. In the specimens mixed with polypropylene fibre, during the entire process of heating to a target temperature, the polypropylene fibre will gradually melt after being heated to a certain temperature, resulting in an increase in the number of capillaries in the specimen, which provides a channel for the moisture and steam in the specimen to escape. The channels also provide a certain space for the expansion stress and heat conduction of the concrete itself, which reduces the possibility of concrete spalling, and finally ensures that the concrete specimens are still relatively intact after being subjected to high temperature.

When loaded at 30% of the ultimate concrete strength, and when the specimen was heated to 200 °C, 300 °C, 400 °C and 500 °C, with the respective constant temperature held for 30-40 min, the reference specimens encountered cracking and were damaged, losing its load-bearing capacity, which may cause collapse of a building made with such concrete, and casualties and property losses. The specimens mixed with polypropylene fibres were still intact and had no cracking. This is the most important result of tests on ultra-high strength and ultra-high performance concrete mixed with polypropylene fibre. It is also the main purpose to improve fire resistance.

10.6.4 Fire resistance test of reinforced specimens

In this test, no fibre was incorporated in the reinforced specimens; under heating to temperature of 300 °C and loading at 30% of ultimate load, the specimen encountered spalling, as shown in Figure 10.17.

(a) Reinforced concrete Φ100×150 mm (b) Reinforced concrete Φ100×300 mm

Figure 10.17 Reinforced concrete specimens under 300 °C constant temperature and load.

Under the same conditions, the damage of the reinforced high-strength high-performance concrete specimens is more severe than that of the benchmark concrete specimens. This is mainly caused by the thermal expansion of the rebars themselves. The reference concrete specimens spalled but they only had to bear the stress caused by their own thermal expansion under high temperature, while the reinforced concrete specimens were not only affected by their own thermal expansion, but also has to bear the stress generated by the thermal expansion of the rebars; therefore, the explosive spalling occurred more severely.

10.7 Durability of microbead ultra-high performance concrete

10.7.1 Resistance to chloride corrosion of microbead concrete

Mortar specimens made with microbead replacement at 10% and 20% of cement respectively, were immersed in saturated salt water (at room temperature) together with the benchmark mortar specimens, and the chlorine ion penetration depth was tested at 3 days, 7 days and 14 days, as shown in Table 10.31.

Table 10.31 Different ages of mortar specimens Cl⁻-penetration depth.

Age	3 days	7 days	14 days
Benchmark mortar	14 mm	17 mm	21 mm
10% beads	7 mm	10 mm	10 mm
20% beads	6 mm	9 mm	10 mm

The mortar specimen made with microbead replacement at 10% and 20% of cement had a Cl⁻ penetration depth of approximately 1/2 of that of the benchmark mortar in saturated salt water.

The Cl⁻ diffusion coefficient of two concretes (strength of about 100 MPa, with and without microbead, respectively) was determined according to the NEL method, established by Tsinghua University. The result for the high-strength and high-performance concrete containing microbead was $0.176598 \times 10\text{-}12$ m^2/S; the result for the high-strength and high-performance concrete without microbead was $0.235791 \times 10\text{-}12$ m^2/S. The former is about 30% lower than the latter.

10.7.2 Sulphate resistance of microbead concrete

Microbead is the smog emitted from the chimney of coal-fired power plants, and is ultra-fine spherical glass beads. It also belongs to the pulverized fly ash system. The evaluation of sulphate corrosion resistance of fly ash is based on R=(C-5)/F≤1.0. When a type of fly ash can meet the requirement, mixing it into concrete can effectively improve the sulphate attack resistance.

Where： C－CaO content in fly ash, %

F – Fe$_2$O$_3$ content in fly ash, %

According to Table 10.1, the chemical composition of the microbead consists of CaO content at 4.8% and Fe$_2$O$_3$ content is 5.3. Substituting this into the equation R=(C-5)/F≤1.0, R has a negative value, therefore the microbead has high resistance to sulphate attack. Experimental results also proved this effect. Microbead HPC was immersed in two solutions such as Na$_2$SO$_4$ 5.0% solution and Na$_2$SO$_4$ 5.0%+NaCl 3% solution, respectively, and soaked for 14 hours, and then taken out to for drying for 2 hours, and oven dried for 6 hours (temperature 80 °C± 2 °C), cooled for 2h, and then put back into the solution to complete one cycle. The results after 50 cycles are shown in Table 10.32.

Table 10.32 Sulphate resistance and combined sulphate and chloride corrosion resistance of concrete specimens

Soaking solution	Na$_2$SO$_4$ 5.0% solution	Na$_2$SO$_4$5.0 %+NaCl 3% solution
No. of cycles	50 nos.	50 nos.
UHPC specimen	No change in appearance and quality	No change in appearance and quality

10.7.3 Freezing resistance test

C120 concrete was subjected to 300 freeze-thaw cycles according to the standard specification, and the results are shown in Table 10.33.

Table 10.33 Test results of freezing resistance of C120 ultra-high performance concrete.

Test content	Test results	Standard requirement
Relative dynamic elastic modulus after 300 freeze-thaw cycles, %	99.9	It can be considered that the freezing resistance of the tested concrete has met the design requirements if after freezing and thawing to a predetermined number of
Rate of loss of mass after 300 freeze-thaw cycles, %	0	cycles, the relative dynamic elastic modulus is not less than 60%, or the mass loss rate is not more than 5%.

10.8 Microstructure of microbead high performance and ultra-high performance concrete

10.8.1 Cement-microbead-silica fume cementitious system

The average particle size of cement is about 20-30 times of that of the microbeads, and the average particle size of the microbeads is about 10 times of that of silica fume. The amount of microbeads that fill in the voids between cement particles is about 30%, changing the particle size distribution and voidage. The average particle size of Portland cement is 10.4 um, the average particle size of microbead is about 1.0 um, and the amount of microbead to fill the voids in cement is about 30%. The optimum amount of silica fume in concrete is about 8%. Therefore, the quantity of cementitious materials of C120UHPC to obtain the densest packing: cement 450-500 kg/m³, microbead 180-200 kg/m³, and silica fume 50-70 kg/m³. When the cementitious material powder is the most densely packed, the porosity of the combination is low, and a dense cement paste structure can be obtained, thereby strength is enhanced.

10.8.2 Microstructure of composite cementitious material paste

The SEM observations of the composite cementitious material paste are shown in Figures 10.18.

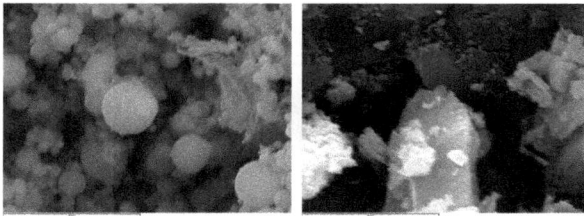

(a) Silica fume + microbead paste (×10000)　　(b) Silica fume + mineral powder paste (×10000)

(c) Silica fume + fly ash paste (×10000)

Figure 10.18 SEM observations of the composite cementitious material paste.

The SEM observations of the cement paste with different microbead replacements are shown in Figure 10.19.

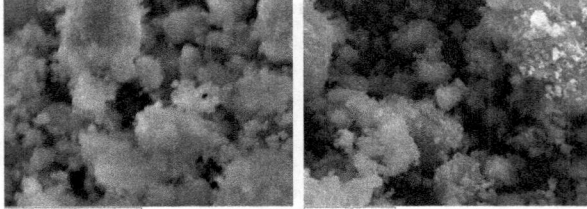

(a) Standard paste   (b) Paste containing 5% microbeads

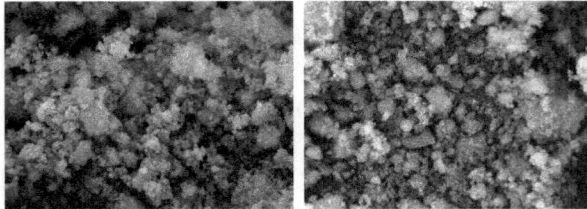

(c) Paste containing 10% microbeads (d) Paste containing 20% microbeads

Paste containing 40% microbeads

Figure 10.19 SEM of paste with different microbeads replacements.

10.8.3 SEM observations of the interface with microbead at different replacement of cement

As shown in Figure 10.20.

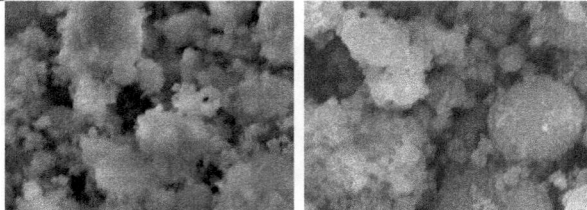

(a) SEM of the reference interface (b) SEM of the 40% microbead interface

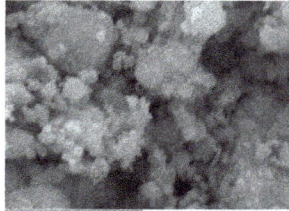

(c) SEM of the 50% microbead interface

Figure 10.20 The interface with microbead at different replacement of cement

10.8.4   XRD Analysis of Cement Paste

As shown in Figures 10.21, no new phases were detected with the increase in the amount of microbeads. Since the microbead is of glass, it does not have any sharp diffraction peak in the XRD spectra.

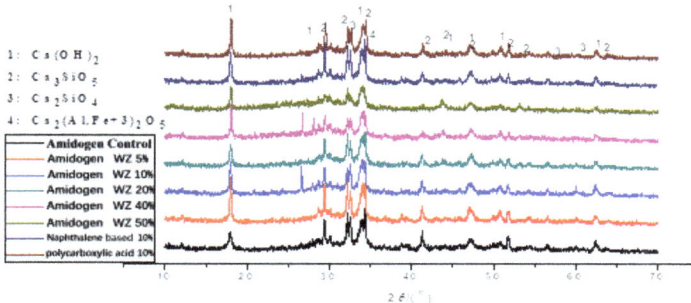

Figure 10.21 XRD patterns of hardened cement pastes.

Taking into consideration that the increase in the content of microbeads may affect the hydration of cement and the amount of hydration products, the samples were analyzed quantitatively. The characteristic peaks of unhydrated C_2S, C_3S, etc. in the cement sample overlap so much that it is simply impossible to propose a quantitative analysis; whereas the characteristic peak intensity of $Ca(OH)_2$ is high and there is no interference of peaks from other phases, thus it can be quantitatively analyzed, and the content of $Ca(OH)_2$ can indirectly reflect the amount of cement hydration degree. Therefore, the content of $Ca(OH)_2$ in the sample was quantitatively analyzed, and the results are shown in Table 10.34. The larger the area of the peak, the higher the content.

Table 10.34 Quantitative analysis results of $Ca(OH)_2$ in the sample.

Sample	$Ca(OH)_2$ Integrated area of characteristic peak
Amino-based benchmark	146727
Amino -based WZ 5%	259706
Amino -based WZ 10%	193434
Amino -based WZ 20%	161451
Amino -based WZ 40%	183036
Amino -based WZ 50%	131065
Napthalene based WZ 10%	202705
Polycarboxylate WZ 10%	226507

It can be seen that in the paste mixed with microbeads, the integrated area of the characteristic peak of $Ca(OH)_2$ is higher than that of the benchmark; especially when the amount of microbeads was 5%, the integrated area of the characteristic peak of $Ca(OH)_2$ is 45% higher than that of the benchmark. It shows that the microbeads promote the hydration of the cement paste.

10.9 Naphthalene-aminosulfonic acid-superfine microbead superplasticizer compound

Aminosulfonic acid superplasticizers have a high water reduction rate (25-28%), and have good plasticity retention effect, but they are prone to causing bleeding and segregation so that the fresh concrete tends to stick on the bottom. Naphthalene-based water-reducing agent has a low water reduction rate, poor plastic retention effect, and fast slump loss. Thus the two can be reasonably mixed to supplement each other's effects, and by adding powder saturated with such mixed superplasticizer, a three-component composite superplasticizer can be formed with some specific effects.

In the paste, the cement particles adsorb the water-reducing agent molecules to form a double electrical layer potential on its surface, and because of the double electrical layer potential, a dispersion effect is produced between the cement particles. The microbead and silica fume that adsorb water-reducing agent molecules on its surface, in addition to having electrostatic repulsion, can also produce a scattering effect. In the voids between cement particles, the microbead and silica fume particles with double electric layer potential on the surface play the role of separator, which makes the cement particles disperse more easily and the paste have better flowability. Additionally, microbead is a spherical glass particle and has very low water absorption, the flowability of the cement paste may be further improved.

The ultrafine powder in the three-component composite superplasticizer releases the water-reducing agent molecules into the cement paste slowly, maintaining the cement particles' capacity of adsorption of the water-reducing agent molecules, thus keeping the cement particles in a dispersed state, i.e. maintaining plasticity of fresh concrete. This is different from adding retarders. The three-component composite superplasticizer can retain the plastic state for a longer time, but there is no retardation.

10.10 Summary of this chapter

Through performance tests of microbead, tests on ultra-high performance mixtures with microbead, durability testing and fracture mechanical properties tests with incorporated fibres; the conclusions of the research work are drawn as follows:

1) Microbead is a kind of ultrafine powder with spherical glass particles. The maximum particle size is 1.45 um and the minimum particle size is 2.38 nm. The combination of cement-microbead-silica fume powders in proper proportion can minimize the voidage of the powders to the lowest; thus under the same water consumption, the concrete can achieve the maximum flowability; or under the same flowability, the lowest water demand can be achieved.

2) With cement-microbead-silica fume as the three main components in the binder, and combined with appropriate types of water-reducing agents, the UHPC prepared can reach a strength of 150 MPa; plasticity retention can reach 3 h, (initial slump 265 mm—265 mm, slump flow 710x740 mm—670x710 mm; inverted slump cone flow time 4"—5") and there is no sign of bleeding or segregation.

3) Compared to the benchmark C120 UHPC, after adding 5% sulpha-aluminate expansion agent, the autogenous shrinkage is reduced from 269×10^{-6} to 138×10^{-6}, which is about 50% lower, and adding a small amount of sulpha-aluminate expansion agent can effectively inhibit the autogenous shrinkage and cracking of UHPC.

4) C120 UHPC has very low chloride ion diffusion coefficient, strong resistance to freeze-thaw damage, and high durability.

5) With 1-2 kg/m^3 polypropylene fibre mixed into C120 UHPC, the load-crack opening displacement curve (P-CMOD) and softening relationship curve (σ-w) of the concrete are similar in shape and the difference is slight. The cracking stress is basically the same. This is because the concrete matrix is the same, and the role of the fibre comes into effect only after cracking of the concrete. With the increase of the fibre content, the fracture energy of the concrete increases, the brittle characteristic length increases, and the toughness increases.

Chapter 11

Characteristics of Superfine Natural Zeolite Powder and its Application in Concrete

11.1 Introduction

Zeolite was discovered by the Swedish mineralogist Cronetedt in 1756. In the lava hollows of the Icelandic rocks, he discovered a vase-like magnificent crystal. The crystal can release water vapor and expand through burning by blowing tube. In Greek, zeolite is named after boiling stone *zeo* (to boil) and *lite* (stone). Moreover, zeolite has the same meaning in Chinese by the name 沸石. Zeolite is a family of aqueous aluminosilicate minerals with a framed structure. The rock with zeolite as the main mineral is called zeolite rock, or also known as natural zeolite. The chemical composition is often expressed as follows:

$$(Na,K)_2(Mg,Ca,Sr,Ba)y〔Al_{(x+2y)}•Si_{(n-(x+2y))}•O_{2n})〕•mH_2O$$

where the number of Al is equal to the total number of cations, and the number of O is twice the number of Al and Si.

Currently, there are 36 types of zeolites found in the world, in which the followings are common and useful: clinoptilolite, mordenite, chabazite, and erionite. In China, clinoptilolite and mordenite are the common types of zeolites.

The application of natural zeolite in China has a history of nearly 50 years. It was originally used in the cement industry, especially for shaft kiln cement, where the use of natural zeolite as an ingredient solves the problem of poor soundness of small cement factory and improve the quality of cement. It is also widely used in concrete and high-performance concrete. The technological application of natural zeolite plays an important role in self-compacting concrete, self-compacting self-curing concrete, and plasticity-retentive concrete. More and more research attentions have been attracted to this aspect of the technology all over the world.

11.2 Natural zeolite resources in China

Since the first zeolite deposit with industrial significance was discovered in Jinyun, Zhejiang province in 1972, zeolite deposits have also been found in Henan, Shandong, Hebei, Heilongjiang, and the Autonomous Region of Inner Mongolia. Currently, more than 120 zeolite deposits have been found in 21 provinces and autonomous regions in China, as shown in Figure 11.1. There are three large zeolite mines in China, i.e. Dushikou zeolite mine in Chicheng County, Hebei Province, as shown in Figure 11.2, zeolite mine in Jinyun County, Zhejiang Province and zeolite mine in Hailin County, Heilongjiang Province. Dushikou zeolite mine has the largest storage among them, which is more than 400 million tons. The zeolite in Dushikou zeolite mine is high-quality clinoptilolite with mineral content of 50-70%.

Figure 11.1 Zeolite deposits in China.

Figure 11.2 Dushikou zeolite deposit.

11.3 Chemical and physical properties of zeolite

11.3.1 Chemical composition

Zeolite is a water-bearing aluminosilicate mineral, and the two types of zeolites commonly used in China are clinoptilolite and mordenite. Their chemical compositions are as follows:

Clinoptilolite $Na_6(Al_6Si_{30}O_{72})\cdot24H_2O$, which usually contains K, Ca, Mg, where the content of Na and K $\geq$ Ca and Mg, Si/Al = 4.25 - 5.25.

Mordenite $Na_8(Al_8Si_{40}O_{96})\cdot24H_2O$, sometimes containing K, Ca, Na, where the content of Na and Ca $\geq$ K, Si/Al = 4.17 - 5.0.

The chemical composition of zeolite is shown in Table 11.1. The SiO_2 content is more than 70%, followed by Al_2O_3 with content more than 10%. The contents of potassium and sodium are also as high as 3–5%. The contents of other components are very low. The chemical composition of natural zeolite in Dushikou of China is similar.

Table 11.1 Chemical composition of zeolite.

Type	SiO_2	Al_2O_3	Fe_2O_3	CaO	MgO	SO_3	Na, K	Loss on ignition
Clinoptilolite	71.65	11.77	0.81	0.88	0.52	0.34	5.24	9.04
Mordenite	71.11	11.79	2.57	2.07	0.15	0.27	2.99	9.50

11.3.2 X-ray diffraction, thermal analysis

The X-ray diffraction diagram of zeolite is shown in Figure 11.3, and the TG-DTA curve is shown in Figure 11.4. Similar to clay minerals, its diffraction peaks appear at low angles.

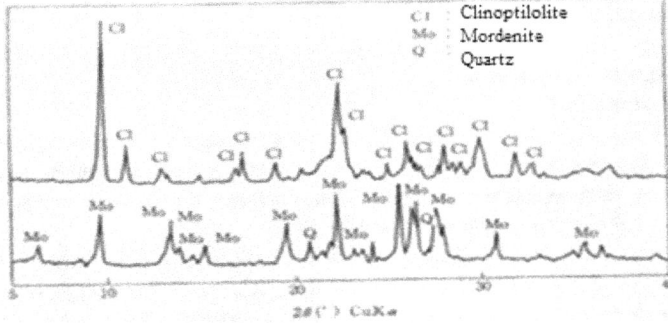

Figure 11.3 X-ray diffraction.

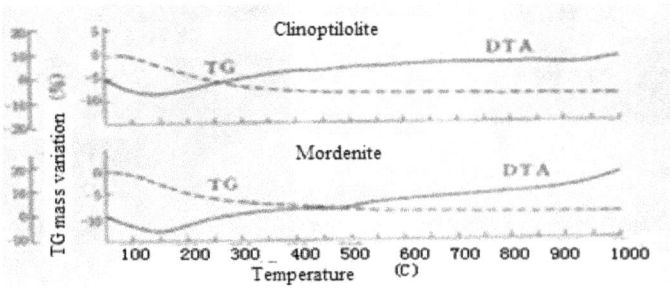

Figure 11.4 Example of thermal analysis results of zeolite (DTA curve).

From the DTA curve in Figure 11.3, there is a slow endothermic peak from 50 °C to 500 °C, which is the dehydration of crystal water in zeolite. From the TG curve in Figure 11.3, the mass of zeolite is reduced by about 10%, which is consistent with the result shown in Table 11.1.

11.3.3 Observation under electron microscope and polarizing microscope

The zeolite observed by using electronic microscope is shown in Figure 11.5.

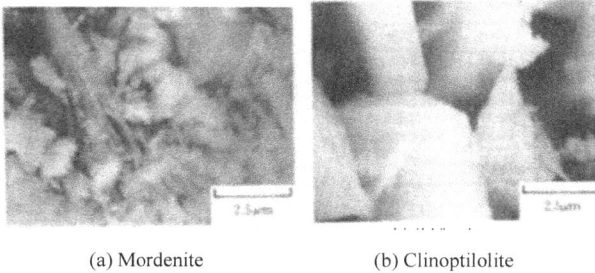

(a) Mordenite   (b) Clinoptilolite

Figure 11.5 Electron microscopic observation of zeolite (in-flight).

The observations by using electron microscope of the natural zeolite from Dushikou and Jinyun in China is shown in Figure 11.6.

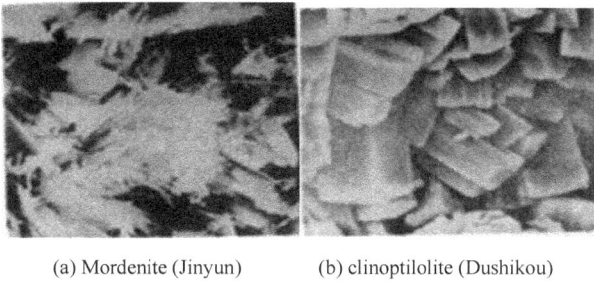

(a) Mordenite (Jinyun)   (b) clinoptilolite (Dushikou)

Figure 11.6 Electron microscopic observation of natural zeolite from China.

It can be seen from the figure that mordenite has hair-like structure, while clinoptilolite has plate-like structure. Figure 11.5 is consistent with Figure 11.6. The concentration of various minerals in the zeolite sample can be seen from the pictures under polarizing microscope in Figure 11.7.

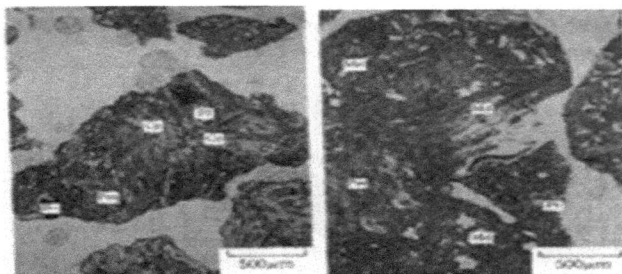

(a) Clinoptilolite (b) Mordenite

Legend: Cp - clinoptilolite, Md - mordenite, Bi - biotite, Pm - pumice, Pl - plagioclase.

Figure 11.7 Polarizing microscopic observation of zeolite.

11.3.4 Apparent density

The apparent density of zeolite is shown in Table 11.2, which is about 2.1-2.3.

Table 11.2 Apparent density of zeolite.

Zeolite types	Apparent density	
	Literature	In-flight tests
Clinoptilolite	/	2.21
Mordenite	2.30	2.24
	2.16	

11.3.5 Mineral properties of zeolite

1. Silicon (aluminum) oxygen tetrahedron

Zeolite is a mineral with silicon (aluminum) acid lattice structure. Its basic unit is a silicon-oxygen tetrahedron with Si as the center and four oxygen ions arranged around it, as shown in Figure 11.8. The silicon ion is at the center of the tetrahedron, and the four oxygen ions are at the corner tips of the tetrahedron. The distance between silicon and oxygen ions is about 1.6 Å. Aluminum-oxygen tetrahedron is formed when the silicon ion in the tetrahedron is replaced by an aluminum ion. In aluminum-oxygen tetrahedron, the distance between aluminum and oxygen ions is about 1.75 Å, and the distance between oxygen ions is about 2.86 Å.

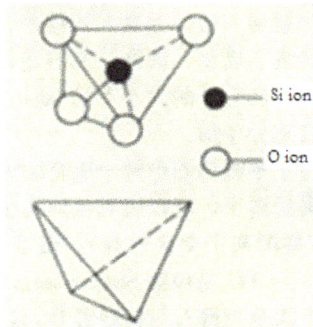

Figure 11.8 Silicon-oxygen tetrahedron in zeolite.

Silicon-oxygen tetrahedrons can only be connected to each other through corner tips to form clusters of silicon-oxygen tetrahedrons. The oxygen ions located at the common corner tips of the silicon-oxygen tetrahedron are shared by four adjacent silicon-oxygen tetrahedrons, and their negative bivalent charges are neutralized by the silicon ions in the center of the two adjacent tetrahedrons. Therefore, the oxygen ions at the corner tips are electrically inactive and become inert oxygen ions. The ratio of silicon to oxygen in each silicon-oxygen tetrahedron is 1:2, and Si^{4+} ion is neutralized by the four oxygen ions at the corner tip of the tetrahedron (each with negative valence), so the total electrical charge is 0. If the silicon in the silicon-oxygen tetrahedron is replaced by aluminum ions, the aluminum-oxygen tetrahedron is formed. Aluminum is positive trivalent, so that one of the four oxygen ions at the corner tips of the tetrahedron cannot be neutralized, resulting in a negative charge. In order to neutralize the electricity, corresponding metal cations join. Therefore, all zeolite skeletons contain alkali metal, alkali earth metal ions. Because the amount of silicon displaced by aluminum ion in zeolite is changed, the contents of alkali metal and alkaline earth metal ions are also different with different ratios of silicon to aluminum.

2. Zeolite water

The silicon-oxygen and aluminum-oxygen tetrahedrons are connected with each other through corner tips, which forms a three-dimensional silicon- (aluminum-) oxygen lattice structure of various shapes, namely zeolite structure. Due to the diversity in connection modes of silicon-(aluminum-) oxygen tetrahedrons, many pores and channels are formed in the zeolite structure, some of which are one dimensional, some are two dimensional, and some are three dimensional. The pores and channels inside the zeolite structure are usually filled with water, where water molecules exist in crystalline form. The water molecules can be removed after heated at a particular temperature, but the removal of the water molecules does not destroy the zeolite structure. The water is called "zeolite water".

3. Zeolite molecular sieve

In the zeolite structure, the removal of zeolite water leaves many pores and channels, which possess the adsorption characteristic. Thus, the molecules smaller than the pores and channels in the zeolite are adsorbed into the pores and channels. However, the molecules larger than the pores and channels in the zeolite cannot be adsorbed. Therefore, zeolite can be used as a molecular sieve to screen out the gas of different molecular sizes. This is called zeolite molecular sieve.

4. Cation exchange capacity

In order to balance the charges in the zeolite structure, alkali metal and alkali earth metal ions enter the crystal structure of zeolite, but these ions can be replaced by other metal ions. The capacity of substituting with a base is called the cation exchange capacity. It is often referred to as CEC (Cation Exchange Capacity). CEC is measured by Shollenberger's ammonium acetate method. CEC is usually expressed as mg equivalent per 100 g sample. Examples of alkali displacement capacity of zeolite and clay minerals are shown in Table 11.3.

Table 11.3 CEC examples of zeolite and clay minerals.

Mineral	CEC	Exchanged cations		
		Ca^{2+}	Na^+	K^+
Zeolite	153.0	15.9	72.1	74.9
Bentonite	97.9	16.4	96.6	5.3

Based on the cation exchange sequence of clinoptilolite, the radioactive ion Cs and the nitrogen in the form of NH4 in the aqueous solution can be removed by ion exchange with natural Ca- and Na- type clinoptilolite. Alkali metals, alkaline earth metal ions and heavy metal ions in zeolite can also be replaced, in the order of $Pb^{2+} › Ca^{2+} › Cd^{2+} › Zn^{2+}$. The effect of replacing different cations on the structure is very small. However, it has a significant effect on cation exchange, catalysis, and adsorption performances of zeolite.

11.3.6 Chemical reactivity of zeolite

Influence of different cation contents in zeolite skeleton on chemical reactivity

An original sample of natural zeolite (Nenjiang) has zeolite content $\geq 60\%$ with fineness of 4-5% screened out with 4900-holes sieve. Through ion exchange, the original natural zeolite turns into NH^{4+}, Ca^{2+}, Na^+, and K^+ type, respectively. All types of zeolite samples are used to replace cement at 10% and prepare standard specimens. The flexural and compressive strengths at an age of 28 days have been measured, which are shown in Table 11.4.

Table 11.4 Strength of natural zeolite cement with different cations (MPa)

Strengths Type	28 d		Note
	Flexural	Compressive	
Original zeolite	7.9	39.2	Strength of pure cement
NH^{4+}	7.7	44.2	paste specimens
Ca^{2+}	8.0	44.3	32.1
Na^+	7.5	42.0	
K^+	7.8	39.5	

It can be seen from Table 11.4 that, when 1% of cement is replaced by various zeolite samples, both compressive and flexural strengths at 28 days are higher than those of benchmark specimens. Among them, the Ca^{2+} and NH^{4+} samples have the best performance, which increases by 38% compared with the strengths of benchmark specimens.

Influence of different fineness

Cement mortar specimens are prepared by grinding Dushikou clinoptilolite to pass through sieves of 100 mesh, 200 mesh, 260 mesh and 360 mesh and replacing 20% and 40% of cement, respectively. The 7 day and 28 days strengths are shown in Table 11.5.

Table 11.5 Strength of zeolite cement with different fineness and replacement rates.

Fineness		100 mesh		200 mesh		260 mesh		360 mesh	
Replacement rate (%)		20	40	20	40	20	40	20	40
Strengths	7 d	8.3	5.6	7.9	6.7	8.5	7.0	8.7	8.6
	28 d	26	21	27	23	30	25	34	28

It can be seen from Table 11.5 that the 28-day strength of zeolite cement increases with the increase of fineness of zeolite at the same replacement content. Regardless of the fineness, the strength of zeolite cement with a replacement content of 20% at 28th day is higher than that of zeolite cement with a replacement content of 40%. Fineness has a significant effect on the chemical reactivity of zeolite.

Zeolite content and lime absorption value

Natural zeolite with different zeolite contents is applied to react with hydrated lime. The higher the reactivity of zeolite, the higher the absorption value of hydrated lime, as shown in Table 11.6.

Table 11.6 Lime absorption value of zeolite (mg CaO/g)

Zeolite type	Zeolite content (%)	30-day lime absorption value
Jinyun Mordenite Xuanhua clinoptilolite	≥ 50	≥ 150
	≥ 50	176.12
Hailin clinoptilolite (1)	≥ 60	196.58
Hailin clinoptilolite (2)	≥ 70	262

According to China's national standard, the 30 days lime absorption value is 50-60 (mg CaO/g). The value of lime absorption of natural zeolite is 4-5 times higher than that of the value specified in the national standard, and it increases with the increase of zeolite content.

Soluble silicon and aluminum in zeolite

Zeolite's lime absorption value is high and zeolite content is high, and its chemical reaction activity is also high. This may be related to the soluble silicon and aluminum in Zeolite as shown in Table 11.7.

Table 11.7 Soluble silicon and aluminum in zeolite.

Types	N_2 adsorption BET (m^2/g)	Pore size (Å)	Average particle size (μm)	Ammonium exchange capacity m mol/100g	Soluble SiO_2 (%)	Soluble Al_2O_3 (%)
Zeolite 1	34.30	25.30	5.90	147.81	12.08	8.93
Zeolite 2	19.54	23.00	5.00	107.82	8.08	8.20

As shown in Table 11.7, zeolite content in Zeolite 1 is about 50%, zeolite content in Zeolite 2 is about 70%. Zeolite contains soluble silicon and aluminum, so after mixing with hydrated lime, chemical reactions occur in the early stage.

In summary, the chemical reactivity of zeolite is related to its ammonium ion exchange capacity (CEC), type of zeolite (Ca^+, NH_4^+), soluble silicon and aluminum contents in zeolite, and fineness.

11.3.7 Chemical reactivity of crystalline zeolite

The chemical reactivity of zeolite mentioned above is tested when the zeolite is in a crystalline state. What will happen about the reactivity of calcined zeolite, if zeolite is calcined and converted to vitrified material?

Thermal treatment of natural zeolite (Oya stone)

After crushing the natural zeolite (Oya stone) to form powders with size smaller than 1.2 mm, heating at 500 °C and 800 °C for 2 hours respectively, tests were conducted by using XRD and SEM, of which the results are shown in Figures 11.9 and 11.10, respectively. After 2 hours of thermal treatment at 800 °C, TG-DTA results of natural zeolite are shown in Figure 11.11.

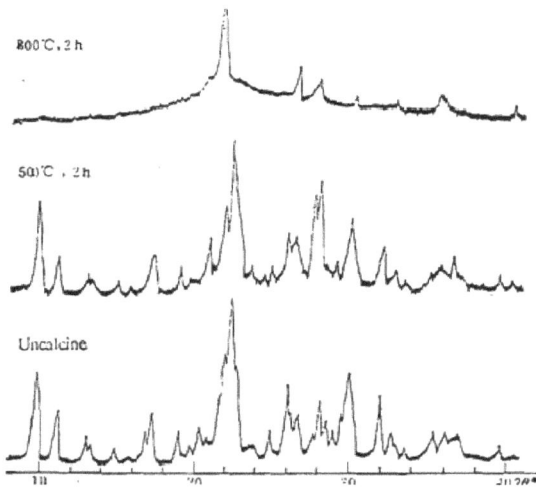

Figure 11.9 XRD patterns of natural zeolite (Oya stone).

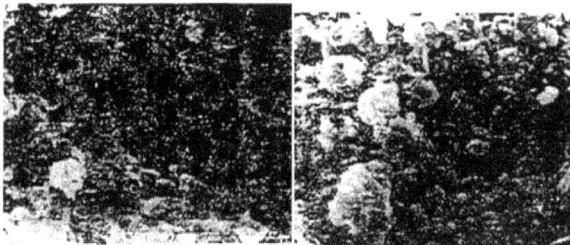

(a) Untreated zeolite   (b) 500 °C 2h treated zeolite

(c) 800 °C 2h treated zeolite

Figure 11.10 Effect of thermal treatment on SEM image of zeolite.

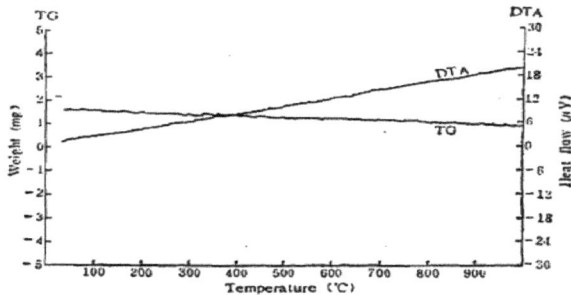

Fig 11.11 TG-DTA of natural zeolite after 2 hours thermal treatment at 800 ℃.

The XRD patterns of Oya stone treated at 500 °C and untreated are basically the same, but Oya stone treated at 800 °C turns into vitrified state, and the characteristic peaks of zeolite disappear. There is no endothermic change on the DTA curve of Oya stone treated at 800 °C; thermogravimetry (TG) curve also has very little weight loss. It can also be seen from the SEM image (Figure 11.10) that most of the Oya stone and zeolite minerals treated at 800 °C have been vitrified and become a coarse agglomerate structure.

Thermal treatment on the fineness of zeolite (Oya stone) and reaction with Ca(OH)₂

After crushing the Oya stone to particles of size < 0.15 mm, and using the powders that have not been untreated, those treated at 500 °C and 800 °C for 2 hours, respectively, mixed with $Ca(OH)_2$ and water, according to the ratio of 3:1:2.6 to conduct tests. After standard curing and autoclave curing (180 °C, 3h), the XRD, TG-DSC, and SEM analysis of each specimen is carried out. The results are shown in Figures 11.12, 11.13, and 11.14. The results are summarized in Table 11.8.

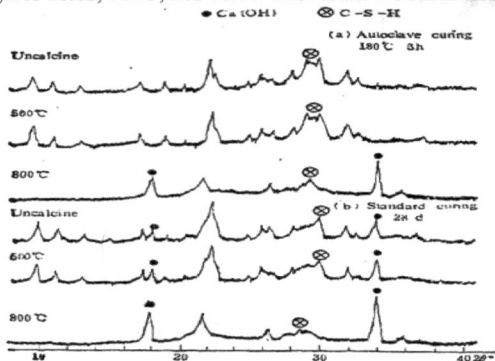

Figure 11.12 XRD showing the reaction between natural zeolite (Oya stone) and Ca(OH)₂.

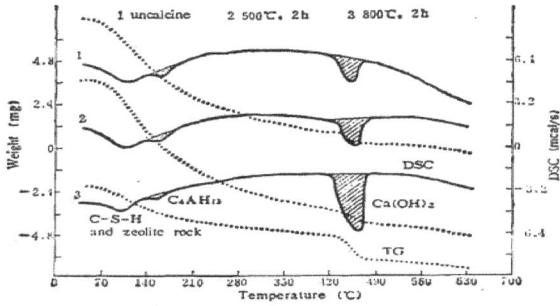

(a) Standard cured specimen, 20 °C, 28 d

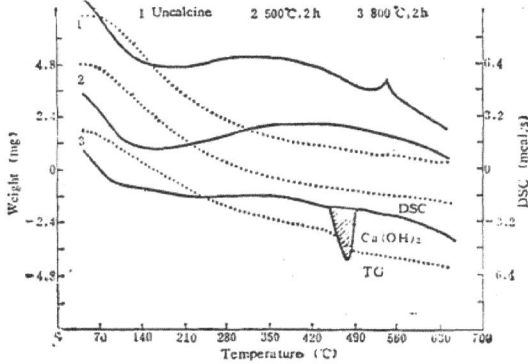

(b) Autoclave cured specimen, 180 °C, 3 h

Figure 11.13 TG-DSC of Autoclave cured specimen.

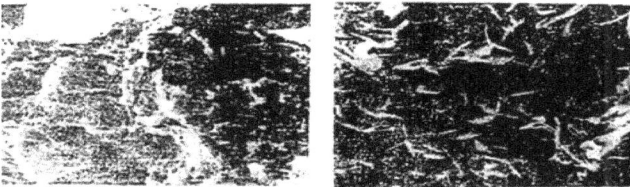

(a) untreated (without calcination)

(b) 500 °C

(c) 800 °C

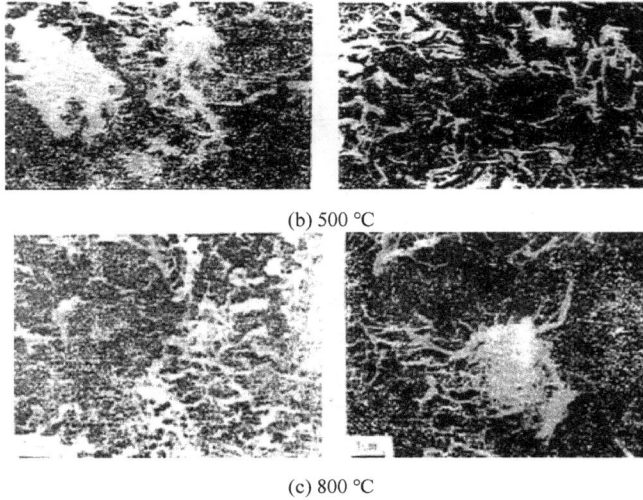

(1) Standard cured specimen   (2) Autoclave cured specimen

Figure 11.14 Specimens with zeolite + Ca(OH)₂+H₂O.

Table 11.8 Reaction between natural zeolite treated at different temperatures and lime.

Curing condition	Standard cured specimen			Autoclave cured specimen 180 °C, 3 h		
Temperature	untreated	500 °C	800 °C	untreated	500 °C	800 °C
XRD	C-S-H CH	C-S-H CH	C-S-H CH	C-S-H	C-S-H	C-S-H CH
TG-DSC	Thin-film C-S-H;28d $Ca(OH)_2$ with ~90% of reaction	Thin film C-S-H;28d $Ca(OH)_2$ with ~90% of reaction	Thin-film $Ca(OH)_2$ with ~75% of reaction	C-S-H CH with full reaction	C-S-H CH with full reaction	CH with ~80% of reaction
SEM	Thin-film C-S-H	Thin-film C-S-H	Thin-film C-S-H	Needle-shaped C-S-H	Needle-shaped C-S-H	Thin-film C-S-H covering particles of Oya stone

Thermal treatment on active silicon and aluminum contents in natural zeolite

Through the analysis of active silicon and aluminum contents in natural zeolites which are untreated and treated at 500 °C and 800 °C, it is further proved that as the treatment temperature of natural zeolite increases, the content of active silicon and aluminum decreases, which means that the contents of active silicon and aluminum decrease, i.e. the ability of reaction with $Ca(OH)_2$ is also reduced, as shown in Table 11.9.

Table 11.9 Thermal treatment and active silicon and aluminum contents in Oya stone.

Treatment Temperature	Oya stone	500 °C	800 °C
Soluble SiO_2 %	12.08	11.81	9.02
Soluble Al_2O_3 %	8.39	8.37	7.37

11.4 Improvement of natural zeolite powder on concrete with different water/cement ratios

By using three types of natural zeolite powder and one type of silica powder to prepare concretes with different w/c ratios for different types of cement with different strengths, with replacing of 10% cement, the change of concrete strength is observed.

11.4.1 Raw materials

Cement: Jidong 52.5 Portland cement, 28-day compressive strength of 60 MPa;

Capital cement factory manufactured slag 425(K4), slag 325(K3);

Natural zeolite powder: zeolite collected from 1. Fangshan (F); 2. Heihe (H); 3. Longhua (L); zeolite content ≥60%, specific surface area = 4000 cm^2/g;

Silica fume :SiO_2≥90%, specific surface area≥200000 cm^2/g.

11.4.2 Concrete mix proportion

Mix proportion for the benchmark concretes is shown in Table 11.10. Mix proportion for the benchmark concretes with replacing 10% natural zeolite powder is shown in Table 11.11.

Table 11.10 Mix design of the benchmark concretes (kg/m^3).

ID	W/C (%)	C	W	S	G	Nf	Age (3,7,28 d)
G-1	30	500	150	650	1200		
G-2	45	400	180	670	1250		
G-3	60	350	210	680	1260		
K 4 -1	35	500	175	650	1200		
K 4 -2	45	400	180	670	1250		
K 4 -3	60	350	210	680	1260		
K3-1	40	500	200	650	1200		
K3-2	50	400	200	670	1250		

Table 11.11 Comparison between concretes containing natural zeolite powder and silica fume.

ID	W/C (%)	C	W	S	G	NF	Age 3 7 28
GF-1			150				
GH-1	30	450+50	150	650	1200		3 d 7 d 28 d
GG-1	30	450+50	150	650	1200		3 d 7 d 28 d
GF-2	45	360+40	180	670	1250		3 d 7 d 28 d
GH-2	45	360+40	180	670	1250		3 d 7 d 28 d
GG-2	45	360+40	180	670	1250		3 d 7 d 28 d
GF-3	60	315+35	210	680	1260		3 d 7 d 28 d
GH-3	60	315+35	210	680	1260		3 d 7 d 28 d
GG-3	60	315+35	210	680	1260		3 d 7 d 28 d
K4H-1	35	450+50	175	650	1200		7 d 28 d
K4F-1	35	450+50	175	650	1200		
K4G-1	35	450+50	175	650	1200		
K4H-2	45	360+40	180	670	1250		
K4F-2	45	360+40	180	670	1250		
K4G-2	45	360+40	180	670	1250		
K4H-3	60	315+35	210	680	1260		
K4F-3	60	315+35	210	680	1260		
K4G-3	60	315+35	210	680	1260		
K3H-1	40	450+50	200	650	1200		
K3F-1	40	450+50	200	650	1200		
K3G-1	40	450+50	200	650	1200		
H3H-2	50	360+40	200	670	1250		
K3F-2	50	360+40	200	670	1250		
K3G-2	50	360+40	200	670	1250		7 d 28 d

11.4.3 Test results

Test results are shown in Table 11.12.

Table 11.12 Summary of test results

ID	W/C (%)	Replacement C	Slump	Compressive strength (MPa) 3 d 7 d 28 d	Relative strength(%) 3 d 7 d 28 d
G-1	30	0	1-5	57 78 79	100 100 100
GF-1	30	10	3-5	61 78 86	107 99.7 108.9
GH-1	30	10	1-3	63 75 89	110. 96 113
GG-1	30	10	1-3	65 83 94	115 107 119
G-2	45	0	7-5	35 47 57	100 100 100
GF-2	45	10	8-12	41 59 66	117 123 116
GH-2	45	10	8-12	40 50 71	113 107 125
GG-2	45	10	5-8	42 59 67	120 124 118
G-3	60	0	18-22	14 29 47	100 100 100
GF-3	60	10	18-22	19 36 49	132 124 106
GH-3	60	10	18-22	15 32 46	109 110 99
GG-3	60	10	12-18	18 37 53	126 127 113
K_4-1	35	0	23	38 41	100 100
K4H-1	35	10	8-12	42 64.3	111 157
K4F-1	35	10	8-12	38 58	100 142
K4G-1	35	10	5-8	46.5 64	122 155
K_4-2	45	0	18-22	23 41.3	100 100
K4H-2	45	10	8-12	26.7 47.7	116 116
K4F-2	45	10	8-12	24 46.3	104 112
K4G-2	45	10	5-8	32.7 53.7	142 130
K_4-3	60	0	18-12	8.3 17	100 100
K4H-3	60	10	18-22	12.3 27	149 159
K4F-3	60	10	18-22	13 23	157 135
K4G-3	60	10	8-12	18.5 31	223 182
K3-1	40	0	8-12	24 32	100 100
K3H-1	40	10	6-8	27,3 48	114 150
K3F-1	40	10	6-8	26 42.7	108 133
K3G-1	40	10	8-11	32 48	122 150
K3-2	50	0	12-15	14.7 23.7	100 100
H3H-2	50	10	8-12	16.5 33	112 139
K3F-2	50	10	8-12	13.7 27	93 114
K3G-2	50	10	6-8	19 34	129 144

11.4.4 Conclusions

According to 76 groups of concrete tests with different types of cement, different admixtures, and different water/cement ratios, the slump and the strength of concrete at different curing ages are measured, and the conclusions are as follows:

Portland cement No. 575

(1) By using the admixture replacing 10% cement, significant increment in 3 day strength of GF (Fangshan zeolite), GH (Heihe zeolite), and GG (silica fume) concrete, at W/C of 0.30, 0.45, and 0.60 respectively were observed. When W/C=0.30, the minimum increment is 7% and the maximum increment is 14.6%; when W/C=0.45, the minimum increment is 13%, and the maximum increment is 20%; when W/C=0.60, the minimum increment is 9.3%, the maximum increment is 32%.

For 28 days strength, when W/C=0.30, the effect of zeolite on increasing strength is significant, where the minimum value of the increase is 8.9%, and the maximum value is 19%; when W/C=60%, the minimum value of the increase is 5.5%, and the maximum value is 13.4%.

(2) As to the effect on concrete slump, after replacing 10% cement in the concrete with natural zeolite powder produced at different regions, the slump is roughly the same or larger than that of the benchmark concrete. When the same amount of silica fume is used to replace 10% of the cement, slump decreases significantly.

For 425 slag cement

(1) When 10% of cement was replaced by admixtures, there was a significant enhancement on the 7 day and 28-day strength of GF (Fangshan zeolite), GH (Heihe zeolite) and GG (silica fume) concrete with W/C of 0.35, 0.45 and 0.60, respectively. For W/C of 0.35, the enhancement on the 28 day strength of the GF, GH, and GG concrete was 56.8%, 41.5% and 55.4%, respectively. For W/C of 0.45, the enhancement on the 28 day strength was 15.5%, 12% and 30%, respectively. For the W/C of 0.60, the enhancement on the 28 day strength was 58.8%, 35.3% and 82.4%, respectively.

11.5 High-strength/high-performance concrete made of natural zeolite powder

High-strength/high-performance concrete was prepared by using natural zeolite powder with a specific surface area ≥6000 cm^2/g and an average particle size ≤ μm to replace 10% of cement by mass.

11.5.1 Raw ingredients

Cement: 525 Portland cement, 425 Slag cement

Aggregate: Coarse aggregate was crushed pebbles produced in Beijing Lishuiqiao, and fine aggregate was river sand produced in Beijing Lishuiqiao. The physical properties of the coarse and fine aggregates are shown in Table 11.13.

Table 11.13 Physical properties of coarse and fine aggregates

Properties / Type	Maximum particle size (mm)	Fineness modulus	Apparent density (g/l)	Water absorption (%)	Bulk density (kg/m^3)	Particle size distribution
Coarse aggregate	20	6.40	2.69	0.5	1.56	qualified
Fine aggregate	≤5mm	3.23	2.61	0.8	1.55	qualified

Ultrafine mineral powder: natural zeolite powder: 1) Oya stone 2) Erjing zeolite and 3) Dushikou; 4) Gaojing fly ash; 5) Shougang slag powder.

High efficiency water reducing agent: Naphthalene-type water reducing agent

11.5.2 Experimental programs

There were four series of test programs: 1. To investigate the effect of four different mineral powders on concrete performance by replacing 10% of cement (by mass) with nature zeolite powders, slag powder and fly ash, respectively. 2. To investigate the influence of different fineness of zeolite powders on concrete performance by replacing 10% of cement with zeolite powders. 3. To investigate the effect of different substitution contents of natural zeolite powder for cement on concrete performance, and the substitution content was 10%, 15% and 20%, respectively. 4. To evaluate the performance of concrete prepared with 525 Portland cement or 425 slag cement mixed with natural zeolite powder. The mix proportions of this series of tests are shown in Tables 11.14, 11.15, 11.16 and 11.17, respectively.

Table 11.14 Replacing 10% of cement (by mass) with different mineral powders (Series 1)

Batch No.	Cement-admixture Type	Particle size	W/B (%)	Sand percentage (%)	Water	Water reducer	Concrete raw ingredients (kg/m^3) Cement	Admixture	Sand	Aggregate
1	Cement	8.0	35	32	175	5.0	500	--	560	1207
2	Oya stone	6.2	35	32	175	5.0	450	50	560	1207
3	Erjing zeolite	6.4	35	32	175	5.0	450	50	560	1207
4	Fly ash	5.6	35	32	175	5.0	450	50	560	1207
5	Slag	6.7	35	32	175	5.0	450	50	560	1207

Table 11.15 Influence of different fineness of zeolite powders on concrete performance (Series 2)

Batch No.	Cement-admixture		W/B (%)	Sand percentage (%)	Water	Water reducer	Concrete raw ingredients (kg/m³)			
	Type	Particle size					Cement	Admixture	Sand	Aggregate
6	Cement	8.0	35	32	175	5.0	500	--	560	1207
7	Oya stone	6.7	35	32	175	5.0	450	50	560	1207
8	Oya stone	6.2	35	32	175	5.0	450	50	560	1207
9	Oya stone	5.6	35	32	175	5.0	450	50	560	1207
10	Erjing zeolite	6.8	35	32	175	5.0	450	50	560	1207
11	Erjing zeolite	6.4	35	32	175	5.0	450	50	560	1207
12	Erjing zeolite	5.6	35	32	175	5.0	450	0	560	1207

Table 11.16 Effect of different substitution contents of natural zeolite powder for cement on concrete performance (Series 3)

Batch No.	Cement-admixture		W/B (%)	Sand percentage (%)	Water	Water reducer	Concrete raw ingredients (kg/m³)			
	Type	Content					Cement	Admixture	Sand	Aggregate
13	Oya stone	10	35	32	175	5.0	450	50	560	1207
14	Oya stone	15	35	32	175	5.0	425	75	560	1207
15	Oya stone	20	35	32	175	5.0	400	100	560	1207
16	Erjing zeolite	10	35	32	175	5.0	450	50	560	1207
17	Erjing zeolite	15	35	32	175	5.0	425	75	560	1207
18	Erjing zeolite	20	35	32	175	5.0	400	100	560	1207
19	Dushikou	10	35	32	175	5.0	450	50	560	1207
20	Dushikou	15	35	32	175	5.0	425	75	560	1207
21	Dushikou	20	35	32	175	5.0	400	100	560	1207

Table 11.17 Performance of concrete prepared with different cement mixed with natural zeolite powder (Series 4)

Type	Batch No.	Natural zeolite powder	W/B (%)	Water (kg/m³)	Water reducer (kg/m³)	Concrete raw ingredients (kg/m³)			
						Cement	Natural zeolite powder	Fine aggregate	Coarse aggregate
Portland cement	22	--	37.8	170	4.0	450	0	560	1207
	23	Yes				390	60	560	1207
	24	--	35.5	160	5.0	450	0	560	1207
	25	Yes				390	60	560	1207
	26	--	33.0	150	5.5	450	0	560	1207
	27	Yes				390	60	560	1207
	28	--	31.0	140	6.0	450	0	560	1207
	29	Yes				390	60	560	1207
Slag cement	30	--	37.8	170	4.0	450	0	560	1207
	31	Yes				390	60	560	1207
	32	--	35.5	160	5.0	450	0	560	1207
	33	Yes				390	60	560	1207
	34	--	33.0	150	5.5	450	0	560	1207
	35	Yes				390	60	560	1207
	36	--	31.0	140	6.0	450	0	560	1207
	37	Yes				390	60	560	1207

11.5.3 Specimen preparation and curing

Specimen size: 10 cm × 10 cm × 10 cm. Raw ingredients were mixed in a concrete mixer, formed on a vibrating table, and demoulded after 24 hours. The specimens were cured in humid air at 20 °C until tested at the specified age.

11.5.4 Test results and analysis

Performance of fresh concrete

In these experiments, the content of fine aggregate, coarse aggregate and water was fixed, and only the fineness, type and content of natural zeolite powder were changed. The slump values of fresh concrete were discrete, ranging from 10 cm to 20 cm. Comprehensive analysis results showed that the slump of the concrete mixed with zeolite powder was about 1.5-3.0 cm lower than that of the reference concrete. Moreover, as the content of zeolite powder increased, the slump decreased more. This was attributed to the porous structure of zeolite powder.

Impact on concrete strength

The test results of each series are summarized in Table 11.18.

Table 11.18 Summary of the test results of each series.

Series	Batch No.	Slump (cm)	Air content (%)	Compressive strength (MPa)		
				3d	7d	28d
Series 1	1	20	3.0	47.5	59.5	70.8
	2	17.4	3.0	52.4	67.0	80.0
	3	16.5	3.0	48.1	61.8	74.6
	4	20.0	3.0	43.1	50.7	66.9
	5	21.0	3.0	41.0	51.6	66.1
Series 2	6	20.0	3.0	47.5	59.5	70.8
	7	17.4	3.5	49.0	65.5	77.9
	8	17.0	3.5	51.3	66.6	77.9
	9	17.0	3.5	48.5	68.4	80.0
	10	17.2	3.0	48.5	62.5	75.7
	11	17.0	3.5	52.2	65.5	76.4
	12	17.0	3.0		64.2	74.3
Series 3	13	17.4	3.0		67.0	80.0
	14	17.0	3.4		59.5	77.8
	15	15.8	3.6		59.0	68.5
	16	17.2	3.5		62.5	75.7
	17	16.8	3.6		61.3	69.4
	18	15.8	3.6		58.3	72.1
	19	17.2	3.2		60.2	77.9
	20	16.5	3.4		54.8	74.3
	21	15.5	3.4		60.5	70.8
Series 4	22	17.0	3.0		52.0	59.3
	23	15.6	3.4		58.2	66.9
	24	14.6	3.6		53.8	66.8
	25	14.0	3.5		59.1	76.8
	26	14.0	3.6		58.7	71.2
	27	13.0	3.6		65.7	84.0
	28	10.0	3.2		61.8	74.2
	29	18.0	3.4		69.8	85.3
	30	15.6	3.4		34.2	45.0
	31	16.0	3.0		36.6	50.4
	32	14.8	3.0		36.2	47.8
	33	15.0	3.0		39.1	54.0
	34	14.0	3.1		39.0	52.0
	35	12.0	3.2		42.1	58.8
	36	10.0	3.0		48.3	56.4
	37	10.0	3.0		53.1	64.3

Series 1

In this series, 10% of the cement was replaced with Oya stone, Erjing zeolite, slag powder and fly ash, respectively. The ratios of their 3 day, 7 day and 28 day strength to the benchmark concrete strength are shown in Figure 11.15.

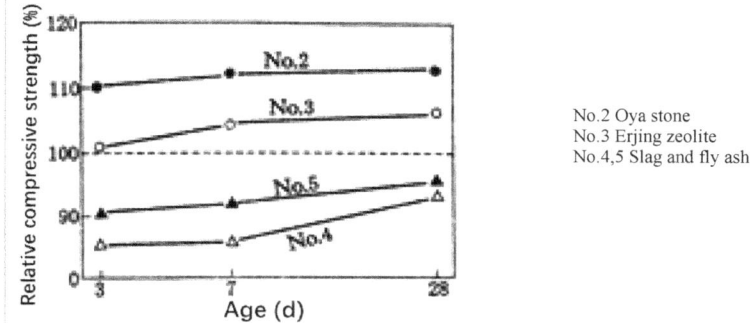

No.2 Oya stone
No.3 Erjing zeolite
No.4,5 Slag and fly ash

Figure 11.15 Influence of different mineral admixtures on concrete strength

It can be seen from Figure 11-15 that, after replacing 10% of cement with Oya stone powder, the 3 day, 7 day and 28 day strength of the concrete was increased by more than 10% compared with the benchmark concrete. The performance of Erjing zeolite was slightly inferior to that of Oya stone, but its concrete strength at all tested ages was still higher than that of the benchmark concrete. However, after replacing 10% of cement with slag and fly ash, the concrete strength was lower than that of the benchmark concrete, especially the 3 day and 7 day strength.

Series 2

In this series, the effect of replacing 10% of cement with zeolite powders of different fineness on concrete performance was studied.

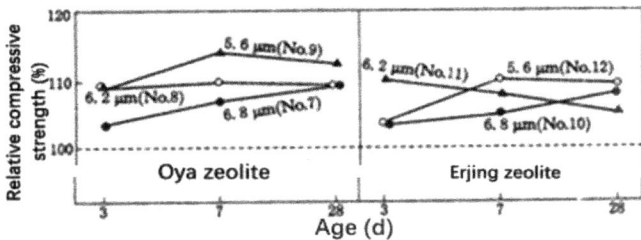

Figure 11.16 Enhancement of natural zeolite powders of different fineness on concrete performance.

As can be seen from Figure 11.16, the 7 day strength of concrete mixed with Oya stone powder with an average particle size of 6.7, 6.2 and 5.6 μm was 10%, 12% and 15% higher than that of the reference concrete, respectively. In addition, the 28 day strength was increased by 10%, 10% and 13%, respectively. Similar results were observed for the concrete mixed with Erjing zeolite powder. In general, when the average particle size of natural zeolite powder was 6 μm, the enhancement effect was more reasonable.

Series 3

In this series, the effect of different substitution contents of natural zeolite powder for cement on the strength of concrete was investigated (Figure 11.17).

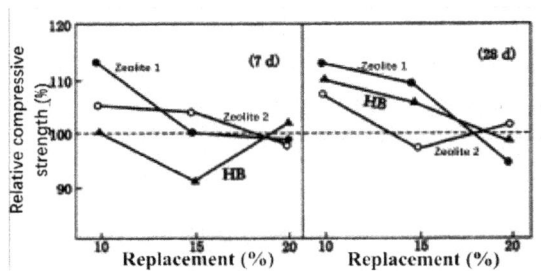

Figure 11.17 Relationship between concrete strength and the substitution contents of natural zeolite powder for cement.

The substitution content of natural zeolite powder for cement was 10%, 15% and 20%, respectively. Three types of natural zeolite powder had different effects on concrete performance. 1) 7 day strength: when the substitution content of Oya stone (zeolite 1) was 10%, the compressive strength was increased by about 15%. But when the substitution content was 15% and 20%, the strength was almost the same as that of the benchmark concrete. For Erjing zeolite, when the substitution content was 10% and 15%, the enhancement on compressive strength was approximately 5%. But when the substitution content was 20%, the strength was almost the same as that of the benchmark concrete. For Hebei Dushikou zeolite, when the substitution content was 10% and 20%, the compressive strength was almost the same as that of the benchmark concrete. When the substitution content was 15%, the strength was about 10% lower than that of the benchmark concrete. 2) 28 day strength: when the substitution content was 10%, the strength of these three types of concrete was all higher than that of the benchmark concrete. When the substitution content was 20%, their strength was close to that that of the benchmark concrete.

Series 4

In this series, different types of cement were used. Portland cement and slag cement were used as benchmarks, and 13.3% of the cement was replaced with natural zeolite powder by mass. The change of concrete strength under different water to cement ratios was evaluated.

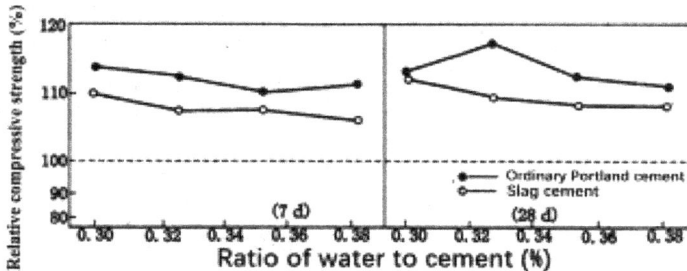

Figure 11.18 Strength of concrete with 13.3% natural zeolite powder and different water/cement ratios.

It can be seen from Figure 11.18 that, for both Portland cement and slag cement-based concrete, the 7 day and 28 day strength of concrete with 13.3% Oya stone powder and water to cement ratios of 0.30-0.38 was higher than that of the reference concrete. These mixes exhibited different degrees of improvement. The strength enhancement of Portland cement-based concrete was more significant, reaching 15.3%.

11.6 The microstructure of concrete with natural zeolite ultrafine powder

When the equivalent amount of natural zeolite powder replaces 5-10% of the cement in the concrete, the strength of the concrete is increased by 10-15% compared to the reference concrete. Zeolite is a kind of crystalline mineral, which is different from ordinary pozzolans. What effect does natural zeolite powder have in cement concrete, and why can it increase the strength of concrete? This is related to the improvement of the microstructure of concrete by natural zeolite ultrafine powder.

11.6.1 Pozzolanic reaction of natural zeolite powder

F.M. Lea suggested that natural pozzolans are some zeolite-like compounds. That is, natural Zeolite belongs to the category of pozzolanic materials. For the lime-volcanic ash reaction mechanism, two main arguments have been proposed: alkali exchange and direct combination.

Alkali exchange: Natural zeolite powder is combined with $Ca(OH)_2$ through ion exchange. Through measurement, when the solution is CaO 0.5 mg/L, the calcium ion exchange capacity of natural zeolite from Dushikou is 0.25 Ca mmol/g raw material. F.M. Lea proposed that it is equivalent 1-2% of volcanic ash. During alkali exchange, the crystal lattice of zeolite minerals does not change at all. One alkali ion is replaced by another ion and replace the same position in the crystal lattice. This reaction may not form a binding material.

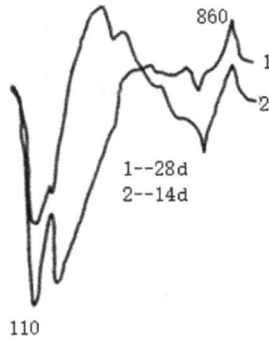

Figure 11.19 DTA of slaked lime-zeolite powder specimen.

Direct combination: The role of natural zeolite powder in the hydration and hardening of cement is mainly the direct combination of SiO_2 in the zeolite powder and $Ca(OH)_2$ to form hydrated calcium silicate, as shown in Figures 11.19 and 11.20.

(a) Standard curing 28 days (b) Autoclave specimen

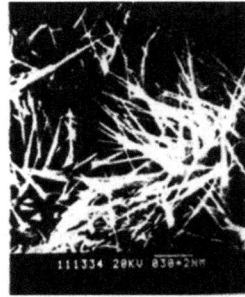

Figure 11.20 SEM of natural zeolite powder + slaked lime specimen.

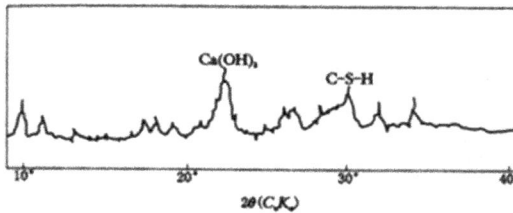

Figure 11.21 XRD pattern of natural zeolite powder + slaked lime.

In Figure 11.21, the C-S-H phase appears on the XRD pattern for age of 28 d. In Figure 11.22, the infrared spectra (IR) of samples at 14 d and 28 d both show the presence of hydrated calcium silicate, and the peak number is 970 cm^{-1}. The characteristic peaks of Ca(OH)$_2$ and zeolite alone are not obvious. This also shows that the natural zeolite powder reacts with slaked lime to form hydrated calcium silicate.

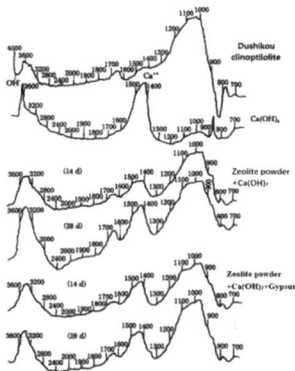

Figure 11.22 Infrared spectra (IR) of natural zeolite powder + lime + gypsum.

Natural zeolite powder reacts with slaked lime: Soluble silicon and aluminum also increase with age. The composition of the test specimen for testing the reaction between natural zeolite powder and slaked lime is shown in Table 11.19 and the results are shown in Table 11.20.

Table 11.19 Composition of natural zeolite ultrafine powder and slaked lime specimens.

Group	Natural zeolite	Slaked lime	Gypsum	Fluorite
A	50%	50%	--	--
B	50%	46%	4%	--
C	50%	46%	--	4%

Table 11.20 Soluble silica alumina in different ages of the reaction between natural zeolite powder and slaked lime.

Group	Curing duration	3 d	7 d	28 d
A	SiO_2	4.88	6.13	9.63
	Al_2O_3	1.66	2.14	3.11
B	SiO_2	8.9	9.4	11.1
	Al_2O_3	2.14	2.29	3.26
C	SiO_2	8.7	7.1	
	Al_2O_3	2.14	2.24	3.11

It can be seen that the natural zeolite powder reacts with slaked lime, and the active silicon and alumina increase with age.

11.6.2 The role of natural zeolite powder in cement concrete

Increase the CSH phase to increase the density of the cement paste. A 2 cm x 2 cm x 2 cm specimen was prepared with Portland cement mixed with 20% natural zeolite powder, W/B=0.25. The results of the samples under standard curing tested at age of 28 d by using SEM and XRD are shown in Figures 11.23 and 11.24. CSH gel is formed, and the density of cement paste is improved.

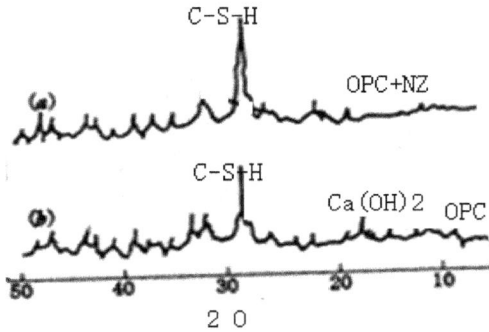

Figure 11.23 XRD with and without zeolite powder in cement stone.

(a) Portland cement stone

(b) Cement stone containing 20% zeolite powder

Figure 11.24 SEM of cement stone with and without NZ powder.

Self-vacuum effect of natural zeolite powder in cement stone.

The zeolite mineral content in natural zeolite is about 60%. There are many cavities and pores in the zeolite mineral, and the internal surface area is very large. The internal surface area of the A-type zeolite molecular sieve is about $1100 \text{ cm}^2/\text{g}$. When zeolite powder is mixed with cement, the zeolite absorbs water, and part of the free water is absorbed by the zeolite powder. During the hydration and hardening process, the zeolite powder releases out the originally adsorbed water, which supplies the water required for cement hydration. The connection between the zeolite powder particles and the cement hydrate is closer. The contact between zeolite powder and hydrate product is much closer through chemical and physical effects.

Improvement of pore structure of cement paste. The sample prepared with Portland cement and 10% zeolite powder, and the reference specimens, were cured under the same conditions. The pore structure was measured at the age of 7 d and 28 d. The results are shown in Figure 11.25.

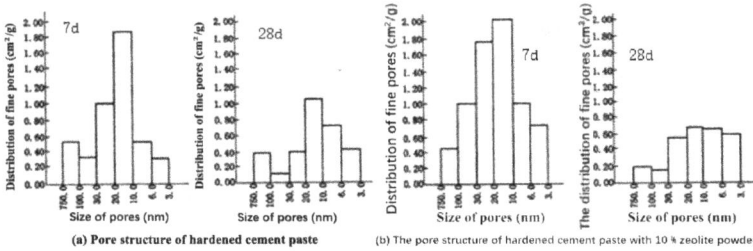

(a) Pore structure of hardened cement paste

(b) The pore structure of hardened cement paste with 10 % zeolite powder

Figure 11.25 Pore structure of cement stone with and without zeolite powder.

Improve the interface structure between coarse aggregate and cement paste in concrete

The concrete specimen was prepared with 10% natural zeolite powder, W/B of 0.30 and cured for 28 d. SEM analysis and EDAX analysis of the cement stone-aggregate interface transition zone was performed. The results are shown in Figures 11.26 and 11.27, respectively.

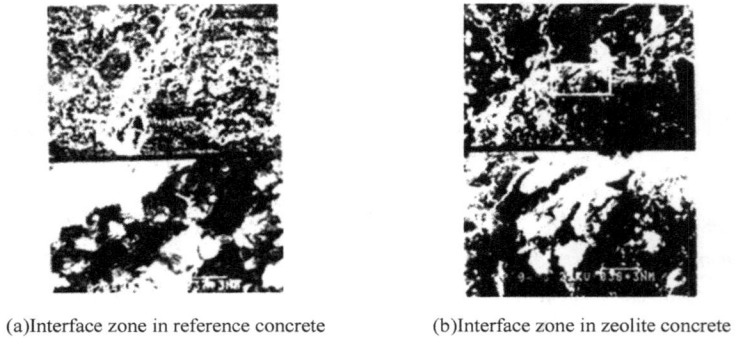

| (a)Interface zone in reference concrete | (b)Interface zone in zeolite concrete |

Figure 11.26 Interface zone in concrete for reference concrete and concrete mixed with 10% natural zeolite powder.

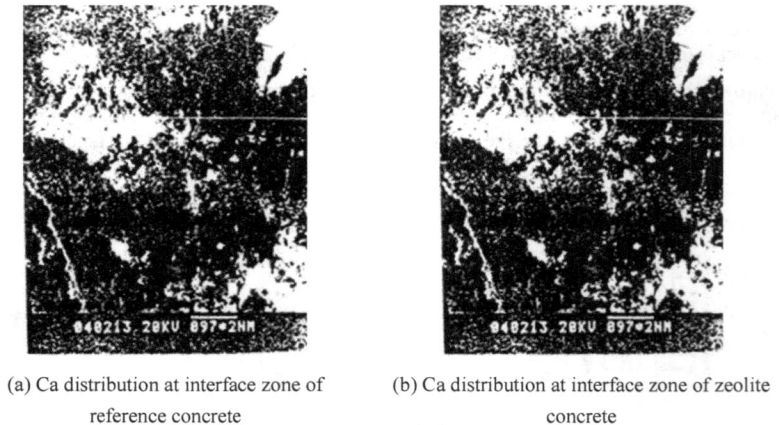

| (a) Ca distribution at interface zone of reference concrete | (b) Ca distribution at interface zone of zeolite concrete |

Figure 11.27 EDAX with Ca element distribution in the interface layer (the left side is aggregate).

In Figure 11.27, the white horizontal line is the selection line for the distribution position of the selected element. The white spots are the distribution of the Ca element along the position line. From the figure (a), it can be seen that in the interface transition zone of the pure cement stone, the Ca element is enriched at the aggregate end, which is caused by the directional enrichment and crystallization of $Ca(OH)_2$ in the water film layer, which is harmful to the mechanical properties of the interface. After adding 10% natural zeolite, due to the high silica content and soluble silica and alumina, the $Ca(OH)_2$ enriched crystallization effect is weakened. In Figure 11.27 (b), the Ca element is evenly distributed, and there is no obvious enrichment phenomenon.

EDAX point analysis of the content of CaO and SiO_2, and calculation of mass ratio of SiO_2/CaO in interface zone, were conducted, of which the results are shown in Figure 11.28. The SRD analysis was conducted on reference concrete and test concrete with 10 % zeolite, of which the results are shown in Figure 11.29.

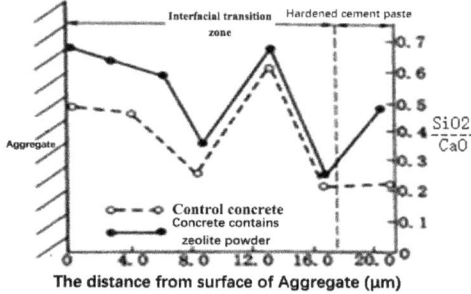

Figure 11.28 SiO₂/Ca O mass ratio of the concrete interface transition layer.

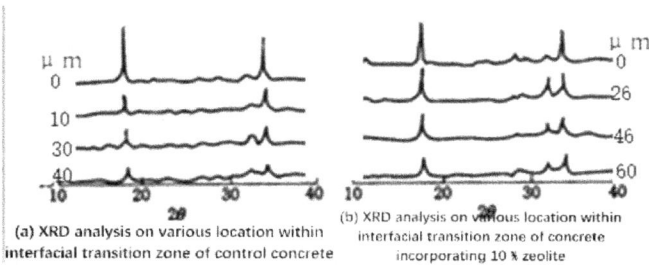

(a) XRD analysis on various location within interfacial transition zone of control concrete

(b) XRD analysis on various location within interfacial transition zone of concrete incorporating 10 % zeolite

Figure 11.29 XRD of the characteristic peak of Ca(OH)₂.

After calculation, the $Ca(OH)_2$ orientation index is shown in Figure 11.30. In the concrete mixed with natural zeolite powder, the $Ca(OH)_2$ crystal orientation in the interface transition layer is obviously weaker than that in the reference concrete. The orientation index is reduced from 50 µm in the reference concrete to about 40 µm in concrete with natural zeolite powder.

The crystal orientation of $Ca(OH)_2$ is weakened and the orientation range is reduced, so that the structural unevenness of the transition layer is improved, and the interface is strengthened.

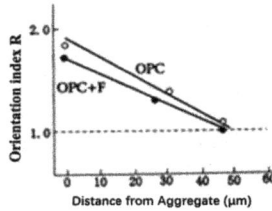

Figure 11.30 The orientation index and range of Ca(OH)$_2$.

11.7 Special functions of natural zeolite powder applied in cement concrete

Natural zeolite powder has a huge internal surface area due to the porous structure of zeolite. As mentioned above, the internal surface area of A-type zeolite molecular sieve is about 1100 cm^2/g. Zeolite powder can absorb a large amount of water and gradually release it out slowly, which means it can be used as a carrier of water and a carrier of water reducing agent. Through ion exchange, it can absorb and solidify heavy metals, and inhibit alkali-aggregate reaction, etc.

11.7.1 Natural zeolite powder can be used as a self-curing agent to inhibit the shrinkage and cracking of concrete

The shrinkage caused by the self-drying of cement stone in concrete is called autogenous shrinkage. When the water-cement ratio of concrete is less than 0.38, the cement in the cement stone cannot be completely hydrated. During the setting and hardening process, when the unhydrated cement further hydrates, it will absorb the moisture in the capillary of the cement stone, make the capillary generate a self-vacuum and generate a negative pressure in the capillary. And this will cause the hardened cement stone to auto-shrink. When the auto-shrinkage stress is greater than the tensile stress of the cement stone, the cement stone (or concrete) will produce cracking. The lower the water/cement ratio and the finer the binders, the more serious the situation. In addition to autogenous shrinkage early shrinkage also includes the shrinkage due to early water loss.

Early shrinkage (including autogenous shrinkage) is different from long-term drying shrinkage. The former is caused by early external and internal water loss. As the strength of cement is very low, cracks often occur. The latter is also due to the evaporation of water in concrete, and the dehydration of the hydrated gel in concrete. At this time, the tensile strength of the concrete is relatively large, and it can resist the shrinkage stress, so it is not easy to crack.

Guse and Hilsdorf observed the surface cracking of high-strength/high-performance concrete with W/C of 0.3 and 0.25. When demoulding after 24 hours, the surface of the concrete showed network cracks, as shown in Figures 11.31 and Figure 11.32, respectively.

Multifunctional Concrete Technology
Materials Research Foundations **127** (2022)

Materials Research Forum LLC
https://doi.org/10.21741/9781644901991

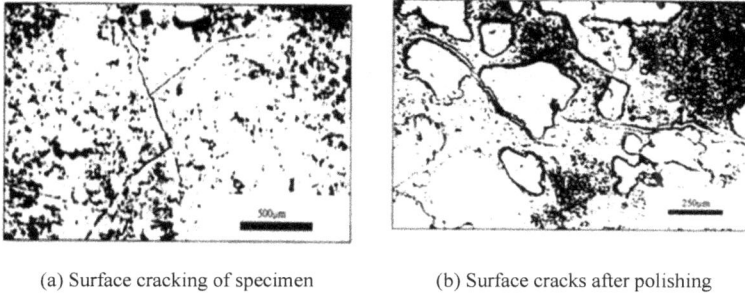

(a) Surface cracking of specimen   (b) Surface cracks after polishing

Figure 11.31 Autogenous shrinkage cracking of low W/C concrete.

Concrete with W/C=0.50 has the lowest autogenous shrinkage-caused cracking, only 3 mm/cm² at 1 d age. Concrete with W/C=0.25 has the largest autogenous shrinkage-caused cracking at 1 d age, 49 mm/cm², while concrete with W/C=0.30 also has larger autogenous shrinkage-caused cracking of 20 mm/cm² at 1 d age and 38 mm/cm² at 7 d age. The autogenous shrinkage of concrete with silica fume is slightly larger, about 39 mm/cm² at 7 d age.

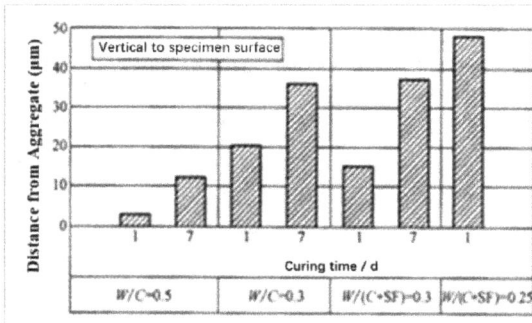

Figure 11.32 Analysis of cracking.

For example, when the C80 concrete shear wall was demolded on the ground floor of a super high-rise building in Guangzhou, there were cracks on the wall, and more than 120 cracks appeared on some walls as shown in Figure 11.33.

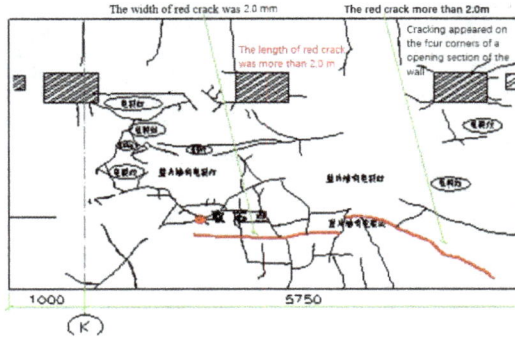

Figure 11.33 Autogenous shrinkage-caused cracks on the C80 concrete shear wall.

After adding natural zeolite powder, the zeolite powder absorbs water during the mixing process, but slowly release out water during the setting and hardening process. The water released will distribute evenly in concrete and be supplied for cement hydration. This will avoid the self-vacuum effect of capillary pores and the autogenous shrinkage cracking will be reduced.

For the case of the shear wall of the C80 concrete, the composition of materials is as shown in Table 11.21. 42.5R cement was used, and the total binding material was 560 kg/m³. The amount of mineral additives was also large with W/B=0.24. After the initial setting, the cement with high early strength absorbed the capillary water for hydration, and the shrinkage of the mineral powder is also high. Therefore, both autogenous shrinkage and cracking occurred.

Table 11.21 Mix design of C80 concrete (kg/m³).

P·II42.5R	Slag S95	Micro beads	River sand	Crushed stone	Water	superplasticizer	W/B
350	150	60	700	1025	135	1.7%	0.24

Through the experiments, the proportion of concrete was adjusted, and the properties of fresh concrete were measured, of which the results are shown in Tables 11.22 and 11.23. The autogenous shrinkage and hydration heat-caused temperature rising of concrete are shown in Table 11.24.

Table 11.22 Adjusted mix design of C80 (kg/m³).

Cement	Micro beads	Fly ash	NZ	Sand	Crushed rock	Water	CFA	admixture
320	80	170	15	800	900	140	1.5%	1.9%

Table 11.23 Fresh properties and strength of concrete.

slump(cm)		Slump flow (mm)		Flow time (s)		strength (MPa)		
Initial	2 h	Initial	2 h	Initial	2 h	3 d	7 d	28 d
25	26	650	680	7	5	56.6	75.4	92.1

Table 11.24 Concrete autogenous shrinkage and hydration heat temperature rise.

24 h	48 h	72 h	Time for starting rise (h)	Time to peak (h)	Initial – peak Temp (°C)
1.0/10k	1.06/10K	0.99/10K	25	35	25-75

After the tests of large plates and actual size plates, no autogenous-shrinkage cracking on the plate surface was found, as shown in Figures 11.34, 11.35 and 11.36.

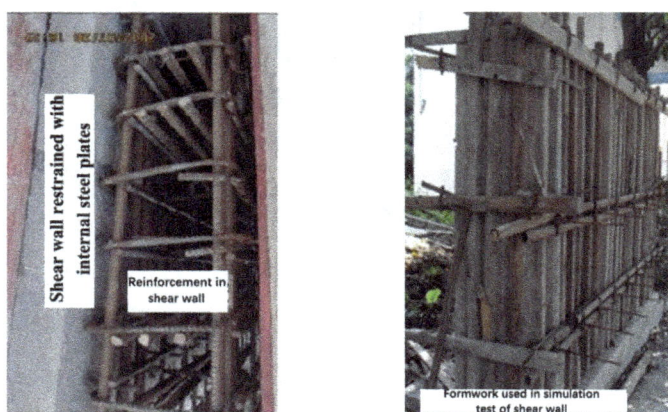

Figure 11.34 Simulated shear wall structure and formwork appearance.

Since natural zeolite powder is used as a self-curing agent, during the mixing process of concrete, the zeolite powder absorbs water. After the initial setting of the concrete, when the cement continues to hydrate and needs water, the zeolite powder releases out water for the needs of cement hydration and inhibits the production of capillary dehydration and the capillary tension. Autogenous shrinkage and cracking of concrete in the shear wall is suppressed.

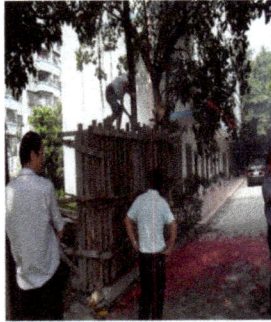

Figure 11.35 Simulating concrete pouring of the shear wall.

Figure 11.36 The surface of the shear wall after demoulding.

11.7.2 Natural zeolite powder used as the carrier of water reducing agent to reduce water and keep plasticity of the concrete

The natural zeolite superfine powder is used to adsorb the superplasticizer to prepare the zeolite water-reducing and plasticizing agent. Zeolite water-reducing and plastic-preserving agent can be used in ordinary concrete, high-performance concrete and ultra-high-performance concrete to increase the fluidity of concrete and control the change of fluidity with time. In ultra-high-performance concrete, it can also reduce the viscosity of fresh concrete, easy to pump for construction.

Zeolite water-reducing and plastic-preserving agent can control the slump loss of fresh concrete over time, mainly due to the slow release of high-efficiency water-reducing agent in concrete, maintaining the adsorption capacity of cement particles to water-reducing agent, so that the surface of cement particles has a certain value Zeta potential to make the cement slurry in a dispersed state.

Zeolite water-reducing and plastic-preserving agent mixed into concrete can also improve the structure of hardened concrete and increase the strength of concrete.

Experiment on the adsorption and desorption of natural zeolite powder on superplasticizer and its influence on the fluidity of cement paste

a) Experiment on the adsorption and desorption of zeolite powder to superplasticizer

Four parts of 100 g of natural zeolite powder each were mixed into the water solution of superplasticizer with 45% solid content. The solution was then filtered, washed, dried, and weighed every 30 min to obtain the weight of the zeolite powder after soaking for 30 min, 60 min, 90 min and 120 min. Calculate the capacity of the zeolite powder adsorbing the water reducing agent, of which the results are given in Table 11.25.

Table 11.25 Isotherm capacity of zeolite powder adsorbing water reducing agent (g/10g).

Time (mins)	30	60	90	120
A Sample	2.5	3.5	5.6	5.7
B Sample	2.6	3.6	5.5	5.8

Table 11.26 Isothermal volume of zeolite desorbing water reducing agent in aqueous solution (g/10g).

Time (mins)	30	60	90	120
A Sample	2.3	3.3	4.1	4.5
B Sample	2.4	3.4	4.0	4.6

After the zeolite powder was saturated with the superplasticizer, 100 g of the obtained zeolite superplasticizers A and B were sampled, put in an aqueous solution, and the amount of desorbing the superplasticizer is measured for 30 min, 60 min, 90 min, and 120 min, respectively, of which the results are given in Table 11.26.

According to Table 11.25 and Table 11.26, the adsorption and desorption curves of natural zeolite powder to water reducers A and B are obtained as shown in Figures 11.37 and 11.38. It can be seen from Figures that the adsorption capacity of natural zeolite powder for water reducer A and B increases almost linearly before 90 minutes, and after 90 minutes, it basically reaches a saturated state, and the saturated adsorption value for water reducer is about 5.6-5.8 g /10g. With time increasing, the adsorption amount increases slightly. The desorption amount of zeolite to water reducing agent in the water keeps increasing with time, but the limit value of its desorption amount is below the saturated adsorption value.

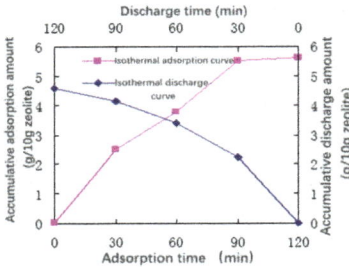

Figure 11.37 Isotherm adsorption and
discharge of zeolite powder to water
reducing agent A.

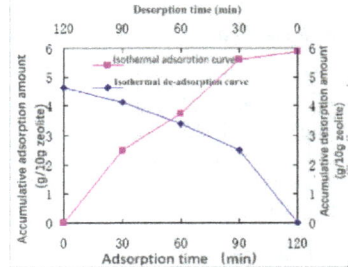

Figure 11.38 Isotherm adsorption and
discharge of zeolite powder to water
reducing agent B.

b) The effect of zeolite water reducer on the fluidity of cement paste

With 500 g of cement, 150 ml of water, and 10 g of zeolite water reducer, the fluidity test of the pure slurry was carried out according to GB/T-8077-2000, and the changes of flow over time were measured. The results are shown in Table 11.27. It can be seen that the zeolite water reducing agent can effectively control the fluidity loss of cement paste. At 120 min, the fluidity of Mitsubishi cement pastes not only did not decrease, but increased by 9.2%; while for Weifang shaft kiln cement, the fluidity decreased by about 20 %, which may be caused by the excessively high content of f-CaO or C_3A in the cement.

Table 11.27 Change of fluidity of cement paste (mm) with time.

Cement	Flow				
	Initial	30 min	60 min	90 min	120 min
Mitsubishi P.O.42.5	238	240	240	270	260
Weifang P.O.32.5	220	200	200	180	175

c) Change of concrete slump with time

The mix design of concrete test is shown in Table 11.28. The changes of concrete slump within 2 hr and the strength of each age are shown in Table 11.29. It can be seen from the test results that after 2 hr, the concrete slump dropped from 20 cm to 16.5 cm, a decrease of 17.5%. But the slump is still more than 16 cm, which can still meet the requirements of pumping concrete.

Table 11.28 Mix design of concrete.

W/C	Raw materials (kg/m³)						Remarks
	Cement	Fly ash	Sand	Crushed rock	Water	Zeolite admixture	Mitsubishi cement
0.40	350	100	750	1100	180	11.25 (2.5%)	

Table 11.29 Changes of slump and strength of concrete

Slump (cm)			Compressive strength （Mpa)		
Initial time 0 (11：20)	70 min (12：30)	120 min (1：20)	3 d	7 d	28 d
20	19	16.5	16.4	25.6	42

d) Mechanism of control on concrete slump loss by zeolite water reducing agent

Experiments have proved that both the slump loss of concrete and the increase in the apparent yield value of cement slurry are related to the decrease in the residual amount of superplasticizer in the liquid phase (actually the amount of superplasticizer adsorbed by cement particles) as shown in Figures 11.39 and 11.40. Comparison of the change of Zeta potential for the cement slurry with W/C=0.40 with and without zeolite water reducing agent was made and the results are shown in Figure 11.41. The absolute value of the Z potential of the cement paste without zeolite water reducing agent decreased rapidly, from 5.5 mv to 4.7 mv in 30 min and to 4.2 mv in 1 hr. But the cement paste containing zeolite water-reducing agent, the initial Z potential is 5.95 mv, and basically maintained at the level of 6 mv within 140 min. The cement paste remained in a good dispersion state.

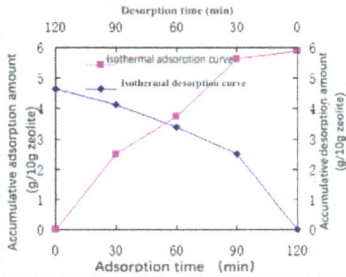

Figure 11.39. Changes in slump with time and the residual concentration of water-reducing agent.

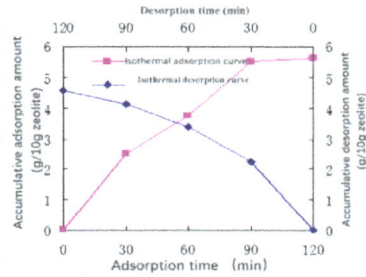

Figure-11.40. Changes in yield value of cement slurry over time and the residual concentration of water-reducing agent.

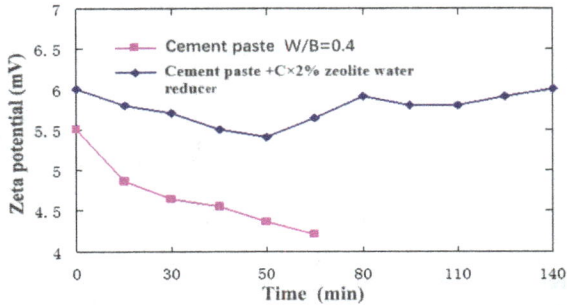

Figure 11.41 Changes in zeta potential on the surface of cement particles in cement paste.

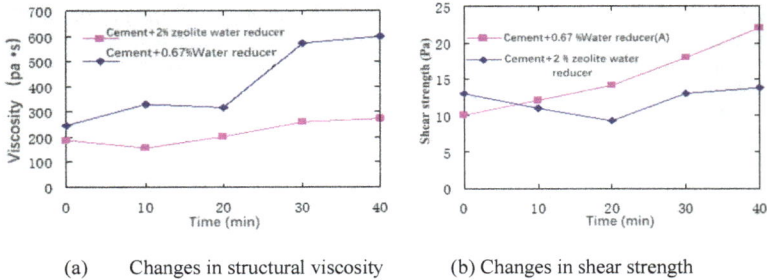

(a)　　Changes in structural viscosity　　　(b) Changes in shear strength

Figure 11.42 Changes in structural viscosity and shear strength of cement Paste with and without zeolite water reducing agent over time.

Zeolite water reducing agent also has a great influence on the shear strength and structural viscosity of cement paste. W/C=0.30. The cement paste was mixed with 0.67% naphthalene-based water-reducing agent and 2.0% zeolite water-reducing agent. Two samples were mixed with a Warming mixer, and the Fanning viscometer was used to determine the viscosity of cement paste. The changes of shear strength and structural viscosity till 40 min are shown in Figure 11.42.

It can be seen from the above that the zeolite water reducing agent can slowly desorb the adsorbed high-efficiency water-reducing agent into the cement paste, maintaining the adsorption amount of the water-reducing agent on the surface of the cement particles, thereby maintaining the zeta potential on the surface of the cement particles, and keeping the cement particles in a dispersed state. The structural viscosity and shear strength of the cement paste are also in a stable state, and the slump loss of the concrete can be controlled.

Application of Zeolite Water-reducing and Plastic-retaining Agent in Ordinary Concrete

On September 14, 2006, the Concrete Company of China Construction Third Engineering Bureau produced C35 ordinary concrete using zeolite water-reducing and plastic-preserving agent. The concrete was used also in the Wuhan Shengshihuating project. Concrete production mix design, the change of concrete slump and flow, and concrete compressive strength are shown in Table 11.30, Table 11.31, and Table 11.32, respectively. The construction site of Shengshihuating and the feeding method of water reducer and plasticizer are shown in Figure 11.43 (a) and Figure 11.43 (b) respectively.

Table 11.30 Concrete mix design (kg/m3).

Material / Concrete type	Cement	Fly ash	Slag	River sand	Crushed stone	Water	FDN	Zeolite admixture
Reference	230	90	100	710	1080	180	9.0	—
Slow slump loss	230	90	100	710	1080	180	—	11.76

Table 11.31 Change of slump and flow with time.

time / concrete	0 h	1 h	2 h
Reference	205 mm/360 mm	160 mm/—	—
Slow slump loss	190 mm/510 mm	195 mm/500 mm	190 mm/480 mm

Table 11.32 Setting and compressive strength of concrete

Concrete type	Setting time （h）		Compressive strength （MPa）		
	Initial	Final	3d	7d	28d
Reference	13.5	18	21.3	29.8	45.1
Slow slump loss	10	14.5	24.8	32.8	51.5

(a)Pumping concrete at site   (b)Feeding of admixture

Figure 11.43 Wuhan Shengshihuating project.

The slump and flow of concrete with zeolite water-reducing and plastic-preserving agent can remain basically unchanged for more than 2 hr, which ensures the requirements of concrete pumping construction. The initial setting and final setting time of concrete is also shorter than that of reference concrete. The concrete strength at 3 d, 7 d and 28 d is 10% higher than that of reference concrete.

Application of zeolite water-reducing and plastic-preserving agent in ultra-high performance concrete

In October 2008, in the Guangzhou Zhujiang New Town West Tower project, a test of ultra-high performance concrete with strength grade C100 was carried out. In order to reduce the viscosity of concrete and control the workability change of concrete with time, 2% zeolite water-reducing plasticizer (containing polycarboxylate superplasticizer with 12% solid content) was added to concrete to improve the workability of fresh concrete. The workability remained unchanged for 3 hr, as shown in Table 11.33.

Table 11.33 Slump retention effect of zeolite water-reducing plasticizer for ultra-high performance concrete.

property time (min)	slump (cm)	Slump flow (cm)	Flow time (s)
initial	24	64/62	6.0
60	25.5	61/67	4.4
120	23	56/57	5.3
180	24	52/54	9.0

For slump flow, the numerator and denominator are the slump flow in two directions respectively.

The use of zeolite water-reducing and plastic-preserving agent for ultra-high performance concrete can also retain flowability for more than 3 hr, ensuring the construction of ultra-high pumping concrete. The West Tower project adopted this technology to pump C100 UHPC and C100 UHP-SCC from the ground to 421 m high for construction application. In the research and development of Shenzhen Jingji Building C120 super high-performance concrete and its super-high pumping test, the zeolite powder plastic preservative was also used with liquid superplasticizer to keep the C120 ultra-high performance concrete plastic for 3 hr, and it is pumped from the ground to a height of 521 m.

11.7.3 Effect of natural zeolite on inhibiting concrete alkali aggregate reaction

According to Industry Standards SD105-82 mortar bar method, the criteria for inhibiting AAR: (1) 14d mortar bar expansion rate <0.02%. (2) The ratio of difference between the 14 d expansion rate of the standard mortar and the comparison mortar to the expansion of the standard mortar ≧75%; and the expansion rate of the 56 d test mortar is ≦0.05%. If the above two requirements can be met, it is considered that the cement will not cause an alkali-aggregate reaction even if the alkali active aggregate is used.

Dam cement, zeolite cement (mixed with 30% zeolite powder) and active aggregate were used to prepare mortar bar specimens, and the expansion was measured, of which the results are shown in Figure 11.44.

Mortar bars were prepared with Zeolite cement (incorporating 30% zeolite powder) and active aggregate or inactive aggregate to test the expansion rate as shown in Figure 11.45.

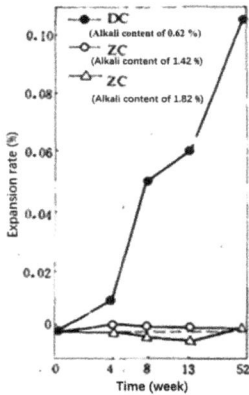

*Figure 11.44. Reaction of dam cement,
zeolite cement with reactive aggregate.*

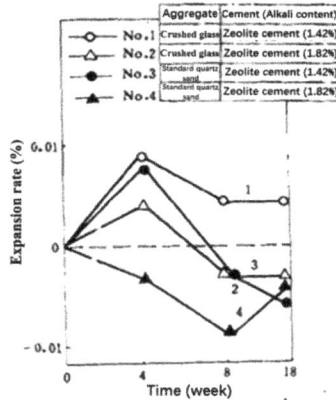

*Figure 11.45. Reaction of zeolite with active
aggregate and non-active aggregate.*

From Figure 11-44 and Figure 11-45, we can see:

(1) The 14-day expansion rate of the test mortar is not more than 0.02%; (2) The ratio of the difference between the 14 d expansion rate of the standard mortar and the test mortar to the expansion rate of the standard mortar is as follows.

The expansion rate of test mortar (at 28 d) age

$$Et\ (ZC1.42\%) = 0.009\%$$

$$(ZC1.82\%) = 0.002\%$$

The expansion rate of standard mortar at 14 days: $Et=0.093\%$

$$R1(ZC1.42\%) = （(0.093-0.009)/0.093）\times(100\%) = 90\%$$

$$R2(ZC1.82\%) = （(0.093-0.002)/0.093）\times(100\%) = 98\%$$

Both are greater than 75%. The expansion rate of both mortar with zeolite at 56 d age is less than 0.05%. It can be concluded that, with 30% zeolite powder mixed into the cement, even if the concrete is prepared with active aggregate, there will be no alkali-aggregate reaction. Studies have shown that natural zeolite inhibits alkali-aggregate reaction. It is the zeolite that makes the alkali ions (K^+, Na^+) in the concrete enter the zeolite skeleton through ion exchange, thereby inhibiting the reaction of active SiO_2 in aggregate to react with the alkali ion in the cement paste.

11.7.4 Porous concrete of gas carried by natural zeolite powder

Natural zeolite powder is used as a gas carrier to bring air or other gases into the cement paste, so that the cement paste expands, hardens, and becomes porous concrete. Moreover, natural zeolite also participates in the hydration reaction of the cement, so that the strength of the porous concrete can increase with age.

1) The natural zeolite gas carrier: zeolite has a large internal surface, as mentioned above, the internal surface area of the type A zeolite molecular sieve is about 1100 cm^2/g. However, in general, the internal surface of the zeolite is filled with zeolite water. If zeolite is used as the carrier for gas, the zeolite must be dehydrated so that the zeolite can absorb air or other gases. The parameters such as heating dehydration of zeolite powder and the amount of carrier gas are shown in Figure 11.46.

Figure 11.46 Porous concrete of gas carried by natural zeolite powder.

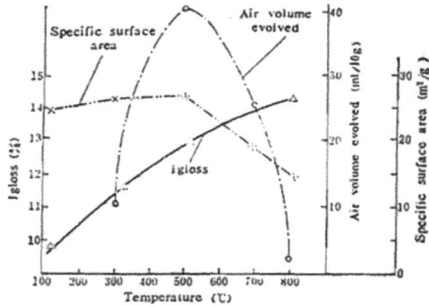

Figure 11.47 Relationship between burning temperature, weight loss, gas emissions and specific surface area.

It can be seen from Figure 11.47 that, after the burning temperature exceeds 500 °C, the dehydration volume continues to increase, but the specific surface area and gas emissions are reduced. Due to the burning under high temperature, the structure of Otani stone is destroyed, the specific surface area is reduced, and the gas emissions are reduced accordingly. Therefore, the optimal dehydration temperature of Otani zeolite as a gas carrier is 500 °C. If the processing temperature is too high, the chemical reactivity of natural zeolite will be reduced, as shown in Table 11.34.

It can be seen that the heat treatment temperature should be as low as possible under the condition that the natural zeolite (Otani stone) obtains the maximum amount of gas generation and natural zeolite can have higher chemical reactivity.

Table 11.34 Heat treatment and active silica and alumina content in natural zeolite (Otani stone).

Temperature	Original Otani Stone	500 °C	800 °C
Soluble SiO_2 %	12.08	11.81	9.02
Soluble Al_2O_3 %	8.39	8.37	7.37

The particle size of natural zeolite powder also affects the amount of gas carried. It can be seen in Figure 11.48 that the generated gas volume of Oya stone with a particle size of 1.2-0.6 mm is slightly higher, while the gas volume of Oya stone with a particle size of less than 0.15 mm is lower. This may be due to the fact that, when the same volume of Oya stone is broken into finer particles, the surface area increases and the area of the internal pores is relatively reduced, and the gas absorbed is relatively reduced. On the other hand, if the fineness is high, the hygroscopicity will also be high. In general, the particle size is smaller, and the chemical reaction performance is better. However, if the particle size is too small, and the amount of gas generated will be reduced. Taking into account the effects of these two aspects, when natural zeolite (Otani stone) is used as a gas carrier, the powder gradation should be as shown by the dotted line in Figure 11.49.

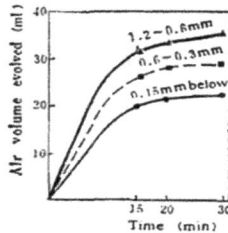

Figure 11.48 Gas volume carried vs particle size.

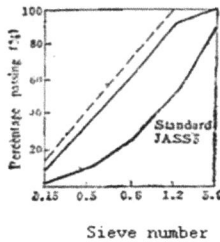

Figure 11.49 Range of particle size for gas carrier.

2) Production and application of natural zeolite gas-carrying porous concrete partition wall

A three-story testing building with floor area of 500 m^2 was built with zeolite new material in Zhangjiakou City, Hebei Province, as shown in Figure 11.50. All internal partition walls were made of zeolite-carrying gas porous concrete. The production process is shown in Figures 11.51-11.53.

Figure 11.50. Testing building with Zeolite.

Figure 11.51. Casting of porous mortar with zeolite carried gas.

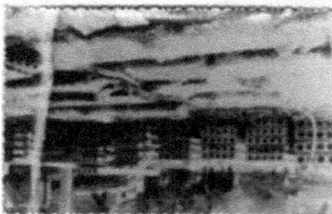

Figure 11.52. Curing of porous partition wall.

Figure 11.53. Demoulding of partition wall.

The zeolite-carrying gas-concrete partition wall panels include: Household partition: 235x59.5x12 cm; 137 kg/piece

Indoor partition board: 263x59.5 x9 cm; 110 kg/piece

Sound insulation: 90 mm thick hollow board, 40.14 Db.

Strength: 7d, 2.6 MPa; 28d, 6.9 MPa.

After 30 years, the test building has good functions and excellent durability.

11.7.5   The solidification of heavy metals by natural zeolite

The ion exchange characteristics of zeolite

The metal ions (generally K^+ and Na^+) that enter the zeolite crystals to balance the negative charge can be replaced by other ions, such as Ca^{++}, Sr^{++}, Ba^{++}, Cu^{++}, Zn^{++}, Ni^{++}, Ag^+, and other Alkaline earth metals, heavy metals, etc. This is called cation exchange capacity. It is commonly known as CEC (Cation Exchange Capacity). Usually, CEC is expressed in milligram equivalents per 100 g sample. Alkali metals, alkaline earth metal ions and heavy metal ions in zeolite can be replaced. The order is $Pb^{++} > Ca^{++} > Cd^{++} > Zn^{++}$.

Isothermal adsorption of natural zeolite to heavy metals

Natural clinoptilolite with 70% zeolite content was crushed and passed through a 0.178 mm sieve. The isotherm adsorption capacity of Cd^{++}, Zn^{++}, Pb^{++}, Cu^{++}, at 25 °C is measured. The maximum adsorption capacity of natural zeolite for Cd^{++}, Zn^{++}, Cu^{++}, Pb^{++} is 909.1 mg/kg, 3333.3 mg/kg, 3639.0 mg/kg and 20000 mg/kg, respectively. Through ion exchange, natural zeolite can effectively fix heavy metal ions.

Chapter 12

Super-Fine Slag – Characteristics and Applications

12.1 Introduction

Fused slag is a by-product of blast furnace ironmaking. Upon water quenching, it becomes granulated blast furnace slag. This process is illustrated in Figure 12.1.

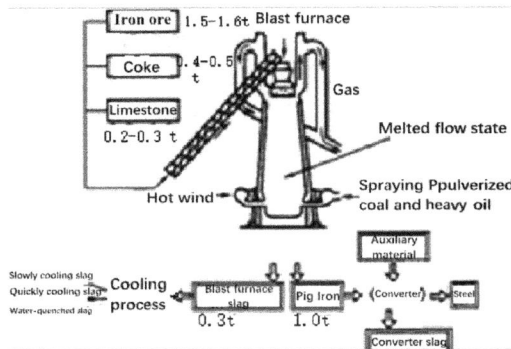

Figure 12.1 An illustration of the ironmaking and blast furnace slag production process.

To produce one ton of cast iron, the raw input composes of 1.5-1.6 ton of iron ore, 0.4-0.5 ton of coke, and 0.2-0.3 ton of limestone. These raw materials are fed into a blast furnace, followed by inlet of hot air. The combustion of coke then releases heat that melts the ore, and reducible gases that turn the oxidized iron to molten iron. Impurities in the iron ore, ash content in the coke, and the limestone in the furnace together become the slag which floats on the surface of the molten iron. Nearly 0.3 ton of slag is produced accompanying the production of one ton of cast iron.

In the year 1862, E. Langen discovered the binding capability of slag upon alkali activation. Since then slag has been developed as a hydraulic binder. In China the slag cement is made by the mix and co-grinding of ordinary cement clinker, natural gypsum and water-quenched slag whose percentage varies from 20% to 80% by mass.

In August 1925, the Japanese Ministry of International Trade and Industry announced the maximum substitution content of slag in cement to be 70%. It was later modified multiple times to an A category of slag cement with slag substitution content to be <30%, and C category with substitution content between 30% and 60%. Since a later announcement in October 1979, up to 5% of slag is allowed in ordinary Portland cement (OPC) to save resources and energy.

With the recent advance in grinding technology, super-fine slag powders with specific surface area (SSA) up to 6000-8000 cm^2/g are produced which, upon mixing with cement and silica fume, are used to manufacture special cement and concrete.

Slag powder (S95 slag) or super-fine slag (P800) has a potential cementitious capability, which is weak by itself but can be activated by alkali. When mixed with OPC, the $Ca(OH)_2$ and sulfate also enhance the cementing property of slag. Concrete made with OPC mixed with slag or super-fine slag (P800) has several advantages: (1) slower hydration heat release which inhibits concrete temperature rise, (2) higher long-term strength, (3) reduced permeability, (4) higher resistance to Cl⁻ ingress and Cl-induced rebar corrosion, (5) higher resistance to sulphate attack, (6) higher resistance to alkali silica reaction (ASR), better effect achieve using P800 super-fine slag, (7) better workability with reduced bleeding, and (8) capable of manufacturing (ultra) high strength concrete.

With a yearly production of ~10 million tons, the S95 slag has been widely used to produce concrete in China. Super-fines slag, meanwhile, is less produced and mainly used in high performance concrete (HPC) and ultra-high-performance concreting (UHPC).

12.2 Super-fine steel slag in general

12.2.1 Characteristics

Super-fine slag is often used in HPC and UHPC. The influence of its fineness and substitution content on the performance of concrete is shown in Table 12.1.

Table 12.1 Influence of the fineness and substitution content to concrete performance

Slag	Fineness (cm²/g)	2750-5500			5500-7500			>7500		
	Substitution content (%)	30	50	70	30	50	70	30	50	70
Concrete performance	Flowability	○	○	○	◎	◎	◎	◎	◎	◎
	Bleeding	○	○	△	◎	◎	◎	◎	◎	◎
	Retarding effect *	◎	◎	◎	◎	◎	◎	◎	◎	◎
	Adiabatic temperature rise	-	-	◎	-	-	◎	-	-	◎
	Reduce hydration heat	○	◎	◎	○	○	◎	○	○	◎
	Early-age strength	○	△	△	○	△	△	○	○	△
	Long-term strength	○	◎	◎	○	◎	◎	◎	◎	○
	High strength	○	△	△	◎	○	○	◎	◎	○
	Drying shrinkage	○	○	○	○	○	○	○	○	○
	Frost attack resistance	○	○	○	○	○	○	○	○	○
	Carbonation	-	-	△	-	-	△	-	-	△
	Water permeability	○	◎	◎	◎	◎	◎	◎	◎	◎
	Cl⁻ penetration	○	◎	◎	◎	◎	◎	◎	◎	◎
	Sulfate attack resistance	○	○	◎	○	○	◎	○	○	◎
	Heat resistance	○	○	○	○	○	○	○	○	○
	Curing improvement	○	○	○	○	○	○	○	○	○
	ASR mitigation	○	◎	◎	○	◎	◎	○	◎	◎

Note: ◎Superior performance is obtained, compared with OPC.
○Better performance is obtained, compared with OPC.
△Cautions are needed when using the slag cement, compared with OPC
* Retarding effect refers to the effect of postponing the setting time.

Concrete performance can be improved by adjusting the substitution content and fineness of slag. The following observations about slag substitution are worth noting. (1) Higher substitution content leads to larger reduction of heat release rate and the adiabatic temperature increase. (2) Concrete made with slag substitution often exhibits higher long-term strength compared with control specimens made with OPC. This effect is more obvious when the substitution content is higher. (3) Better resistance to sulfate attack is achieved at larger slag substitution. (4) A proper slag substitution could reduce the (free) alkali content and mitigate ASR. (5) Finer slag powder leads to less bleeding and improved consistency. (6) Substituting OPC with super-fine slag (SSA ≥ 7500 cm^2/g) can facilitate high early-age strength concreting, and high strength concreting in general. (7) Slag substitution densifies the concrete (pore structure), reduces the permeability and improves the resistance to Cl$^-$ ingress.

12.2.2 Application

The aforementioned characteristics of slag substitution justify its application in structural concreting for general construction. Examples of application scenarios are summarized in Table 12.2. Slag with either SSA larger than 5500 cm^2/g or substitution content larger than 30% can improve the flowability and reduce the bleeding of fresh concrete, along with a retarding effect. After hardening, the concrete has a low permeability and Cl$^-$ ingress rate. Substitution content higher than 50% can further mitigate ASR damage. At substitution content $\geq 70\%$, the adiabatic temperature rise can be significantly reduced, which helps minimizing drying shrinkage.

Table 12.2 Effect of slag and the corresponding application

Effect	Application
High consistency	High workability concrete (labor-saving, better quality control)
High retarding effect	Concreting in hot weather; large quantity continuous casting
Low heat release	Massive concrete
High 28-day strength	Lower cement content in the mix design
High long-term strength	Improved structural durability; lower cement content in the mix design
High strength	High-rise reinforced concrete; underground infrastructure
Low permeability	Underground infrastructure; infrastructure in (sea)water
High resistance to salt ingress	On-shore infrastructure; marine infrastructure
High resistance to seawater	Infrastructure on/in seawater
High resistance to sulfuric acid	Concrete in the vicinity of chemical engineering factory, hot spring, and area exposed to acid rain
Mitigating ASR	Enhanced durability

12.2.3 Significance

The significance of using slag substitution are two folds. (1) Compared with OPC, slag substitution better satisfies the requirement on mix-design from concrete performance, quality, and construction conditions. (2) Slag substitution is also environment-beneficial as it lowers the need of energy, natural resources, and net CO$_2$ emission.

12.3 The characteristics of steel slag

12.3.1 Chemical properties

The chemical properties of slag mainly determine its hydraulic cementing capability, which is the most crucial characteristics when used as an admixture in concrete. Alkalis such as calcium hydroxide or sulfate are often used as activators, as OH^- and/or SO_4^{2-} can accelerate the hydration of slag. The type and dosage of activators depend on the chemical composition and content of glass component of the slag. A typical chemical content of slag is shown in Table 12.3.

Table 12.3 Chemical content of slag, %

SiO_2	Al_2O_3	CaO	MgO	Fe_2O_3	TiO	MnO	S	K_2O, Na_2O	Cl
27-40	5-33	30-50	1-21	<1	<3	<2	<3	1-3	0.19-0.26

Note: SiO_2, Al_2O_3, CaO and MgO are the main component; Fe_2O_3, TiO, MnO, S, $K(Na)_2O$ and Cl are the minor contents.

One way to quantify the reactivity of slag is by calculating its basicity b from the chemical composition.

$$b = (CaO + MgO + Al_2O_3) / SiO_2$$

The reactivity is considered high when b is larger than 1.4.

$$CaO/SiO_2 > 1.0$$

12.3.2 Contents of the glassy and crystalline phases

The content of glassy phase is the most relevant index to assess the cementing capability of slag. The internal energy of the glassy phase can be up to 200 J/g, thus making it more active than crystalline phases. The glassy content in slag is a function of the chemical composition, the temperature when cooling starts, and the manner the cooling is conducted. When $(CaO + MgO) / (SiO_2 + Al_2O_3) > 1.15$, the glassy content will decline.

By measuring the content of crystalline phases using X-ray diffraction (XRD), the content of the glassy phases can be calculated.

glassy content = (1- crystalline content) × 100%

In China, the glassy content in slag is typically larger than 98%, thus the reactivity of the slag is high. The slag-OPC cement in China can have slag substitution up to 70%.

12.3.3 Chemistry and structure of the glassy phase

Electron probe micro-analyzer (EPMA) data suggest that the chemical compositions of crystalline and glassy phases are similar. With the reduction of glassy content, the glassy phases tend to contain less Al_2O_3 but more MgO. This is due to the localized chemical reaction from the crystalline region. According to Pu and Xu, the glassy region often contains two phases despite the alkalinity of the slag. One is continuously distributed, within which the other phase homogeneously exists as isolated spheres and/or pillars, as shown in Figure 12.2. EPMA element analysis suggests that the continuous phase is richer in Ca while the isolated phase richer in Si.

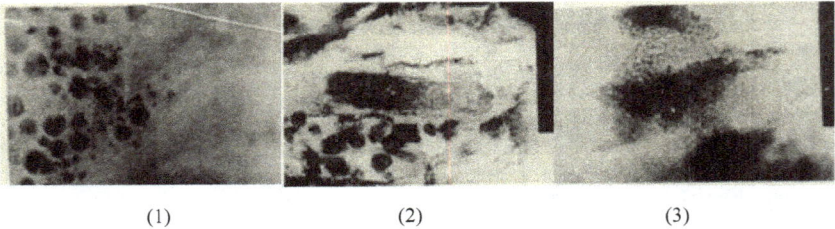

(1)   (2)   (3)

Figure 12.2 Dual-phase nature of glass phases in (1) basic slag, (2) neutral slag, and (3) acidic slag.

The microstructure of glassy region in slag can be summarized as follows. (1) In basic slag, the content of the Si-rich phase is low. It exists as small grains within the continuous Ca-rich phase. (2) In neutral slag, the Si-rich phase is more abundant, and exists as larger particles. (3) In acidic slag, there is much more Si-rich phase such that it inter-connects into a pillar or rod morphology.

The continuous Ca-rich phase forms the matrix of the glassy content. It dissolves rapidly when exposed to aqueous condition with OH⁻, which dismantle the glassy content. Si-rich phase then gradually exposes to the alkali environment and react with $Ca(OH)_2$ to form calcium silicate hydrate (C-S-H) gel.

12.3.4   The activity index of slag

The reactivity of slag is usually indexed through its basicity b and glassy content. In China, the water-quenched slag usually has a b above 1.8 and glassy content above 98 %. This high reactivity makes it very suitable mineral admixtures for cement. In US, the ASTM 989 classifies slag into three grades: grade 80, grade 100 and grade 120, as shown in Table 12.4. The strength test is conducted on mortar specimens with slag/OPC=1:1, binder/sand=1:2.75, and at a mortar flow of 110±5 % mm. The measured strength is compared with that of a control mortar made with pure OPC which is set as 100. Such assessment framework considers not only the chemical and mineral compositions, but also the fineness of the slag. It is holistic and relevant.

In Japan, the classification of slag reactivity is similar as in the US. A 50% slag substitution was employed in the mortar specimen preparation according to the Japanese industrial standard JISR5201. The strength of the specimens is bench-marked with standard specimens made with OPC, to yield the activity index of slag = (strength of slag cement mortar / strength of OPC mortar) × 100%. Figure 12.3 displays the activity index of slag with different fineness.

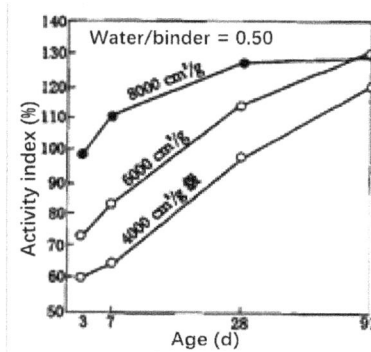

Figure 12.3 Activity index of slag with different fineness.

Reactivity of slag at an early-age is significantly increased when the powder is finer, whereas such effect of fineness is less substantial at late-age. For example, at 91 days the activity indices of slag samples with SSA = 6000 cm^2/g and 8000 cm^2/g are both around 128% (Figure 12.3).

The influence of chemical compositions to the reactivity of slag is illustrated in Figure 12.4. Slag with higher basicity usually has higher activity index. Alternatively speaking, higher (CaO + MgO + Al$_2$O$_3$) content leads to higher reactivity; higher SiO$_2$ content leads to lower reactivity.

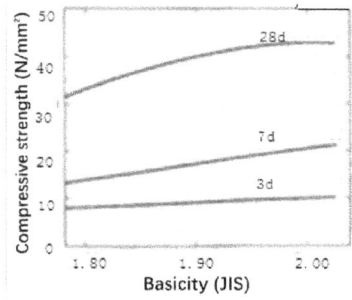

Figure 12.4 Compressive strength of slag cement mortar as a function of the slag basicity.

12.4 Technical standards of slag

12.4.1 Chinese standard

The Chinese standard GB/T 18046-2008 classifies slag into three grades: S 105, S 95 and S 75, based on their 28-day activity index. Table 12.4 shows other specifications for each grade, such as the flow value ratio (100%), moisture content (mass percentage), SO$_3$ content (mass percentage),

Cl content (mass percentage), loss on ignition (mass percentage), glass content (mass percentage), radioactive content, etc.

Table 12.4 Specifications of slag used in concrete (GB/T 18046-2008).

Items		Grade		
		S 105	S 95	S75
Density (g/cm^3) $\geq$		2.8		
Specific surface area (m^2/kg) $\geq$		500	400	300
Activity index (%)	7 days	95	75	55
	28 days	105	95	75
Flow value ratio (%) $\geq$		95		
moisture content (mass %) $\leq$		1.0		
SO$_3$ content (mass %) $\leq$		4.0		
Cl content (mass %) $\leq$		0.06		
Loss on ignition (mass %) $\leq$		3.0		
Glass content (mass %) $\geq$		85		
radioactive content		Satisfactory		

12.4.2 Japanese standard

In Japan the standard for slag used in concrete is JIS A 6206:1997. This standard uses the SSA as index for classification, as shown in Figure 12.6. For concrete designed for higher strength and higher workability, slag with larger SSA is preferred. With the development of grinding technology, super-fine slag powder with SSA larger than 600-800 m^2/kg is included in the standard.

In Table 12.5, the SO$_3$ is considered to originate from gypsum, since slag forms in a reductive condition in the furnace where SO$_3$ cannot be produced. The basicity of each grade in Table 12.5 is higher than 1.6.

Table 12.5 Specification of slag (JIS A 6206:1997).

Items		SSA 400	SSA 600	SSA 800
Density (g/cm^3)		$\geq$ 2.8	$\geq$ 2.8	$\geq$ 2.8
Specific surface area (m^2/kg)		300-500	500-700	700-1000
Activity index (%)	7 days	$\geq$ 55	$\geq$ 75	$\geq$ 95
	28 days	$\geq$ 75	$\geq$ 95	$\geq$ 105
	91 days	$\geq$ 95	$\geq$ 105	$\geq$ 105
Flow value ratio (%)		$\geq$ 95	$\geq$ 90	$\geq$ 85
MgO content (%)		$\leq$ 10	$\leq$ 10	$\leq$ 10
SO$_3$ content (%)		$\leq$ 4.0	$\leq$ 4.0	$\leq$ 4.0
Loss on ignition (%)		$\leq$ 3.0	$\leq$ 3.0	$\leq$ 3.0
Cl content (%)		$\leq$ 0.02	$\leq$ 0.02	$\leq$ 0.02

12.4.3 US, UK and Canadian standard

Specifications of slag in US, UK and Canada are shown in Table 12-6.

Table 12.6 Specification of slag (USA, UK and Canada).

Items	ASTM-C989	BS-6699	CSA-A363
Basicity	/	≥ 1.0	/
Loss on ignition (%)	/	≤ 3.0	/
SO$_3$ content (%)	≤ 4.0	≤ 2.0	≤ 2.5
45 μm sieve remain (%)	≤ 20	SSA 275m^2/kg	≤ 20
Activity index (%), 28 day	≥ 75 (S 80); ≥ 95 (S 100); ≥ 115(S 120)	/	≥ 80

12.5 Performance of slag concrete

The performance of slag concrete is significantly affected by the fineness and substitution content of the slag. A careful choose of type and substitution content of slag is crucial to reach the designed performance of concrete.

12.5.1 Performance of fresh concrete

When using normal slag (with low SSA) at high substitution content, the required water and/or water reducer content are lower, to reach the same workability of OPC concrete. However when using slag with high SSA, e.g. super-fine slag with SSA ~800 m^2/kg, the fresh concrete becomes more sticky (than OPC concrete). More water reducer or air-entraining agent is thus needed to reduce the stickiness or to obtain the same amount of air entrainment. On the other hand, using slag with SSA ~600-800 m^2/kg help reduce the bleeding. The adiabatic temperature rise is also related to the substitution content of slag.

12.5.2 Early-age strength of slag concrete

The early-age strength of slag concrete containing with slag SSA ~400 m^2/kg is usually lower than that of OPC concrete. This phenomenon is more significant at higher slag substitution content. Meanwhile the strength after 28 days are higher than that of OPC concrete. The early-age strength of slag concrete can be improved by using slag with SSA ~600-800 m^2/kg.

The relationship between slag SSA, curing age and strength of concrete is shown in Figure 12.5. After 1-day curing, the strength of slag concrete is lower than OPC concrete despite the SSA of the slag. After 3-day or 7-day curing, slag SSA ~400 m^2/kg leads to lower strength, whereas slag SSA ~600 and 800 m^2/kg results in higher strength than the OPC counterpart. In general, slag with SSA ~800 m^2/kg often results in good early-age strength. As shown in Figure 12.5b, after 6 months all slag substitution contents result in higher strength than OPC concrete.

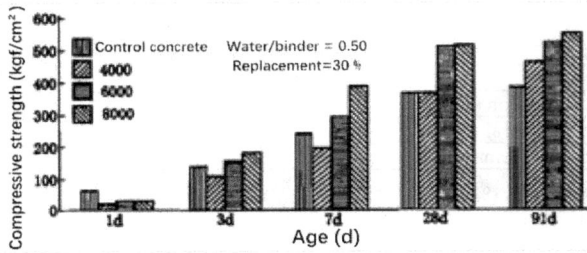

(a) Slag substitution content 30%, W/B=50%

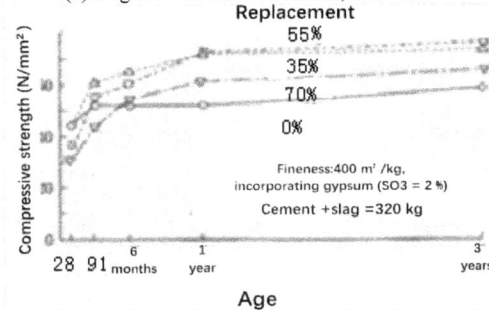

(b) same fineness, different substitution content.

Figure 12.5 Compressive strength growth of slag concrete.

The relationship between curing temperature (water curing at 20 °C, 15 °C, 10 °C and 5 °C) after demolding) and 28-day strength is shown in Figure 12.6. (1) When water-cured at 15 °C the slag concrete is much weaker than OPC concrete at 28 days. (2) Compared with 20 °C, the water-curing at 15 °C, 10 °C and 5 °C reduces the strength by 2-7 MPa when slag substitute content is 30%, and by 3-10 MPa when substitution content is 50-70%. The strength reduction is more significant at lower curing temperature. (3) The slag with SSA~800 m²/kg results in a clearly higher 28-day strength than the OPC concrete. (4). Given sufficient curing, concrete containing slag can continuously develop its strength. It is feasible and economical to design based on 56-day or 91-day strength.

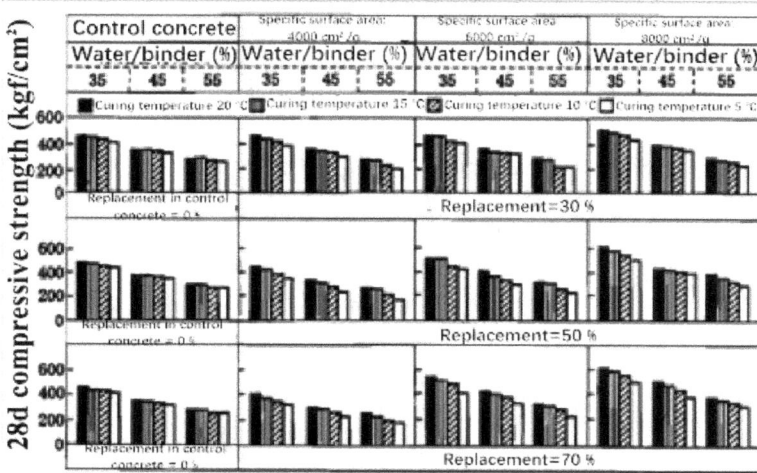

Figure 12.6 28-day strength of slag concrete under different curing temperature.

The relative strength development of slag concrete is shown in Figure 12.7, where the SSA of slag is 4000, 6000 and 8000 cm^2/g (note, 10 cm^2/g = 1 m^2/kg), and substitution content is 30%, 50% and 70%. Larger substitution content and larger SSA of slag results in larger relative strength at long-term. For W/B=0.30, the 91-day strength relative to 28-day strength is 120% for OPC concrete; 130%, 128% and 170% for slag SSA=4000 cm^2/g and substitution content of 30%, 50% and 70%, respectively; 120%, 122% and 170% for slag SSA=8000 cm^2/g and substitution content of 30%, 50% and 70%, respectively. It can be concluded that the later-age relative strength development is not improved when using slag with larger SSA. When the slag SSA is larger than 558 m^2/kg, the slag concrete strength at all ages are higher than that of the OPC concrete.

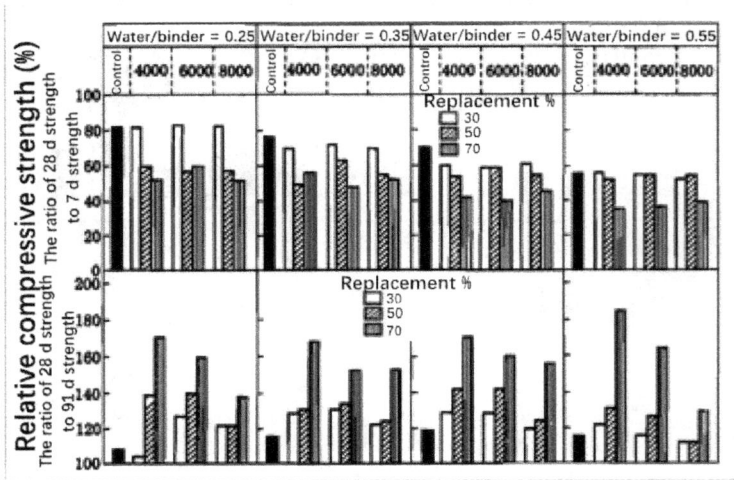

Figure 12.7 Relative strength as a function of slag fineness and substitution content.

The compressive strength of slag concrete is also affected by the post-mixing and curing temperature, and the influence is more significant than OPC concrete. Using 20 °C as a standard, lower post-mixing temperature results in slower strength development and lower strength, and vice versa. In particular, coarse slag further decreases the temperature. However, despite the slow strength development at lower post-mixing temperature, it may recover if a higher curing temperature is later used, as shown in Figure 12.8 (a) and (b).

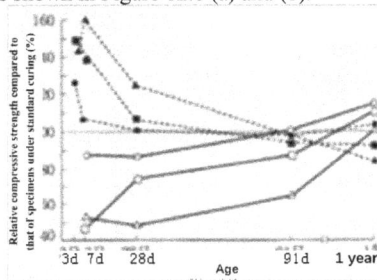

(a) Relative compressive strength with respect to OPC concrete. Slag from source A and B (SSA = 400 m²/kg) are used at substitution content of 55%, at curing temperature 5 °C and 30 °C. Notation: OPC concrete ○ (5 °C), ● (30 °C); Source A △ (5 °C), ▲ (30 °C); Source B □ (5 °C), ■ (30 °C).

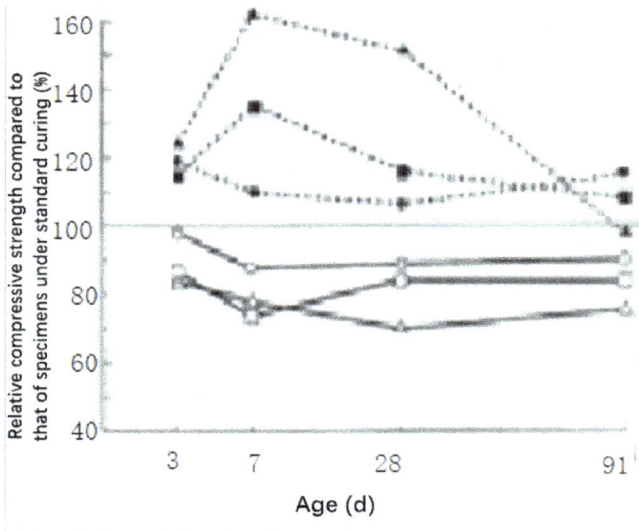

(b) Relative compressive strength with respect to OPC concrete. W/B=0.3, slag substitution content 50%, curing temperature 5 °C and 35 °C. Notation: OPC concrete ○ (5 °C), ● (35 °C); SSA = 600 m²/kg △ (5 °C), ▲ (35 °C); SSA = 800 m²/kg □ (5 °C), ■ (30 °C).

Figure 12.8 Compressive strength of slag concrete compared with OPC concrete.

As shown in Figure 12.8 (a), for OPC and concrete with 55% of slag of Source A or Source B, when cured at 5 °C, their strength is lower than that of the 20 °C-cured control samples at 7 d, 28 d and 91 d (OPC sample is slightly higher at 91 d), but all higher after one year.

As shown in Figure 12.8 (b), when sealing-cured at 5 °C and 35 °C, the strength development is clearly different for all samples. At all ages, 35 °C curing results in a strength relatively higher than 100%, while the 5 °C results in a value lower than 100%, and the 91 d strength is almost the same as the 7 d and 28 d strength.

The elastic modulus of slag concrete is comparable to that of OPC concrete of similar compressive strength. As shown in Figure 12.9, the empirical equation $E = 1400w^{1.5}\sqrt[3]{f'c}$ (w=2.3) generally fits the measured data, where the SSA = 300, 400, 500 and 800 m²/kg, at curing temperature of 20 °C, 5 °C and 30 °C. When strength is 40 MPa, the elastic modulus of both OPC concrete and slag concrete is around 32×10^3 N/mm².

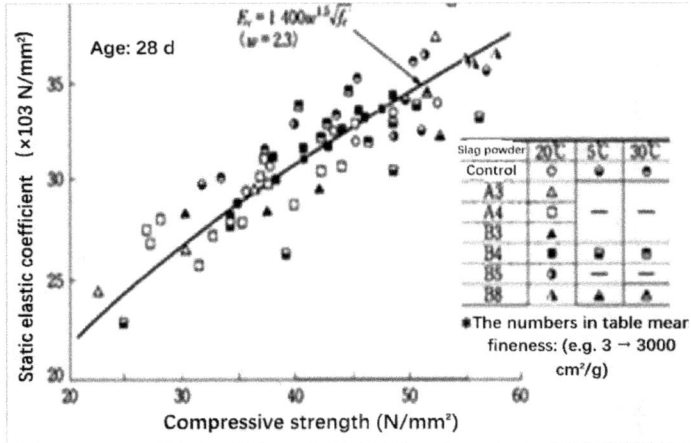

Figure 12.9 Relationship between the compressive strength and elastic modulus of slag concrete.

The hydration heat release of slag-substituted cement is shown in Figure 12.10, as measured by heat of dissolution. When substitution content is less than 30%, the hydration heat is not clearly decreased, but rather often increased (compared with OPC). At substitution content higher than 30%, the early hydration heat is reduced despite the fineness of slag. The reduction is proportional to the substitution content.

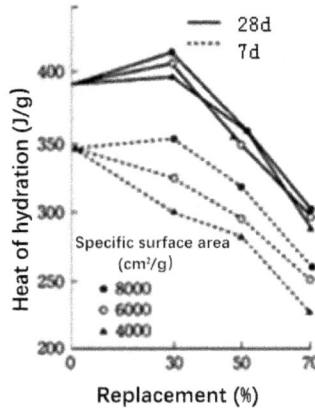

Figure 12.10 Hydration heat as a function of slag substitution content.

Adiabatic temperature rise: for slag with large SSA, the ultimate adiabatic temperature rise of slag concrete is lower than that of OPC concrete at slag substitution content of 70%, but higher than OPC concrete at substitution content of 30-50%. The early age heat release is generally lower at a higher substitution content, as shown in Figure 12.11. Smaller slag SSA and higher substitution content both reduce the early-age heat release. Alternative opinions suggest that the adiabatic temperature rise of slag concrete is smaller than that of OPC concrete at slag substitution content of 30-60%. In general, the hydration heat release of slag concrete is higher than that of fly ash concrete, thus the latter is preferred for temperature control in massive concreting. Experience from a project in Guangzhou (China) suggests that 30% slag concrete is not helpful with temperature control, whereas 30% fly ash concrete substantially reduces the heat release and risk of cracking.

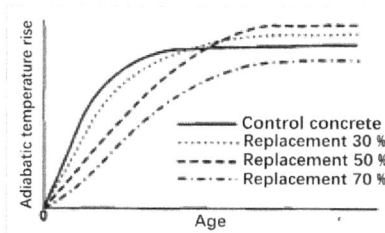

Figure 12.11 Adiabatic temperature rise as a function of slag substitution content.

(1) Shrinkage and creep

Shrinkage of slag concrete is affected by the age when drying starts. Drying shrinkage at 20 °C and 50% indoor relative humidity (R.H.) is shown in Figure 12.12. The drying shrinkage of slag concrete is generally comparable with or slightly lower than that of OPC concrete. Drying shrinkage can be reduced given sufficient long-term curing.

Autogenous shrinkage of slag concrete is also affected by the SSA and substitution content of slag, as shown in Figure 12.13. Coarse slag (i.e. small SSA) results in reduced autogenous shrinkage even at high substitution content. However, slag with large SSA increases the autogenous shrinkage, particularly when at substitution content is 50-70%. In general, smaller water/cement ratio leads to larger autogenous shrinkage. In case of slag concrete, with the decrease of water/cement ratio and increase of slag fineness, the pore structure of the hydrated paste is refined, thus results in a larger capillary force and a larger autogenous shrinkage.

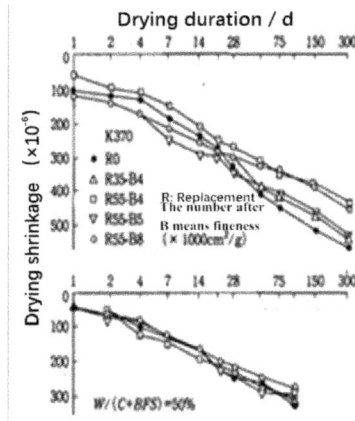

Figure 12.12 Drying shrinkage of slag concrete.

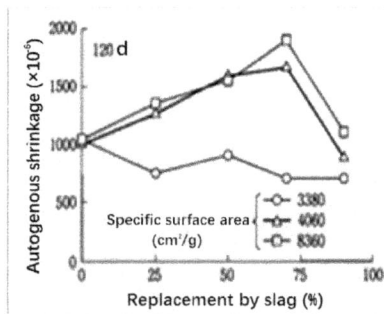

Figure 12.13 Autogenous shrinkage of slag concrete (effect of substitution content and slag fineness).

12.6 Applications of slag and issues of concern

12.6.1 The humidity and temperature of curing slag concrete

(1) Prolonged humid condition is key to ensure the strength development and durability of slag concrete. When slag is coarse and/or at large substitution content, the influence of curing temperature becomes substantial. Therefore, during construction of slag concrete, special attention should be paid to the curing temperature and humidity at early age. A recommendation is given in Table 12.7.

Table 12.7 Curing temperature and duration as a function of slag fineness and substitution content.

Substitution content (%)	30-40	50			55-70
Type of slag	SSA (m^2/kg)	SSA (m^2/kg)			SSA (m^2/kg)
Average daily temperature (°C)	400	400	600	800	400
17	≥ 6 d	≥ 7 d	≥ 7 d	≥ 6 d	≥ 8 d
10	≥ 9 d	≥ 10 d	≥ 9 d	≥ 8 d	≥ 11 d
5	≥ 12 d	≥ 13 d	≥ 12 d	≥ 10 d	≥ 14 d

In addition, the casting temperature should be in principle above 10 °C, and the surface temperature during curing is recommended to be above 10 °C. If the casting temperature is below 10 °C, and not likely to be increased by the later hydration heat evolved, it is necessary to guarantee the humidity, temperature and duration of curing.

(2) Bleeding and segregation.

When the water content or the dosage of water reducer is too higher in slag concrete, water is often separated from the fresh concrete and floats to the surface of the fresh mixture. This phenomenon is called bleeding. Bleeding increases the stickiness of the bottom of the mixture, causing it to settle to the base of the mold, which requires larger force to mix and hence making casting more difficult. Bleeding and settlement often accompany each other, resulting in a watery surface and a solid sediment at the base, which is called segregation.

Since 98% of the slag is glassy phase, coarse slag powder at high substitution ration often increase the tendency of bleeding and segregation. Apart from properly reducing the water content in the fresh mixture, a slight dosage of air-entraining agent also helps reduce the bleeding risk. Homogeneously distributed small bubbles break the upward channel of capillary, thus reduce the bleeding and increase the consistency of fresh mixture. In addition, introducing small amount of super-fine powders, such as silica fume, super-fine natural zeolites, can help reduce segregation as well.

(3) Decrease the environmental impact.

Slag is obtained by grinding the by-product of blast furnace iron-making. Compared with OPC, slag needs no calcination, thus reducing the environmental impact of cement production. Using slag can reduce the consumption of natural resources such as limestone, the consumption of energy, and the emission of CO_2. Table 12.8 illustrates the CO_2 emission of cement with 45% slag substitution, compared with the CO_2 emission of producing 1 ton of OPC.

Table 12.8 Fuel consumption and CO_2 emission for 1 producing 1 ton of cement.

		OPC	45% slag cement
Limestone consumption (kg)		1113	612 (45)
Energy consumption	Fuel (kg)	104	57 (45)
	Electricity (kW·h)	99	84 (15)
CO_2 emission (kg)		730	412 (44)

Chapter 13

Silica Fume (SF)

13.1 Introduction

Micro silica also known as silica fume, is powder collected as a by-product of the silicon and ferrosilicon alloy production. The diagram of Smelting electric furnace and silica fume collection is shown in Figure 13.1(a) and (b) respectively.

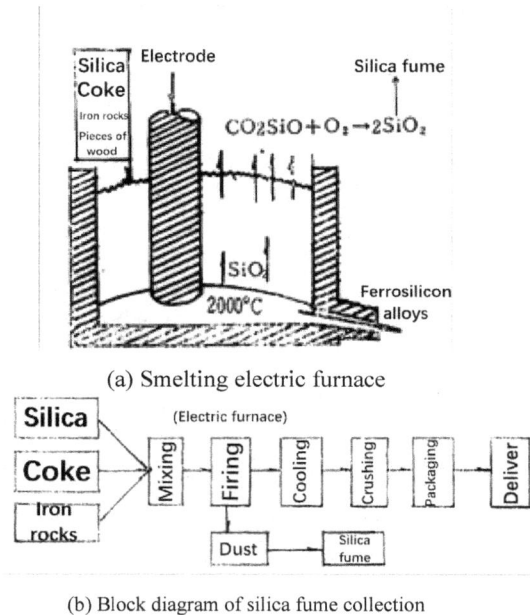

(a) Smelting electric furnace

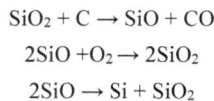

(b) Block diagram of silica fume collection

Figure 13.1 Electric furnace for smelting and silica fume collection.

Silica fume is produced by the reduction of quartz to SiO and Si at high temperature in an electric furnace during the smelting of ferrosilicon alloys or silicon alloys, and then oxidized to SiO_2. The reaction is:

$$SiO_2 + C \rightarrow SiO + CO$$

$$2SiO + O_2 \rightarrow 2SiO_2$$

$$2SiO \rightarrow Si + SiO_2$$

Unlike SiO_2 in quartz, this SiO_2 is an amorphous, with amorphous structure of silicon dioxide. It is a highly reactive pozzolanic ash material, of which mass is not in an energy levelling state. Thus it is chemically unstable. It is a glassy spherical particle with a diameter of 0.1-0.2 μm; it may be used as a mineral admixture in concrete, which can effectively improve the strength and durability of concrete. Silica fume has been used in a large number of high-strength concrete structures in super high-rise buildings in China. Silica fume can not only improve the strength and durability of concrete, but also improve the construction performance of concrete with low water/cement ratio.

In 1950, the Norwegian Institute of Industrial Technology first began to study the use of silica fume in concrete, and two years later, it was used in the concrete of a tunnel in Norwey. This was the first time that silica fume was applied to concrete in the world. As an admixture, silica fume has the same technical specification as that for other mineral powder in a cement standard in Norway in 1967. It is clear that its content should be below 10%. In 1981, it was further specified in technical standards that its content should be below 7.5%. In 1983, high-strength light aggregate concrete mixed with silica fume was used in construction of the oil drilling platform in Arctic Ocean. In Norway, silica fume was added to concrete to obtain a high strength concrete with a compressive strength of 70-100 N/mm², while in Denmark, ultrafine particles such as silica fume and a large number of efficient water reducing agents were mixed together to significantly reduce the water consumption, but the fresh concrete still had excellent fluidity, and the ultra-high performance concrete technology with a compressive strength of about 200 MPa has been developed. Japan has introduced this technology and has studied the use of this material (UHPC) for replacing metal materials. Japan began to use silica fume in concrete in the field of civil engineering to improve durability. In the construction field, the Japanese Ministry of Construction has implemented a comprehensive technology development plan "Technical development of ultra-lightweight and ultra-high-rise buildings for reinforced concrete buildings"; and silica fume was added in to concrete with strength of 80-100 MPa. In 1992, silica fume was mixed into concrete-filled steel tube with a design strength of 60 MPa, and since then, high-strength concrete with silica fume has been used in super high-rise reinforced concrete structures. More recently, silica fume and fibers have been mixed into mortar to develop concrete with high flexural strength and high toughness to produce prefabricated parts for use on bridges. In 2000, the Japanese Industrial Standards (JIS) also proposed technical standards for silica fume. In the United States, the first concrete project using silica fume was the Kinzua Dam in 1983. The cement content used in the concrete was 386 kg/m³ and the silica fume content was 70 kg/m³, and compressive strength of the concrete was 70 MPa at 7 d and 86 MPa at 28 d. The compressive strength increased to 110 MPa after 3 years. Since then, silica fume concrete has been developed and applied in the United States. In 1984, in high-rise and super high-rise buildings, concrete was prepared with cement content of 390 kg/m³, silica fume content of 15%, slump 18 cm, and 28 d compressive strength of 84 MPa. The purpose of using this concrete is to reduce the cross-sectional size of columns, increase the available using area, and reduce construction cost. In the United States, silica fume was also mixed in the panel concrete of heavy-duty traffic bridges with 450 kg/m³ of cement content, and10 % of silica fume dosage, resulting in 28 d strength of 98 MPa. In France, silica fume was used in a 530 m long span cable-stayed bridge with 28 d strength over 100 MPa in 1991.

In China, a 40 m-span post-tensioned pre-stressed concrete beam produced by a railway engineering company used C80 silica fume concrete. The bottom columns of Shanghai Pudong Bridge and some high-rise buildings have also used C60-C80 silica fume concrete. In Guangzhou West Tower Project, C100 silica fume UHPC and C100 self-compacting concrete have also been used. In the East Tower project in Guangzhou, the C120 multi-functional concrete prepared with silica fume and other raw materials was pumped to a height of 510 m. The silica fume produced in Zunyi, China, is also packaged as Elkem brand and sold all over the world.

Silica fume concrete is usually used in concrete projects with specific requirements, such as high strength, high impermeability, high durability, corrosion resistance, abrasion resistance and anti-corrosion of steel bars.

After further processing and purification, silica fume is called white carbon black which is used as a mineral filler or suspending agent in the paint, coating or printing industries. Silica fume is also used in the rubber, resin and other polymer material industries as a filler; Improve elongation, tensile strength and crack resistance of product. It can also be used to produce refractory materials, improve production efficiency, reduce costs, etc. In short, silica fume has a broad development and application prospect in the civil construction, metallurgy, chemical, and other sectors.

13.2 Production, types and output of silica fume

As mentioned above, silica fume is a by-product recovered during the smelting of ferrosilicon alloys. There are two processes for recovering silica fume from the fume produced by ferrosilicon: one is without heat recovery system, and the other is with heat recovery system, as shown in Figure 13.2 (a) and (b). The silica fumes produced through these two systems are of different carbon content and colour.

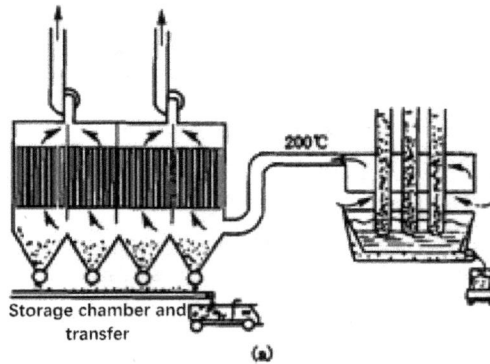

(a) Silica fume recovery process without heat recovery and utilization

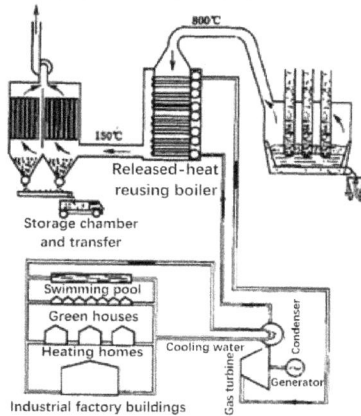

(b) Silica fume recovery with heat recovery and utilization

Figure 13.2 Process flow chart of heat recovery and utilization in during ferrosilicon smelting in ferroalloy plants.

(a) In the silica fume production process without heat recovery device, the gas temperature at the upper part of the load surface is about 200-400 °C, and there is still unburned carbon in the exhaust gas, and the colour of the collected silica fume is dark gray. (b) In the silica fume production process with heat recovery device, the gas temperature on the upper part of the load surface is about 800 °C, and most of the carbon can be burned, and the collected silica fume has low carbon content and the colour of it is white or grey. The form of silica fume products can be divided into dry and wet, as shown in Figure 13.3.

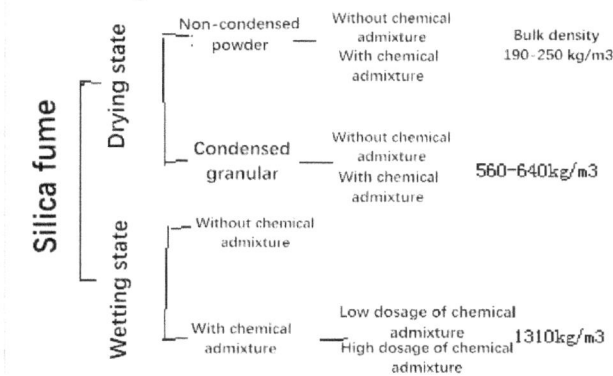

Figure 13.3 Form of silica fume products.

Non-condensed silica fume is a kind of dust, which is difficult to transport, expensive and uneconomical, while condensed silica fume is dust-free, has no consolidation, is easy to flow, is more economical to transport, and is convenient to use. The output of silica fume is shown in Table 13.1. Various countries have different smelting capabilities for ferrosilicon alloys, and also different outputs of recycled silica fume.

Table 13.1 Production capacity and output of silica fume in different countries.

Country	Production capacity	Output	Country	Production capacity	Output
China	487	80	Canada	32	30
Norway	300	65	Brazil	202	30
United States America	170	50	Russia	200	20
South Africa	68	45	Egypt	25	17
Iceland	46	45	Germany	21	12
France	83	40	Australia	13	8
Spain	40	40	India	69	5

(The unit of the data in the table is $x10^3$ t; quoted from the new cement concrete mixture edited by Yoshio Kasai)

According to Table 13.1, the production capacity of silica fume is about $1766x10^3$ t/year, and the actual production capacity is about $487x10^3$ t/year, which means that about $1279x10^3$ t/year of silica fume is discharged into the atmosphere, which may cause severe pollution to the ecological environment.

At present, more than 70% of the application of silica fume in China is in concrete and its corresponding technical fields.

13.3 The physical and chemical properties of silica fume

13.3.1 Shape, density and bulk density

The shape of silica fume is shown in Figure 13.4, where almost all particles are spherical. The specific surface area measured by the BET method is about 15-25 m^2/g, which is 20-30 times of that of ordinary Portland cement. The average particle size is 0.1-0.2 μm. In the particle cluster, between particles, or between particle clusters and particle clusters, there is a loose contact accumulation, with the voids filled with the adsorbed air layer. Therefore, although the density of silica fume is 2.2 g/cm^3, the bulk density is only 0.18-0.23 g/cm^3. Its voidage is higher than 90%. According to calculations, the clear distance between particles of silica fume is 0.103 μm, which is close to the average particle size of silica fume itself.

13.3.2 Chemical composition and amorphous structure

The chemical composition of different silica fume is shown in Table 13.2. The main chemical components of silica fume, especially those can participate in pozzolanic reaction, are different from those of cement, as shown in the triangular coordinate chart in Figure 13.5.

Figure 13.4   SEM of silica fume.

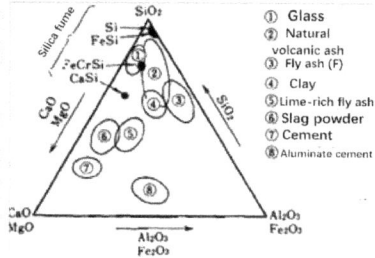

Figure 13.5 Composition of silica fume and related substances.

Table 13.2 Chemical composition of different types of silica fume.

Elemet \ Type	Si	FeSi-90%	FeSi-75%	White SF FeSi-75%	FeSi-50%
SiO$_2$	94-98	90-96	86-90	90	84.1
Fe$_2$O$_3$	0,02-0.15	0.2-0.8	0.3-0.5	2.0	8.0
Al2O$_3$	0.1-0.4	0.5-3.0	0.2-1.7	1.0	0.8
CaO	0.1-0.3	0.1-0.5	0.2-0.5	0.1	1.0
MgO	0.2-0.9	0.5-1.5	1.0-3.5	0.2	0.8
Na$_2$O	0.1-0.4	0.2-0.7	0.3-1.8	0.9	/
K$_2$O	0.2-0.7	0.4-1.0	0.5-3,5	1.3	/
C	0.2-1.3	0.5-1.4	0.8-2.3	0.6	1.8
S	0.1-0.3	0.1-0.4	0.2-0.4	0.1	/
MnO	0.1	0.1-0.2	0-0.2	/	/
L,O.l	0.8-1.5	0.7-2.5	2.0-4.0	/	3.9

SiO$_2$ is amorphous in silica fume. On the X-ray diffraction pattern, a relatively broad peak is shown near the position of the first peak of cristobalite, its range is a broad diffusion peak centred at 0.44 nm, as shown in Figure 13.6. Crystalline minerals, such as quartz, silicon, silicon carbide, sodium sulfate, hematite, and magnetite, etc., can be detected in a diffraction pattern.

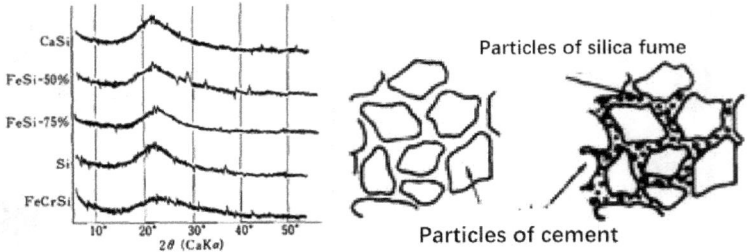

Figure 13.6. XRD pattern of silica fume.   *Figure 13.7. Cement Paste structure.*

13.3.3 Pozzolanic reaction

Silica fume is a very fine amorphous silicon. It has high pozzolanic activity. Because of its fine particles and high activity, it reacts with $Ca(OH)_2$ at the early stage of cement hydration. As the particle size of silica fume is about 1/25 of cement particles, it fills the voids between cement particles, which is also called micro-filling effect, as shown in Figure 13.7. Silica fume can react with $Ca(OH)_2$ released by cement hydration to produce C-S-H gel, which is well known. When the mixing ratio of silica fume and $Ca(OH)_2$ was changed, and the reaction amount of $Ca(OH)_2$ was measured, the results showed that, either when the mixing ratio was 2:1, the age was 2 weeks, or when the mixing ratio was 1:1, and the age was 5 weeks, all $Ca(OH)_2$ had been consumed by the reaction.

In the cement paste with 0.28 water/cement ratio, the replacement amount of silica fume is 15%, and all $Ca(OH)_2$ disappears after 28 days of age. In order to confirm the pozzolanic reaction of silica fume and other admixtures, 30% of various pozzolanic materials were mixed with Portland cement to prepare test specimens. After curing at 80 °C for 40 hours, the XRD pattern is shown in Figure 13.8.

It can be seen from Figure 13.8 that the diffraction peak of $Ca(OH)_2$ is still obvious for cement specimens and specimens mixed with pozzolanic sand and natural siliceous mineral materials, but for specimens containing silica fume, the peaks disappeared completely, which have been consumed to form C-S-H gel.

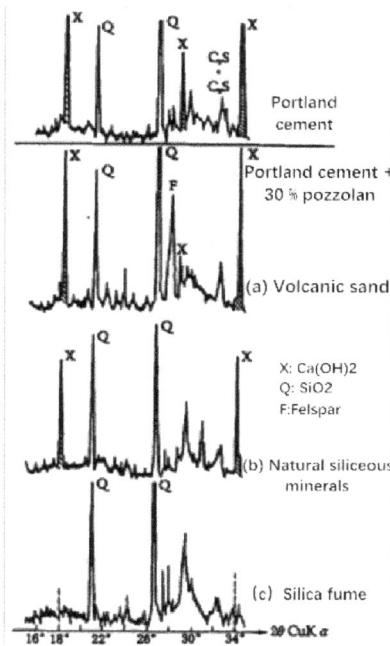

Figure 13.8 XRD patterns of Portland cement mixed with various pozzolanic materials.

13.3.4 Particle size distribution and specific surface area

The particle size distribution of silica fume is related to the manufacturing method and the operating conditions of the electric furnace. The specific surface area determined by the nitrogen adsorption method is about 15-25 m²/g, with an average of 20 m²/g. The specific surface area calculated by Aitcin et al. is shown in Table 13.3.

Table 13.3 Specific surface area (BET method) and average particle size of various silica fume.

Silica Fume	Calculated value of specific surface area (m²/kg)	Measured specific surface area (m²/kg)	Measured specific surface area (μm)
Si System	20000	18500	0.18
FeCrSi System	16000	/	0.18
FeSi—50%	15000	/	0.21
FeSi—75%	13000 (Heat recovery)	15000	0.23
FeSi—75%	13000	15000	0.26

13.3.5 Pore structure

Silica fume can reduce the pores in the cement paste, and can improve the strength, reduce the water permeability and air permeability. Figure 13.9 shows the pore size distribution of cement paste with W/B=0.40 mixed with different mineral admixtures at different ages.

Compared with normal silica cement, slag cement and fly ash cement, concrete with the cement mixed with 10% silica fume at age of 3 days had less pores with size over 50 nm (especially above 100 nm), and at the age of 180 days, there were almost no pores over 50 nm. This is due to the formation of dense CSH gel from the pozzolanic reaction, and the silica fume particles fill the pores between the cement particles to form a dense paste structure, so that strength can be improved.

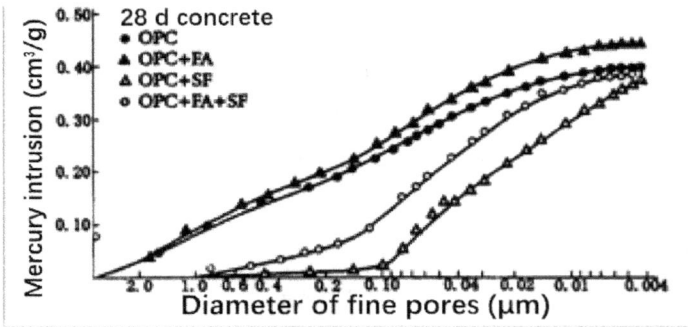

Figure 13.9 Pore distribution of cement paste with various admixtures.

13.4 Technical standards for silica fume

The SiO_2 content is the main parameter specified in the silica fume standards. The specification in Japanese JIS A 6207 is shown in Table 13.4. The quality standards (or drafts) of silica fume in other countries are shown in Table 13.5.

Table 13.4 Quality standard of silica fume (JIS A 6207:2006).

Standard Item Quality	Content (%)	Standard Item Quality	Content (%)
Silicon Dioxide	≥ 85	MgO	≤ 5.0
Sulfur Trioxide	≤ 3.0	Free calcium oxide	≤ 1.0
Free Silicon	≤ 0.4	Chloride Ion	≤ 0.10
LOI	≤ 5.0	Moisture	≤ 3.0
Specific Surface Area (BET method)	$\geq 15 m^2/g$	Activity index (%)	7d >95 28d >105

Table 13.5 Quality standards for silica fume in other countries (or draft).

	Performance / Country	Canada	Norway NS	Denmark DS411	RILEM	Australia	Germany DIN	United States ASTM	Sweden PFS
Score	$SO_3 \leq$ %	1.0	/	4.0	/	/	2.0	4.0	4.0
	LOI ≤ (%)	6.0	/	5.0	6.0	6.0	2.0	10	5.0
	SiO_2(%)≥	85	85	/	85	85	/	70	/
	Moisture content ≤(%)	3.0	/	1.5	3.0	/	/	3.0	/
	MgO≤ %	/	/	5.0	/	/	/	5.0	5.0
	Soluble ≤%	/	/	1.5	/	/	/	1.5	/
	$Cl^- \leq$ %	/	/	0.1	/	/	0.1	/	0.2
	Pozzolanic Activity Index (%)	≥ 85	/	/	≥ 95%	/	≥95%	≥100%	≥ 90%
	Autoclaved length change (%)	≤0.2	/	/	/	0.2	/	0.80	/
	45µm wet sieve residue (%)	10	/	40	/	10	1	34	
	Uniformity, Density Change	±5%	/	/	/	/	±5%	/	±5%
	45µm sieve residual % change	±5%	/	/	/	/	/	/	±5%
	Dry Shrinkage %	≤0.03	/	/	/	/	/	/	0.03
	% Water reducing agent content change	≤20	/	/	/	/	/	/	20
	Suppression of ASR (Min. %)	(80)	/	/	/	/	/	/	75
	Standard soft water consumption	/	/	/	/	/	/	/	115%

There is no separate technical standard for silica fume in China. Instead, fly ash, mineral powder, natural zeolite powder and silica fume are integrated together to be mineral admixtures, specified in the China standard GB/T 18736-2009, in which the technical requirements for silica fume are as follows.

Ignition loss $\leq$6%, Cl$^-\leq$0.02%, SiO$_2\geq$85%, specific surface area $\geq$15000 m^2/kg,

Water content $\leq$3.0%, water demand ratio $\leq$125%, activity index: 3 d/, 7 d/, 28 d$\geq$85%.

Its technical requirements are similar to those of other countries.

13.5 Silica fume concrete

13.5.1 Different applications and replacement content of silica fume

Generally, silica fume is used in high-strength concrete and high durability concrete to improve the construction performance of concrete at low water/cement ratio. According to the purpose of application, the replacement content of silica fume varies depending on the quality of silica fume. This is shown in Table 13.6.

Table 13.6 Replacement content and purpose of different applications of silica fume.

Application		Silicon fume Replacement Content (%)	Purpose of use							
			Strength	Durability			Construction Performance			
			Improve Strength	Reduce Permeable	Abrasion resistant	Prevent salt damage	Improve Durable	Reduced peeling	Fiber Bonding	Pumping
Japan	Tunnel Spit	5-9	o				o			o
	Marine Structures	9	o		o	o				o
	Bridge Beam	8-15	o							o
	Building Structures	10	o							o
Other Country	Tunnel Construction	5-10	o				o	o		o
	Marine Structures	8-10	o			o				o
	Road Paving	5-15	o	o	o	o	o			
	Bridge Beam	4-9	o	o	o		o			o
	Parking	8-10	o	o		o				
	Building Structure	7-8	o							o
	Prefabricated components	4-8	o	o						

Note: o - indicates that the corresponding effect can be obtained.

13.5.2 Performance of fresh concrete

In general, high-strength concrete with compressive strength of 60 MPa or higher can be easily prepared by using superplasticizers and medium-heat or low-heat Portland cement. However, the viscosity is rather high. To further reduce the water/cement ratio and prepare higher strength concrete, it is very difficult to ensure the required fluidity for construction. The concrete is preferably mixed with silica fume or other ultra-fine powder. The results of such application are shown in Figure 13.10. (SFCS- cement mixed with SF).

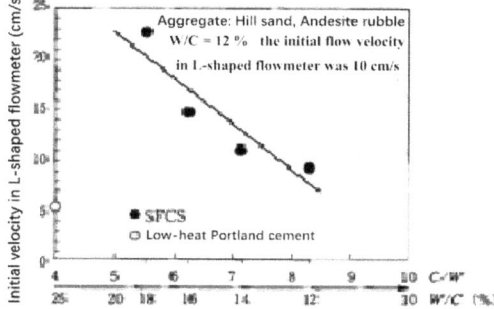

Figure 13.10 Flow rate of SFCS cement concrete in L-meter.

It can be seen from Figure 13.10 that the initial flow velocity of low-heat Portland cement concrete (symbol ○) was 5 cm/s. However, the initial flow velocity of SFCS cement concrete with W/C=0.12 was about 10 cm/s, which had excellent liquidity. R&D centre of China State Construction Commercial Concrete Co., Ltd. casted 60-80 MPa concrete with 5% SF, the fluidity (slump) of the concrete increased by 30-40% compared with that of the reference; the micro-filling effect of the superfine powder was fully utilized.

From Figure 13.10, it can be seen that the concrete with slump 23±2 cm, when mixed with SF powder, the original slump could be maintained even if the water/cement ratio was reduced and the amount of superplasticizer did not increase.

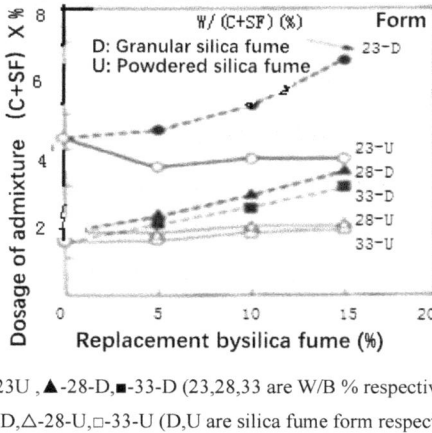

○-23U ,▲-28-D,■-33-D (23,28,33 are W/B % respectively)

●-23D,△-28-U,□-33-U (D,U are silica fume form respectively)

Figure 13.11 Relationship between superplasticizer content and replacement rate in concrete.

As shown in Figure 13.11, the effect of powdered SF is better than that of granular SF for concrete at W/B of 0.23. In concrete at W/B of 0.28 and 0.33, respectively, the replacement content of powdered SF increases, but the amount of admixtures does not increase, and the original fluidity is still maintained, which also proves the effect of powdered SF.

13.5.3 Performance of hardened concrete

1. Strength of concrete

C-S-H gel is formed due to pozzolanic reaction of SF and a dense structure is formed. As the performance of the interfacial transition zone between aggregate and cement paste is improved, the bonding strength at the interface is improved. Generally, the interfacial transition zone in concrete has a loose structure and low strength due to the enrichment and orientation of $Ca(OH)_2$. Figure 13.12 shows the relationship between different replacement contents of silica fume to cement and concrete strength. In the Figure 13.12 (a), W/(C+SF) = 0.28, and in (b), W/(C+SF) = 0.33, the results were similar, that is, the strength at the 3 d age was lower than that of the reference concrete. However, the strength of concrete with a replacement content less than 10% after 7 d age was higher than that of the reference concrete. Even if the replacement content of silica fume to cement reaches 15%, the concrete strength after 28 d was higher than the reference concrete, and highest level was obtained.

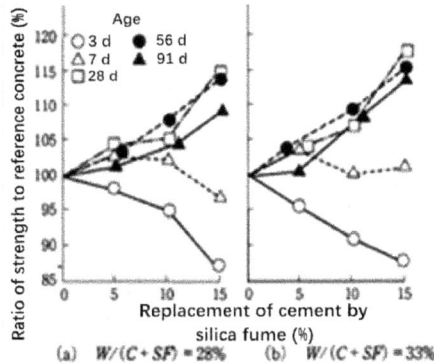

Figure 13.12 Relationship between replacement rate and concrete compressive strength.

2. Impact of aggregates on strength

As to high-strength concrete, aggregate has a significant impact on strength, as shown in Figure 13.13. When W/C=0.14, the strength of concrete mixed with andesite B gravel and gravel sand was 190 MPa, while the maximum strength of concrete mixed with pit sand and sandstone gravel was only 160 MPa; when W/C=0.12, the strength of concrete mixed with andesite B gravel and gravel sand was as high as 210 N/mm^2. Thus, the counterattack-broken crushed andesite and crushed sand are preferably used to prepare high-strength and ultra-high-strength concrete.

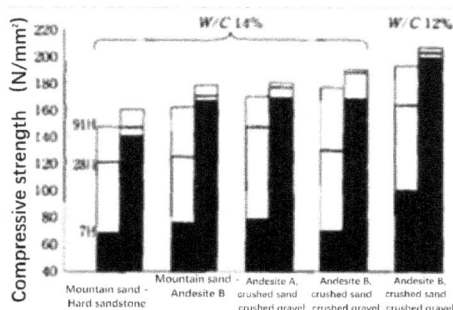

Figure 13.13 Strength of concrete with different coarse and fine aggregates.

3. Durability

Concrete mixed with silica fume can effectively improve the durability. Especially the resistance to neutralization. It has good effect on Cl^- penetration resistance, sulphate corrosion resistance, alkali-aggregate reaction inhibition and penetration resistance, etc.

Resistance to neutralization (carbonation)

With same water/cement ratio, the resistance to neutralization of the concrete containing silica fume is almost unchanged compared with the reference concrete. However, the concrete mixed with silica fume generally has relatively low water/cement ratio, so the neutralization resistance is improved.

Vennestand et al. measured the neutralization performance of concrete and found that concrete mixed with SF slightly reduced the neutralization depth. However, if the early curing is not good, the neutralization depth will increase.

Permeability of silica fume concrete

In Figure 13.14, the test results of concrete immersed in saturated salt water after 28 days of standard curing at W/B of 0.43, i.e., the penetration depths of Cl^- in the concrete measured at 13 weeks, are shown, with various replacement contents of silica fume.

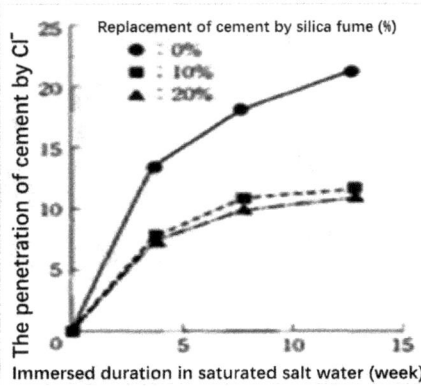

Figure 13.14 Time-dependent change of the replacement rate of silica fume and the penetration depth of Cl⁻.

From Figure 13.14, it can be seen that the penetration depth of Cl^- in concrete can be suppressed by about half when the replacement content of silica fume is 10%.

Figure 13.15 shows the relationship between the replacement content of various mineral powders and the water permeability in concrete with W/B of 0.55, at ages of 2 weeks, 4 weeks and 13 weeks. Concrete mixed with silica fume had the lowest water permeability regardless of its age. This is due to the pozzolanic reaction of concrete mixed with silica fume and has a dense microstructure, which improves the compactness of the concrete and improves the impermeability and Cl^- permeability.

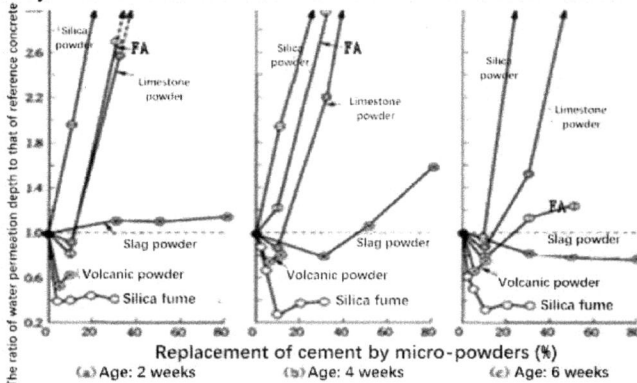

Figure 13.15 Relationship between the replacement rate of various mineral powders and the permeation depth ratio of water.

Inhibition of alkali aggregate reaction

The effect of silica fume and other mineral powders can be used to inhibit alkali-aggregate reactions. By using andesite as alkali-active aggregate and the alkali content in concrete of 7.0 kg/m³ converted from Na_2O, concrete was prepared at W/C of 0.55 with different replacement contents of various mineral powders and the expansion rate at one year of age is shown in Figure 11.16.

5% replacement content of silica fume can effectively inhibit the alkali-aggregate reaction, which is the most effective in inhibiting the alkali-silica reaction among various powders.

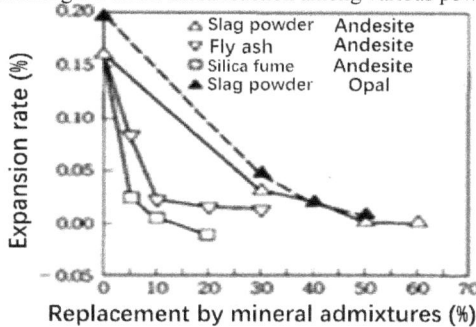

Figure 13.16 Effect of mineral powders to inhibit the reaction of alkali aggregates.

Frost resistance

In order to ensure that the concrete has sufficient frost resistance, the spacing of air bubbles should be less than 200 μm, and it is recommended that the amount of silica fume should be less than 15%. In the concrete, the durability coefficient is improved due to the incorporation of silica fume and the use of superplasticizer.

Fire resistance

Because the silica fume concrete is dense, it is easy to result in explosive spalling in a fire. so that the fire resistance of the concrete mixed with SF must be confirmed first, or mixed with organic fibres to prevent it from bursting.

Shrinkage

The shrinkage of concrete with silica fume increases with increasing amount of cement paste. This is caused by dry shrinkage. However, compared with the reference concrete without silica fume, the shrinkage is not different, as shown in Figure 13.17. If necessary, low heat cement may be used, or expansion agent is added appropriately.

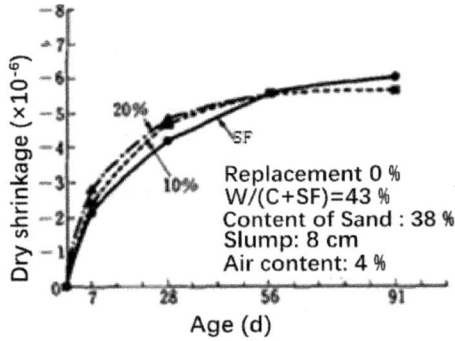

Figure 13.17 Dry shrinkage and length change of concrete

Creep

According to the unit creep, silica fume concrete has the same creep as that of the reference concrete cured in water, but larger than that of the reference concrete in the dry state in the atmosphere, as shown in Figure 13.18.

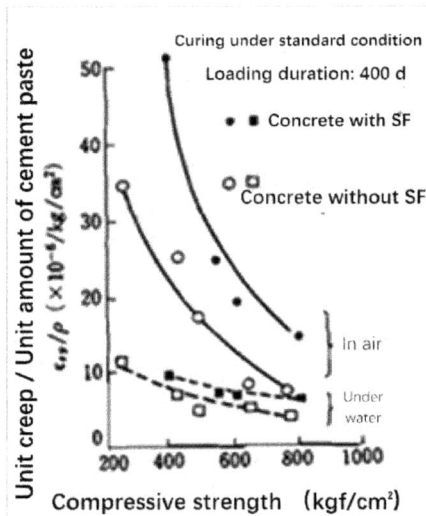

Figure 13.18 Creep of silica fume concrete.

13.6 Microstructure of silica fume concrete

Compared with ordinary concrete, one of the main characteristics of silica fume concrete is that it has a more uniform microstructure.

At low water/cement ratio, the microstructure of the cement paste is mainly composed of poorly crystallized hydrates forming a low porosity and denser matrix when SF is added. With the increase of SF content, the amount of $Ca(OH)_2$ transformed into C-S-H gel increases, i.e. the $Ca(OH)_2$ content in cement paste decreases with the increase of SF content, as shown in Figure 13.19.

It can be seen from Table 13.7 that, when ordinary Portland cement was mixed with SF, the calcium/silicon ratio in the hydrate was reduced, the hydrate could combine other ions (such as alkali ions and alumina ions), so that the resistance of cement paste to ion invasion and inhibition of alkali-aggregate reaction was improved, and the electric resistivity was improved.

Table 13.7 Effect of silica fume on calcium-silica ratio of hydride.

Cementitious Material	Ca/Si ratio
Ordinary Portland Cement (OPC)	1.6
OPC+ 13%SF	1.3
OPC+ 28%SF	0.9

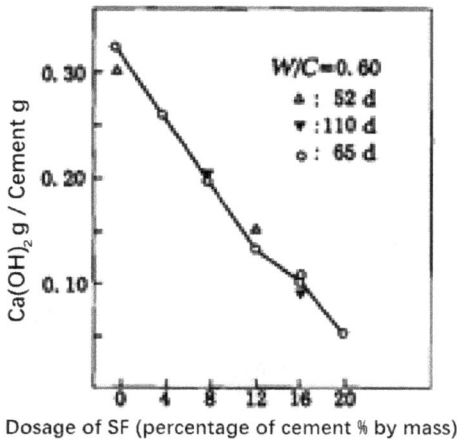

Figure 13.19 Effect of silica fume on content of $Ca(OH)_2$ in ordinary Portland cement paste.

The silica fume in the concrete improves the microstructure of the transition zone between coarse aggregate and cement paste. Research by Rogourd et al. proved that high-strength concrete mixed with silica fume is filled with amorphous and dense C-S-H phases around the aggregate. Due to coarse aggregate is in direct contact with the C-S-H phase, the arrangement of $Ca(OH)_2$ crystalline formation is different from that in ordinary concrete. The pore structure of the interface zone in concrete is quantitatively represented in Figure 13.20. Scivener et al. concluded that, in the interface transition zone, the porosity of SF concrete is significantly reduced, and no porosity gradient has been observed. Khayat et al. replaced the cement with 15% SF in concrete with W/(C+SF) of 0.33 and the porosity of the interface transition zone was significantly reduced. The decrease in the porosity of the interface transition zone and the decrease of the crystal concentration of the primary CH phase are shown in Figure 13.21.

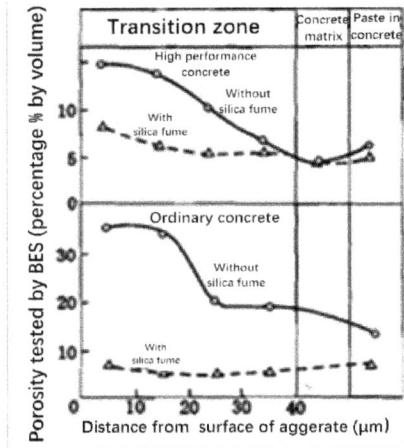

Figure 13.20 Effect of silica fume on porosity of transition zone.

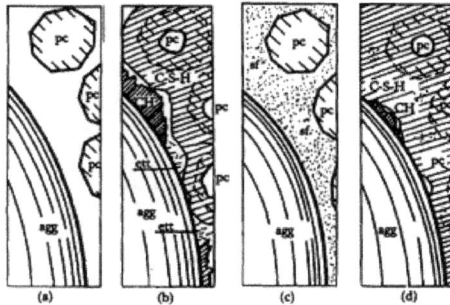

Figure 13.21 Transition zone between cement paste and aggregate interface in concrete with and without silica fume.

In Figure 13.21 (a), the fresh concrete without silica fume, due to bleeding, forms water voids around the coarse aggregate, and the cement particles at the interface connection are also insufficient.

In Figure 13.21 (b), the transition zone in (a) is shown. After hydration, condensation and hardening, there are CH phase, C-S-H phase and a lot of pores in which some needle-like materials filled. In Figure 13.21 (c), fresh concrete mixed with silica fume is shown. SF fills the space around the coarse aggregate, with no water film and bleeding. In Figure 13.21 (d), the transition zone in (c) is shown, where exist C-S-H, a small amount of CH and cement particles, etc., and the porosity is very low.

This reveals that the imbalance in the transition zone is greatly eliminated, and the structural improvement improves the performance of the concrete.

The effect of silica fume in concrete is different from that in mortar in terms of strengthening. The relationship between the amount of silica fume and the compressive strength at the age of 28 d when silica fume is mixed into mortar and concrete with W/B of 0.33, respectively, is shown in Figure 13.22.

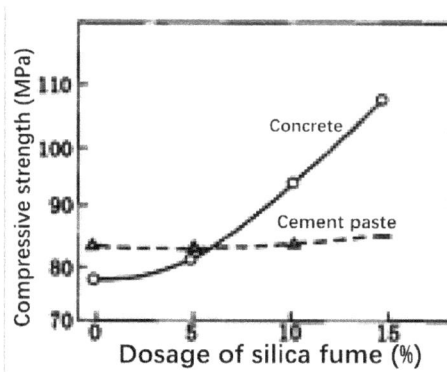

Figure 13.22 Effect of silica fume on the Strength of Mortar and Concrete.

Without silica fume, the strength of mortar (about 82 MPa) is higher than that of concrete (about 78 MPa). When mixed with 5% silica fume, the strength of both is roughly the same. when the silica fume content is more than 8%, the strength of concrete is higher than that of mortar.

It shows the contribution of aggregates to the strength of high-strength high-performance concrete and the effect of the improvement of the interface structure on the strength.

When silica fume is incorporated in concrete, the interfacial structure is improved and the bonding strength between cement paste and aggregate is improved, which also improves the performance of concrete.

The stress-strain curve of silica fume concrete is linear within loading to a stress at 80% of the strength, while that of ordinary concrete is only 40%. During unloaded, the extension of the hysteresis loop is also smaller than that of ordinary concrete.

Similarly, the bonding between reinforcement and concrete is also improved due to the addition of silica fume. The improvement of the microstructure of the transition zone between the reinforcement and the cement paste enhances the pull-out strength of embedded reinforcement.

13.7 Technology of silica fume applied to concrete

13.7.1 C100 ultra-high strength concrete and self-compacting concrete

Mix design of silica fume for UHPC and UHP-SCC are shown in Table 13.8. The properties of silica fume applied to C100 ultra-high strength concrete and self-compacting concrete are shown in Table 13.9 for fresh concrete and Table 13.10 for concrete strength.

Table 13.8 Mix design of silica fume for UHPC and UHP-SCC.

Quantity of raw materials in 1 m³ concrete (kg)									
NO.	W/B	C	BFS	SF	CEA	NZP	S	G	W	HWRA
1	0.20	500	190	60	/	/	750	900	150	18.00
2	0.22	400	190	60	14	28	750	850	154	15.40
3	0.33	415	135 (FA)	/	/	10	750	1000	181	11.0

Table 13.9 Performance of fresh concrete.

NO.	W/B	Slump (mm)		Flow time (s)		Spread (mm)	
		Origin	120 min	Origin	120 min	Origin	120 min
1	0.20	265	265	3.44	3.59	710	690
2	0.22	280	250	3.00	4.80	680	590
3	0.33	225	185	4.97	16.96	550	410

Table 13.10 Strength of hardened concrete (MPa).

NO.	3 d	7 d	28 d	56 d	Strength Grade
UHPC	87.0	108.7	130.8	130.8	C100
UHP-SCC	91.2	106.5	117.3	118.0	C100
SCC	36.0	48.5	68.0	69.2	C50

HPC, UHP-SCC and C50 SCC all have excellent self-compacting performance and construction performance. C50 SCC was used for construction of Wuchang high-speed railway station and UHPC and UHP-SCC were used for Guangzhou West Tower Project.

Chapter 14

Fly Ash

14.1 Introduction

Fly ash is a byproduct in coal-fired power stations, as shown in Fig. 14.1. Usually, fly ash is made up of fine and glassy particles which are even smaller than cement particles.

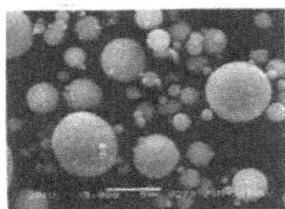

(a) SEM of fly ash

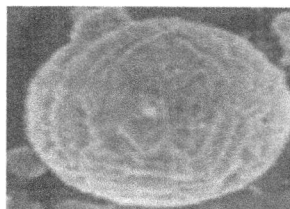

(b) Magnetite and hematite particles in fly ash

Figure 14.1 SEM of fly ash and magnetite and hematite particles in fly ash.

When coal fines combust, fly ash, bottom ash and gas are produced. Fly ash particles are the finest among these flue gas particles in the chimney. Bottom ash is the larger particles which will be separated, as sediments that either deposit in the chimney or are mixed with slag. They travel from the combustion zone to the bottom of the incinerator, as shown in Fig. 14.2.

Part of the water vapor evaporated from the pulverized coal and the fractionated gas is discharged into the atmosphere, and part of it is condensed on the surface of the fly ash. In the flue gas exhaust, there is a lot of SO_x gas, especially when burning coal with high sulfur content. In order to reduce environmental pollution, limestone slurry or limestone powder is often added into the flue gas to capture SO_x gas in the exhaust gas before the flue gas is discharged. Of the dust emitted by coal-fired power plants, 75-85% becomes fly ash, and the remainder is bottom ash and furnace slag.

The performance of fly ash is related to many factors: such as coal variety, quality, fineness of pulverized coal, ignition point, oxidation conditions, pretreatment and desulfurization before burning, and the collection and storage methods of fly ash. Fly ash is used as a mineral admixture for cement concrete, and most countries have corresponding technical standards, such as China standard GB1596-2005, American standard ASTMC618-89, Japanese industrial standard JIS A 6201, etc.

The world's output of fly ash is about 40 billion tons, most of which are used to build wharves, dams, roads, and as additives for concrete and etc. Fly ash is also a raw material and additives for the production of cement. Fly ash is also used to produce bricks, blocks and ceramsite. It is a low-carbon, green and energy-saving material.

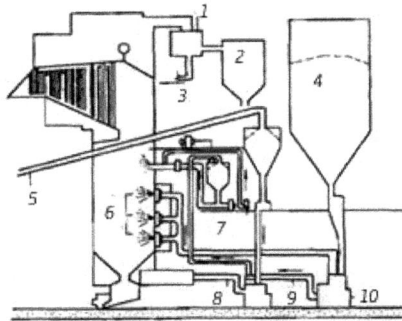

1-Electric dust collection equipment; 2- fly ash hopper; 3-discharge; 4- coal silo; 5-limestone powder; 6-burning kiln; 7-compressor; 8- hot air; 9-air; 10-coal powder spray.

Figure 14.2 Pulverized coal transportation and fly ash collection.

At present, the use of fly ash in different countries in the world is very different, the highest can reach 70-80%, and under normal circumstances, it is 30-40%. Even in a country, the application of fly ash in different regions is not the same. For example, in Guangzhou and Shenzhen, the application rate of fly ash reaches 100%. Since all the local fly ash is used up, additional fly ash has to be transported in.

In China, the annual output of fly ash is about 200 million tons. It is mainly used for building materials, backfilling, road construction, construction and agricultural planting. Compared with other countries in the world, China's utilization of fly ash ranks among the top in the world.

Fly ash also contains toxic metals, such as arsenic (As), chromium (Cr), selenium (Se), titanium (Ti), vanadium (Va), etc. Fly ash may also contain radon (Rn), it is a radioactive element, also known as radioactive gas in the early years. When fly ash containing these harmful metals is piled on the ground for a long time or in a landfill pond, these toxic metal ions will dissolve in the water and increase the harmful cations in the water. Radon (Rn) released in fly ash products is also very low. The US Geological Survey report shows that the radioactivity of fly ash products is not significantly different from traditional building materials such as concrete or red brick. In other words, the amount of radon released in fly ash will not harm human health; it can be used in buildings like building blocks or other building materials.

14.2 Chemical composition of fly ash

Coal ash is a type of pozzolanic material, and its chemical composition is determined by the composition of raw coal and combustion conditions. In the CaO-SiO_2-Al_2O_3 ternary phase diagram, its composition and position are similar to those of pozzolanic materials, as it is an aluminosilicate mineral additive. According to statistical data, the chemical composition of China's fly ash varies, as shown in Table 14.1.

Table 14.1 Chemical composition of China's fly ash.

Compounds	SiO_2	Al_2O_3	Fe_2O_3	Ca O	Mg O	SO_3	loss in mass
variation range	20-62	10-40	3-19	1-45	0.2-5	0.02-4	0.6-51

From the table, the CaO content of fly ash from the Yunnan Kaiyuan Power Plant is as high as 45%. The loss of ignition for fly ash from the Urumqi Power Plant is as high as 51%. The chemical composition of fly ash in other countries, except for the low loss of ignition, is also roughly within the range of Table 14.1. SiO_2 and Al_2O_3 are the main components in fly ash. Most fly ash from China power plants has $SiO_2 + Al_2O_3 \geq 60\%$. The United States ASTM C618 requires $SiO_2 + Al_2O_3 + Fe_2O_3 \geq 70\%$; Japan JIS A6201 requires $SiO_2 \geq 45\%$; the former Soviet Union GOCT6269 requires $SiO_2 \geq 40\%$; some Soviet scholars believe that: Al_2O_3 content of 20-30% belongs to high activity fly ash and Al_2O_3 content $\leq 20\%$ is low activity fly ash. Some Soviet scholars proposed to use the coefficient K to represent the activity of fly ash. $K = (Al_2O_3 + CaO)/SiO_2$. According to the K value, fly ash is divided into four categories, as shown in Table 14.2.

Table 14.2 Relationship between K Value and fly ash activity.

K Value	0.8-1.0	0.6-0.8	0.4-0.6	<0.4
activity	High	Medium	Relatively low	Low

The harmful component in fly ash is the unburned coal particles. The loss of ignition of coal ash is mainly the carbon content. Fly ash with high carbon content has high water absorption rates, low strength, easy to efflorescence, so it is a harmful component. The carbon content of fly ash varies from country to country. China National Standards requires 5-15%, Japan Standards $\leq 5\%$, the American ASTM standards $\leq 10\%$, and the U.S. Bureau of Reclamation standard requires $\leq 5\%$.

14.3 Mineral composition of fly ash

X-ray diffraction is used to identify the mineral composition of fly ash. The XRD pattern is shown in Figure 14.3.

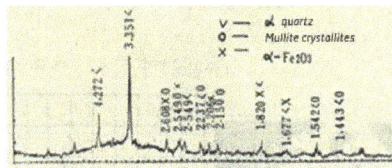

Figure 14.3 XRD pattern of fly ash.

From Figure 14.3, the mineral composition of fly ash is: α Quartz (high temperature type, symbol V.), the characteristic peaks are 3.35, 4.27, 1.92, 1.81, 2.46, 2.28A. Mullite crystallites ($3Al_2O_3$. $2SiO_2$, symbol O.), the characteristic peaks are 5.40, 3.39, 2.69, 2.52, 2.20A. α -Fe_2O_3 (hematite melt, symbol X), the characteristic peaks are 2.69, 1.69, 2.51, 1.84, 1.68A. Quartz crystal minerals account for about 7-13% in fly ash, and mullite crystals account for 8-13%. And the glass content is about 72-80%. According to research, it is agreed that the main components of fly ash are: glassy amorphous materials, mullite, quartz, and small quantities of other minerals. According to data from Japan, the mineral composition of fly ash is as shown in Table 14.3.

Table 14.3 Mineral composition and glass body percentage of coal ash.

	Quartz	Mullite	Magnetite	Hematite	Glass content
Range (%)	2.5~30.8	2.6~63.8	0~3.8	0~6.1	21.6~78
Average	13.9	28.7	0.6	0.5	52.4

The mineral and glass content varies greatly with different types of fly ash. Quartz is a mineral that is not present in the original coal mines, but is derived from the modification of Si-Al clay minerals. Amorphous glass is formed due to the lowered melting point of coal by potassium and sodium. Magnetite and hematite have low content of amorphous particles, and the particles are spheres with uneven surfaces.

As shown in Figure 14.3, glassy amorphous materials are made up of about 60-65% SiO_2, 12-20% of Al_2O_3, and about 9-11% of Fe_2O_3. This indicates that the glassy amorphous materials have high silicon content and low aluminum content. In addition, fly ash also contains calcite, anorthite, $\beta - C2S$, hematite, and a smaller amount of sulfate and phosphate minerals.

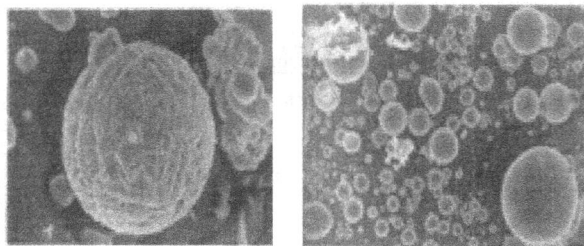

Figure 14.4 Particles of magnetite and fly ash

14.4 Physical properties

The physical properties of fly ash are shown in Table 14.4. The specific surface area is measured according to the GB287-63 specific surface area measurement method for cement.

Table 14.4 Physical properties of fly ash.

Fly Ash	Density (g/cm3)	45 μm (Retained%)	20 μm (Retained%)	10 μm (Retained%)	specific surface area (cm²/g)	glass content (%)
Original fly ash	2.11	35.2	79	/	/	75
Grade 10 μm	2.84	/	/	1.4	7850	90
Grade 25 μm	2.43	/	1.0	22.6	5690	85-90
Grade 45 μm	2.31	5.7	41	/	3290	80
GradeII	2.15	19.1	70	/	4200	35

The density and specific surface area of graded fly ash decrease as particle size increases. The density and specific surface area of grade 10 μm and 20 μm fly ash are significantly larger than that of grade II fly ash. The specific surface area of grade 10 μm fly ash is similar to microbeads, as high as 7850 cm²/g. The particle size distribution curve is shown in Figure 14.5. In the figure, the curves for M-II grade fly ash, S-silica fume, F-grade fly ash (F＜10 μm, F＜25 μm, F＜ 45 μm) are shown. The particle size of silica fume is between 0.2-1.0 μm, and the cumulative percentage is 50-10%. The particle size of graded ash F<10 μm, F<25 μm is between 1.0-20 μm. Grade II ash has a similar particle size distribution as Grade F<45 μm ash, with a range of 20-100 μm. However, for Grade F<45 μm ash, even though most of the <25 μm particles have been extracted, it is still finer than Grade II ash.

The microbeads are also extracted from the fly ash by drying and separating and belong to Grade F<10 μm fly ash.

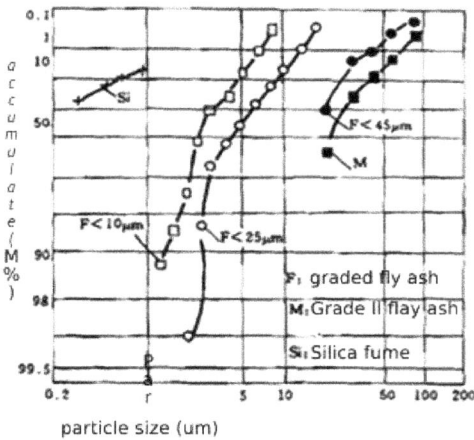

Figure 14.5 Particle size distribution of silica fume, graded fly ash and grade II fly ash

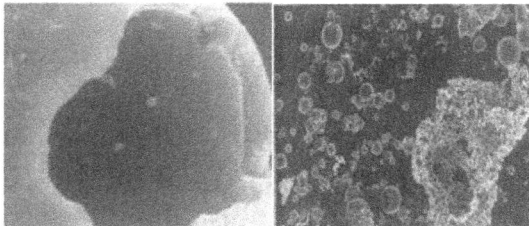

(1) Hollow structure of fly ash   (2) Porous unburnt particles

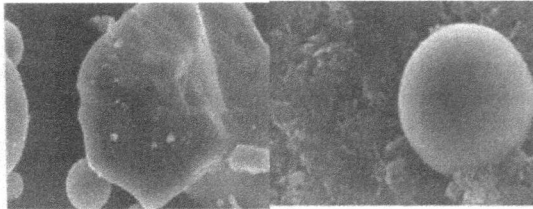

(3) Hydrate on the surface of fly ash particles   (4) Incompletely melted particles

Figure14.6 Different morphologies of fly ash particles (Kasai Yoshio).

As shown in Figure 14.6, fly ash particles are mostly spherical with a smooth surface. And they have a hollow structure. Porous unburnt particles are mixed with these particles. When burning in a furnace, quartz particles are wrapped in clay minerals and become particles that do not melt fully.

14.5 Pozzolanic activity of coal ash

There are two traditional methods to evaluate the pozzolanic activity of fly ash: (1) lime absorption value; (2) slack lime strength test. In Chinese Standards, a comparative test is carried out by mixing 30% of fly ash with cement, and this mixture is compared to cement with the same fluidity but without the addition of fly ash. The compressive strength of both mixtures is then evaluated over 28 days. The results of the pozzolanic activity for some fly ash are shown in Table 14.5.

Table14.5 Test results for pozzolanic activity of fly ash

Fly ash	Replacement of cement (%)	Sand ratio	Water (%)	Flexural strength (%)	Compressive strength (%)
Control	0	S/C=2.5	100	100	100
GradeII Ash	30	S/(C+F)=2.5	98	/	/
10 μm	30	S/(C+F)=2.5	92	110	94
25 μm	30	S/(C+F)=2.5	95	109	92
45 μm	30	S/(C+F)=2.5	99	87	74

From Table 14.5, it can be seen that: (1) The finer the fly ash, the lower the amount of water is needed. Graded 10 μm ash requires 92% water. Graded 25 μm ash requires 95% water. Grade 45 μm ash requires 99% which is similar to the water requirement of 98% for Grade II ash. (2) The compressive and flexural strength increase as the fly ash gets finer. This shows that the dry separation technology can effectively improve the quality and pozzolan activity of fly ash. Microbeads separated from coal ash is one example. In addition, for mortars containing graded ash with diameter ≤25 μm, its flexural strength is significantly higher than the compressive strength. The influence of fly ash fineness on its flexural and compressive strength follows the same trend.

14.6 The mechanism behind volcanic ash reaction

The so-called pozzolan reaction, refers to a unique hydrate that is formed around the fly ash particles when the Portland cement hydration reaction progresses, as shown in Figure 14.6 (3). The pozzolanic reaction model is shown in Figure 14.7.

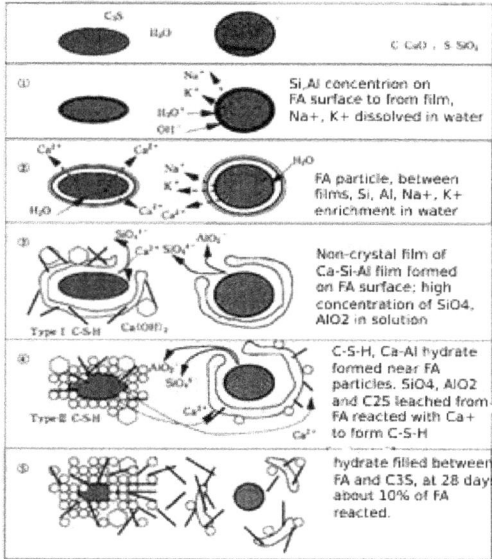

Figure14.7 Pozzolanic reaction model between C3S-Coal ash.

It can be seen from the model of the reaction between C_3S and fly ash: the alkali component (OH-) cuts off the amorphous siloxane (Si-O-Si) bond, water molecules enter and become silanol groups (Si-OH ...HO-Si), a short molecule of negative divalent $H_2SiO_4^-$ is generated. At this time, due to the presence of Ca_2^+ and silanol groups, Ca2+ is adsorbed, connecting H2SiO4- (pH above 11.2 at room temperature) to form a Ca-Si polymer. After adding fly ash to cement, the Ca/Si ratio decreases in the hydrate formed by the pozzolanic reaction, and its value is related to the percentage composition of fly ash as well as the content of CaO. As the pozzolan reaction progresses, the pore structure of the mortar specimen changes. As the duration increases, the pores of 20-100 nm decrease, and the pores of 10-20 nm increase as shown in Figure 14.8(a). The strength is also improved correspondingly

Figure14.8 (b) is an enlargement of the middle part of (a). The thickness is about 15 nm, plate-like crystals, and 10-100 nm pores are formed inside.

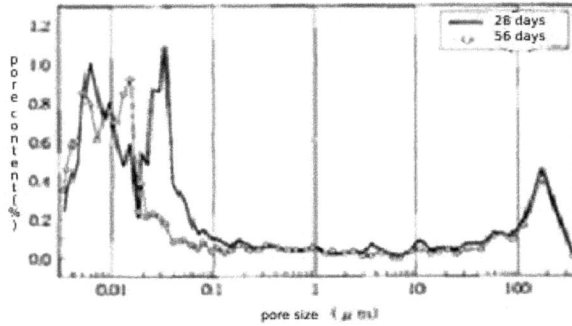

(a) Pore structure changes upon pozzolanic reaction

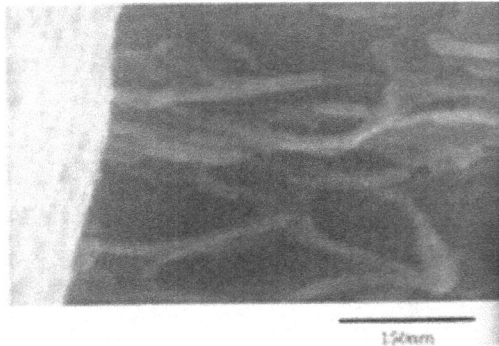

(b) Hydrate formed during pozzolanic reaction

Figure 14.8 Hydrate formed and pore structure changes due to pozzolanic reaction

14.7 Engineering standards of coal ash

14.7.1 China's national technical standard

According to GB 1596-2005 (Use of fly ash in cement and concrete), the quality requirement for fly ash are shown in Table 14.6. Grade II fly ash is finely grounded ash, used for ordinary reinforced concrete structures and lightweight aggregate concrete structures. Grade III fly ash is undisturbed ash, mainly used for unreinforced concrete and mortar. Grade I ash is obtained by electric dust collection and is of good quality. It can be used for prestressed reinforced concrete structures. Various countries have different coal types and technical levels, thus different requirements for the quality of fly ash.

Table 14.6 Quality requirements of Coal Ash.

Fly Ash Grade	Grade 1 Fly Ash	Grade II Fly Ash	Grade III Fly Ash
Source	electric dust collection	finely grounded ash	undisturbed ash
LOI（%）	＜5	＜8	＜15
45μm sieve residue（%）	＜15	＜25	＜45
water demand ratio（%）	＜95	＜105	＜115

14.7.2 Japanese technical standards for fly ash

Japanese technical standards for fly ash are shown in Table 14.7.

Table 14.7 Classification of different grades of fly ash by JIS A 6201-1998.

Grade		Grade	Grade II	Grade III	Grade IV
SiO$_2$ content（%）		＞45			
Water content （%）		＜1.0			
LOI（%）		＜3.0	＜5.0	＜8.0	＜5.0
density (g/cm3)		＞1.95			
fineness	＞45 μm(%)	＜10	＜40	＜40	＜70
	Specific surface area	＞5000	＞2500	＞2500	＞1500
Flowability ration (%)		＞105	＞95	＞85	＞75
Activity index	28 d	＞90	＞80	＞80	＞60
	91 d	＞100	＞90	＞90	＞70

Grade I fly ash is obtained by air separation of undisturbed ash, and has relatively high added value. Grade II fly ash is ordinary fly ash available in large volumes. Grade I and Grade II fly ash are the same before the revision of the standard and are used as an additive for concrete. Grade III and IV fly ash can only be used as concrete additives after the test has been verified.

14.7.3 American technical standards

In American ASTM C 618 fly ash is divided into N, F, C and S grades according to pozzolanic activity. N grade fly ash is used as a concrete additive, and its technical requirements are shown in Table 14.8.

14.8 Fly ash concrete

14.8.1 The properties of fresh concrete

1) Air content of concrete

Due to the difference in the delivery time of the fly ash and the change of LOI within different batches of fly ash, it is difficult to manage the performance of the fresh concrete with fly ash For the application of fly ash concrete in constructions. In the summer of 2012, Guangzhou Tianda Concrete Co., Ltd. encountered such an example. C80 concrete was produced with the same mix design but a new batch of fly ash. But the slump of the concrete was lost quickly and construction could not be carried out. Due to higher water demand, more water reducing agent than the original mix design was required. Figure 14.9 is the relationship between LOI of fly ash, air content, and slump.

Table 14.8 Technical requirements of Grade N fly ash.

SiO2+Al2O3+Fe2O3 (%)	≥70
SO2 (%)	≤4.0
Max water content (%)	≤3.0
LOI (%)	≤10
45 µm retained (%)	≤34
Pozzolanic activity (28d activity%)	≥75
Compressive strength with lime (MPa)	≥55
Water demand (%)	≤115
Max expansion or shrinkage in autoclave (%)	0.08
Relative density (difference between average and Max) (%)	5.00
45 µm retained (difference between average and max) (%)	5.00
Optional requirements	
Mortar shrinkage at 28 d (%)	≤0.03
The loss of expansion of mortar in alkali solution for 14 days (%)	≥75
The expansion of mortar in alkali solution for 14 days (%)	≤0.02

Under the same dosage of AE superplasticizer, with the increase of LOI of fly ash, the slump and air content of concrete decrease accordingly. This is due to the high carbon content in fly ash, which increases the adsorption capacity of the water reducing agent.

2) Concrete flowability

Most of the particles in fly ash are spherical glass. Compared with amorphous cement particles, the amount of water adsorbed on the surface is less. Therefore, after adding fly ash, the water consumption of concrete in order to obtain the required slump is correspondingly reduced. Especially for fly ash with low LOI and fine particle size, the water consumption of concrete is reduced more obviously; as shown in Figure 14.10.

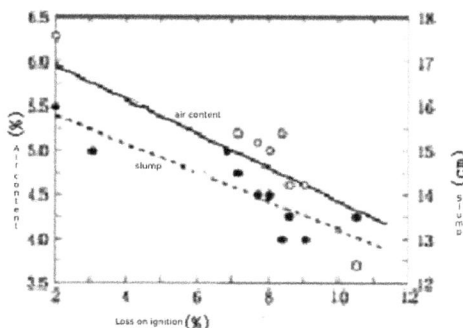

Figure 14.9 the relationship between LOI of fly ash, air content, and slump.

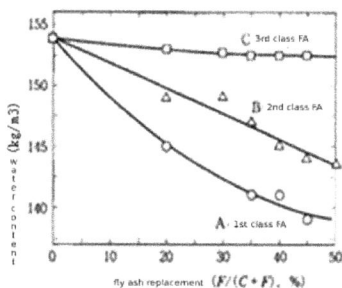

Figure 14.10 The relationship between fly ash dosage and water demand.

Curve C (grade III fly ash) has a high carbon content in the ash and the ash is relatively coarse, so even if the amount of fly ash to replace cement increases, the water consumption of unilateral concrete remains unchanged. Curve B (grade II fly ash) , although the specific surface area of fly ash is the same as that of grade III fly ash, both 2500 cm^2/g, but the carbon content is low, so the amount of substitution increases, and the water consumption of unilateral concrete decreases. Curve A (grade 1 fly ash), fly ash has a large specific surface area and a low carbon content, so the amount of cement substitution increases and the water consumption of concrete decreases more significantly.

3) Bleeding

Fly ash is to replace cement into concrete, and the replacement rate is in the range of 15-45%. As the replacement rate of fly ash to cement increases, the amount of bleeding increases. When fly ash replaces part of the fine aggregate with the same cement content, the replacement rate is 10-30%, no bleeding.

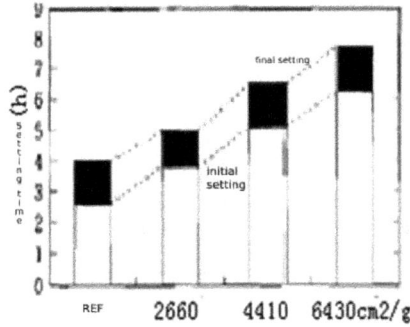

Figure 14.11 The relationship between surface area of fly ash and setting time.

4) Setting time

As fly ash has a large specific surface area, the setting time of concrete with fly ash is longer. This is because the ions dissolved in the cement hydration reaction are adsorbed on the surface of the fly ash particles, reducing the ion concentration in the sub-pore liquid as shown in Figure 14.11.

14.8.2 Heat of hydration

Fly ash is the most effective additive for suppressing the heat of hydration of concrete. The adiabatic temperature rise curves of cement mortar with different fineness of fly ash are shown in Fig. 14.12.

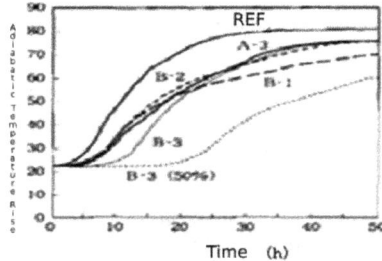

A-3 :7360cm2/g ; B-1: 2640 cm2/g; B-2: 4490 cm2/g ; B-3:9010 cm2/g.

Figure 14.12 The adiabatic temperature rise of cement mortar.

It can be seen from Figure 14.12 that: (1) the adiabatic temperature rise of the control mortar is very high, and after 30 h to 80°C, (2) the adiabatic temperature rise of the mortar mixed with fly ash with different specific surface areas is lower than that of the control mortar; (3). The adiabatic temperature rise of the mortar with B-3 (50%) is the lowest, and the adiabatic temperature rise is only 20 °C at 25 h.

Kasai et al. measured the hydration exothermic development of various cementitious materials after adding water for 24 hours by a conduction calorimeter; expressed in terms of activation energy during hydration: Portland cement 39.0 Kj/mol, mixed with 17% fly ash 26.7 Kj/mol; mixed with 7.5% silica fume 30.4 Kj/mol; It is 49.3 K j/mol when 70 GGBS powder is added. It can be seen that the addition of fly ash is very effective to suppress the heat of hydration. Therefore, in order to reduce the heat of hydration, mass concrete or high-strength concrete is often mixed with fly ash, because the addition of fly ash is more effective in suppressing the heat of hydration than the addition of GGBS.

14.8.3 Hardened concrete

The strength development of fly ash concrete is related to its proportion, performance, temperature and humidity, curing conditions, admixtures and types of fly ash.

1) The strength of concrete with different replacement of cement by fly ash (1:1)

4 kinds of fly ash were used to replace 10%, 20% and 30% of cement respectively to prepare concrete and compare its strength and strength development, as shown in Figure 14.13.

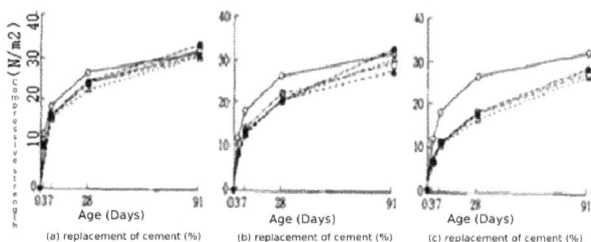

O- OPC, control; Δ-A-20 Japan F; □-E-2-20 Overseas FA ; ▲-K-20Overseas FA; ●-H-1-20Overseas FA .

Slump 12±1cm, W/(C+F)=60%, F/(C+F)= 10, 20, 30%

Figure 14.13 Compressive strength vs age (Kasai).

The replacement rate of fly ash to cement is below 20%, and the compressive strength at 91 d age is basically the same as that of the control concrete.

The strength of fly ash cement concrete is also related to the amount of cement in the concrete; in the rich mix with a large amount of cement, the effect of fly ash on the strength of concrete is more significant Figure 9.17.

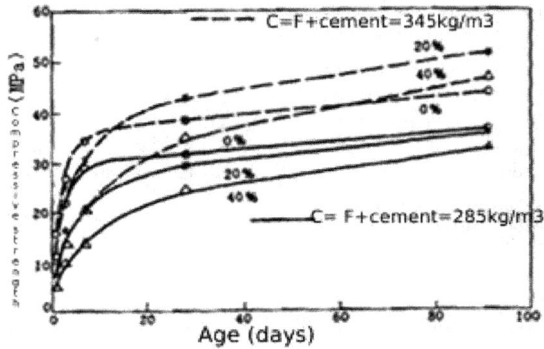

Curing temperature 28±2°C, relative humidity 75±15% (15cmx15cmx15cm specimen)

Figure 14.14 The effect of fly ash on strength in rich concrete mix.

It can be seen from Figure 14.14 that under the same replacement amount of fly ash to cement, the strength with high cement content develops faster. In the concrete with a cementitious material of 345 kg/m³. The strength of the concrete containing 20% fly ash at 28 days is higher than that of the control concrete the concrete is 28 days old. This may be because the concrete with high cement content has a higher concentration of alkali, and the early hydration heat release is larger, which is conducive to the hydration of fly ash, thereby improving its strength.

2)   Strength of concrete with replacement of fine aggregate by fly ash

In order to improve the early strength of fly ash concrete, fly ash is often used to replace part of fine aggregate, so as to increase the total amount of cementitious materials without changing the water demand of concrete. And the amount of fly ash added can be as high as about 50% of the cement content. The strength of concrete does not decrease. If the fly ash content is changed with F/C=100%, the 3 d age strength is the same as the reference concrete strength. As shown in Figure 14.15.

It can be seen from Fig. 14.15 that in the concrete of Portland cement and early-strength cement when the amount of fly ash and cement is the same, the strength of the concrete is one time higher than that of the control concrete (NO1, and NO7) at the age of 180 d.

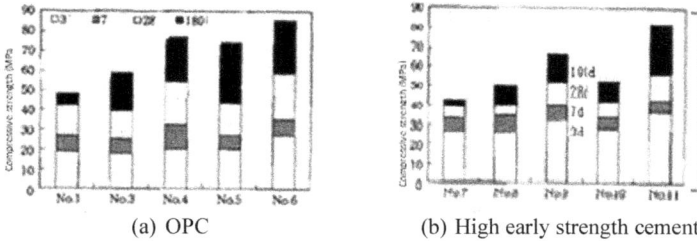

(a) OPC

(b) High early strength cement

(1)OPC: NO1 W/C=53.3%, F/C=0; NO3 W/C=60% F/C=40%; NO4 W/C=60%, F/C=60% ; NO5 W/C=60% , F/C=50%; （with SP） ; NO6 W/C=60%, F/C=100%（with SP）.
(2)High early strength cement: NO7 W/C=54.7% F/C=0; NO8 W/C=60%F/C=25%; NO9 W/C=60% F/C=50% ; NO10 W/C=60% F/C=50% （with SP） ; NO11 W/C =60% F/C=100%（with SP.

Figure 14.15 Concrete strength with additional fly ash.

3) Effect of early temperature on strength of concrete containing fly ash

Portland cement concrete, when cured at a temperature higher than 30 °C, has an increase in early strength, but the later (28 days) strength is significantly lower than that cured under standard conditions. However, the concrete containing fly ash is obviously different. The strength under the early heating curing is obviously improved compared with the curing strength under standard conditions for 28 days, as shown in Figure 14.16.

It can be seen that increasing the early curing temperature is beneficial to the development of fly ash concrete strength. However, in cold temperatures, attention should be paid to early maintenance to avoid freezing damage due to low strength at low temperatures.

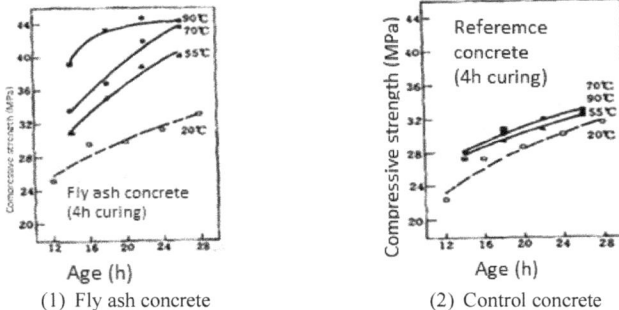

(1) Fly ash concrete

(2) Control concrete

Figure 14.16 Relationship between curing temperature and strength of fly ash concrete.

4) Elastic modulus

The influence of fly ash on the elastic modulus of concrete is similar to that of compressive strength. In general, it is low in the early stage and gradually increased in the later stage. Adding a certain amount of fly ash into the concrete will increase its elastic modulus. Due to the pozzolanic reaction, it is carried out throughout the hydration process, resulting in the formation of hydrated calcium silicate gel, which makes the concrete more compact.

5) Drying shrinkage and autogenous shrinkage

Haque et al. used 40-70% of ASTM C grade fly ash to replace the corresponding cement, and conducted experiments on concrete, and find out that the drying shrinkage of concrete decreased with the increase of fly ash content. The experiments of Kasai et al. of Japan suggested that the slump of the concrete mixed with fly ash was kept the same as that of the control concrete, which could reduce the water consumption and reduced the drying shrinkage of the concrete. Compared with the original fly ash, the fly ash with a large specific surface area has a more obvious effect of reducing the water consumption, and the corresponding drying shrinkage of the concrete is also reduced. Figure 14.17. For original fly ash 3310, FA20 5960, FA10 9700 the higher the replacement rate of fly ash, the lower the autogenous shrinkage of concrete.

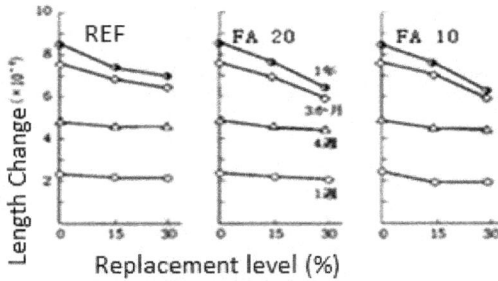

Figure 14.17 Relationship between fly ash replacement rate and drying shrinkage.

14.9 The durability of fly ash concrete

After adding fly ash to concrete, the effect on durability is summarized as follows:

14.9.1 The permeability of concrete

Berry and Malhotra summarized the effect of fly ash on the permeability of concrete as below:

1) Increased water permeability. For concrete containing fly ash with a strength of 40-60 MPa, the original surface water adsorption is 0.09-0.05 ml/m^2s, while the control concrete is 0.08-0.03 ml/m^2s.

2) Air permeability increases. Since the relative humidity of the concrete surface decreases after the addition of fly ash, the air permeability increases. If the concrete is dried prematurely, more pores will be formed inside, which will become channels for air flow, permeability will increase.

3) The chloride ion diffusion coefficient decreases. Li and Roy suggested that the diffusion coefficient of Portland cement slurry with 30% F-class fly ash and the W/C=0.3 and 0.35 at the temperature 38°C was:

W/C 0.3   15.6X 10^{-12} m^2/s ; with fly ash 1.35 X 10^{-12} m^2/s

W/C 0.35   8.7 X 10^{-12} m^2/s ; with fly ash 1.34 X 10^{-12} m^2/s

When the replacement rate of fly ash to cement is 15-45%, the lower the water-binder ratio of concrete, the lower the penetration depth of chloride ions as shown in Figure 14.18.

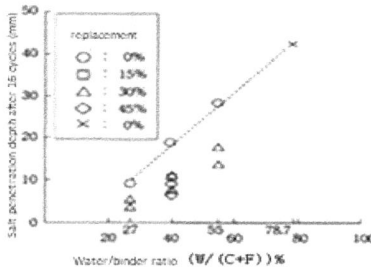

Figure 14.18 Relationship between water-binder ratio and salt penetration (Kasai).

14.9.2 Anti-neutralization properties

In general, the concrete mixed with fly ash consumes a part of the CH phase due to the reaction of the pozzolan, which is not good for the neutralization resistance. And with the increase of the cement replacement rate, the depth of the neutralization of the concrete also increases. For the concrete with 330 kg/m^3(C+FA), 15 cm slump, compared with the control concrete, in the first two years, the neutralization depth is larger. But after 20 years, the neutralization of the concrete mixed with fly ash is still the same as it was in the first 2 years compared to that of the control concrete. This is due to the slow strength development of the early FA-containing concrete.

14.9.3 Autogenous shrinkage

The autogenous shrinkage of the concrete mixed with fly ash is reduced. The stress of the cement slurry is measured by the hollow circular specimen. At W/B=30%, the cement slurry of ordinary Portland cement and slag cement has large shrinkage stress, while the shrinkage stress of fly ash cement is small. That is due to the addition of fly ash which can effectively inhibit the autogenous shrinkage of concrete.

Compared with the control concrete, the concrete mixed with fly ash can appropriately reduce the water consumption under the condition of the same flowability, so the drying shrinkage will also be reduced.

14.9.4   Inhibit alkali-aggregate reaction

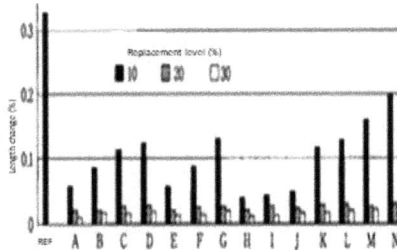

Figure 14.19 Inhibition of alkali-aggregate reaction (relationship between fly ash content and ASR).

Replacing part of the cement with fly ash can reduce the alkali content of the concrete on the one hand, and on the other hand, due to the activity of the pozzolan, the pH value in the pore solution is lowered, which can inhibit the alkali-aggregate reaction (ASR), as shown in the figure 14.19. For the control specimen without fly ash, the length change rate exceeds 0.3%; while for the specimens replaced with 20% and 30% fly ash, the length change rate is within the range of 0.1%. The more amorphous components in fly ash, the greater the fineness of the fly ash and the better the effect. In general, 30% fly ash content can effectively inhibit the occurrence of ASR.

14.10 High performance alkali-fly ash concrete

A new type of concrete can be obtained by using fly ash as the main raw material, sodium silicate as the activator, adding coarse and fine aggregates, stirring to form, and then curing at room temperature or steam. The hydrate generated by this concrete is analcite-type water Sodium silicate. It is a kind of concrete with early strength, high strength, energy saving, durability and high chemical resistance.

1)   The raw materials of concrete

Fly ash: Class I fly ash or Class II fly ash,

Sodium silicate: $Na_2O. nSiO_3$, also known as water glass, commonly known as Paohua alkali, the alkaline water glass with modulus less than 3.

Fine aggregate: river sand or manufactured sand, in line with national standard requirements.

Coarse aggregate: limestone crushed stone, in line with national standard requirements.

Regulator: CaO; Na_2CO_3.

2)   Choice of cementitious material

Using grade I fly ash or grade II fly ash as the main raw material, adding an appropriate amount of CaO to adjust its chemical composition to 40-50% for SiO_2, 12-15% for Al_2O_3, and 10-12% for CaO.

I.   Choice of the modulus of the water glass

The content of water glass is 5-15% of the cementitious material, and the modulus of water glass is 0.8, 1.0, 1.5, 2.0. With appropriate amount of water a pure paste sample was prepared and cured under different conditions. The strengths of 7 d and 28 d were measured, the modulus of the water glass is selected according to the one with the highest strength.

Experiments showed that the modulus of water glass of 1.0, and the dosage of 10% could achieve better result.

II.   Setting time adjustment agent

When the modulus of water glass is 1.0 and the dosage is 10%, the prepared cementitious material has very short initial setting and final setting. Therefore, it is necessary to add a certain amount of soluble carbonate to adjust the setting time. With 0.1% Na_2CO_3 the initial setting time is extended from 8 min to 22 min and the final setting is extended from 38 min to 58 min.

By selecting the modulus of water glass and adding a certain amount of soluble carbonate, the setting time can be adjusted. The proportion of concrete can be selected according to the principle of concrete mix design.

3)   Reaction Mechanism of Alkali-Fly Ash Cementitious Materials and Main Phases Produced

The fly ash cementitious material contains CaO, and when it is mixed with sodium silicate Na_2O. $nSiO_3$, there is a certain amount of Na+ in the system, and the reaction can be studied as a dispersion system of alkali-alkaline earth aluminosilicate. In the whole system, there are alkali metal hydroxide (NaOH), alkaline earth metal hydrogen oxides ($Ca(OH)_2$), acidic oxides (SiO_2), and amphoteric oxides (Al_2O_3).

In the system, the silicate aqueous solution has a negative charge, and the polyvalent metal aqueous solution has a positive charge. The interaction between the two causes the particles to agglomerate. The surface of the gel particles adsorbs Na^+ in the system. Finally, the synthesis of basic compounds occurs, and the system crystallization process.

In the system, aluminum cations have a strong coagulation effect on the alkaline silicate aqueous solution, and then agglomerate into water-resistant alkaline compounds. Each mol Al_2O_3 can capture 1.0-1.5 mol of alkali metal oxide, thus becoming insoluble new hydrate, as the following formula:

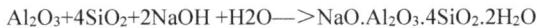

$Al_2O_3+4SiO_2+2NaOH +H2O—>NaO.Al_2O_3.4SiO_2.2H_2O$

Alkaline earth metal oxide $Ca(OH)_2$ can also coagulate with acidic silicic acid aqueous solution and alumina to form gel. The gel contains alkali metal and alkaline earth metal oxides as synthesis of five-component minerals ($R_2O.RO.Al_2O_3.SiO_2.H_2O$).

This type of hydration product is like analcite ($Na_2O.Al_2O_3.4SiO_2.2H_2O$) in nature. Alkali-fly ash cementing material not only produces low alkalinity calcium silicate hydrate, but also produces zeolite-like Hydrated aluminosilicate. This is consistent with the formation of minerals in nature.

4) Durability of alkali-fly ash cementitious material concrete

I. The development and stability of strength

The strength of alkali-fly ash cementitious material concrete can reach more than 10 MPa in 1 d age, and more than 55 Mpa in 28 d. After 1 year, 2 years and 5 years, the strength is slightly improved. It shows that the hydrate formed is stable.

II. Resistance to carbonation

The alkali-fly ash cementitious material concrete specimen was divided into two groups, one group was under carbonation, and the other group was placed in the air. The carbonation resistance can be assessed based on the comparison of the strength of carbonated specimens and noncarbonated specimens, as shown in Table 14.3.

The carbonation coefficient was close to 1.0, which means after complete carbonation, the strength was basically not reduced, and its carbonation resistance is excellent.

Table 14.3 Alkali-fly ash cementitious material concrete carbonation resistance.

NO.	Curing	Strength after carbonation	Control strength	Coefficient of carbonation
1	Steam	54.4 Mpa	56.1 Mpa	0.97
2	Natural	50.4	49.3	1.02
3	Steam	65.1	66.3	0.98
4	Natural	62.7	62.3	1.01

III. Freeze-thaw resistance

The alkali-fly ash cementitious concrete specimen was divided into two groups, one group was frozen and thawed, and the other group was placed in normal temperature water. After certain cycles of freezing and thawing, the compressive strength was carried out. And the comparison of strength to the unfrozen specimen was calculated to evaluate the performance of freeze-thaw resistance. The results are shown in Table 14.4. The frost resistance is excellent.

Table 14.4 The testing result of freeze-thaw test.

NO	NO of cycles	Strength after frozen	Control strength	appearance	coefficient
5	250	50.6 Mpa	52.9	angular peeling	0.96
6	250	63.3	66.3	Good	0.95
7	250	59.0	59.6	Good	0.99

IV. Strong acid corrosion resistance

The specimens were Immersed in different acids, and the strength after immersion was tested. The comparison of the strength to that of the reference specimen was made to evaluate the corrosion resistance coefficient. The results are shown in Table 14.5.

Table 14.5 The testing results of strong acid corrosion.

Type of acid		Immersion time	Test after immersion		Strength of control specimen	Anti-corrosion factor
			strength	Appearance		
H2SO4	1%	210d	50.3 Mpa	Few cracks	56.8 Mpa	0.86
	.5%	210d	56.2	good	56.8	0.99
HNO3	5%	210	55.0	good	56.8	0.97
	1%	210	58.5	good	56.8	1.03
H C l	5%	210	47.3	good	56.8	0.83
	1%	210	58.3	good	56.8	1.03
Mixed acid with 1:1:1	5%	210	46.5	Microcracks	56.8	0.82
	1%	210	55.5	good	56.8	0.98

The corrosion resistance of the test specimens to the higher concentration of sulfuric acid and composite acid was slightly worse.

V. Alkali corrosion resistant

The specimens were immersed in NaOH and saturated lime solution, and after one year, the compression strength was carried out, and the strength was compared with that of the comparative specimens, as shown in Table 14.6.

Table 14.6 The testing results of alkali corrosion.

Type of alkali	Immersion duration (year)	Specimens after immersion		Control strength MPa	Coefficient of Alkali resistance
		Strength MPa	Apperance		
5%Na OH	1	62.3	good	64.3	0.97
Saturated lime	1	63.7	good	64.3	0.99

The results of the alkali corrosion test showed that the alkali corrosion resistance is excellent.

Table 14.7 The resistance to salt corrosion.

Type of salt	Immersion duration	Specimens after immersion		Control strength MPa	Coefficient of salt resistance
		Strength MPa	Appearance		
Sea water (man-made)	1 years	39.5	good	38.4	1.03
	3 years	41.4	good	39.0	1.06
MgCl2 solution 25-30 PPM	1 years	38.5	good	38.4	1.00
	3 years	39.2	good	39.0	1.00
Na2SO4.10H (2500mg/l)	1 years	53.1	good	52.5	1.01
	3 years	59.8	good	54.6	1.09

VI. Salt corrosion resistance

The specimens were immersed in seawater, in $MgCl_2$ solution and sodium sulfate solution. After 1 year and 3 years, the results were tested and are shown in Table 14.7. The resistance to salt corrosion is good.

Materials Research Forum LLC

https://doi.org/10.21741/9781644901991

VII. Economic effects and prospects

The unit material cost of alkali-fly ash concrete is 15% lower than that of ordinary concrete of the same strength grade. And its durability is better than that of ordinary concrete. It is a kind of concrete that saves resources, saves energy, and has a long service life. It has a wide range of development prospect.

Chapter 15

Research and Engineering Application of Multifunctional Concrete Technology

15.1 Introduction

Concrete is a widely used material in engineering. In the past, the technical requirements for concrete materials were mainly based on the construction performance, the bearing capacity (that is, strength) under long-term working conditions, and the resistance performance (durability) under various deterioration factors in the working environment. Concrete is designed to meet the performance of workability, strength, and durability. Due to the continuous development and improvement of engineering technology, the technical requirement of concrete is also diversified and for high-performance. Multifunctional concrete can have a variety of functions according to the construction conditions, use conditions, and environmental conditions of the concrete to meet these various technical requirements.

The most basic properties of multifunctional concrete are as follows: self-compacting, self-curing, low heat of hydration, low shrinkage, high strength, and high durability. The large-scale construction of large-span bridges and high-rise and super high-rise buildings also requires high-strength and ultra-high-strength concrete. As the intensity of steel bars in the structure has also increased, it brings difficulties to the construction and compaction of concrete. If high-strength and ultra-high-strength concrete can be formed by vibration-free and self-compacting, this is suitable for large-span, super high-rise, high-strength, and high-strength concrete. Construction will bring great convenience. If a large-span super high-rise building, high-strength and super-strength concrete construction do not require manual watering and maintenance, it can not only save water resources but also save labor and facilitate management. More importantly, self-curing can inhibit auto-shrinkage cracking and early cracking. The low heat of hydration is to make the temperature difference between the concrete pouring and the ambient temperature $\leq 25°C$ so that the concrete will not produce thermal cracks. This is very difficult for high-strength ultra-high-strength concrete to have a low heat of hydration because of the large amount of cement and low water-cement ratio. How to overcome this challenge for achieving high strength, ultra-high-strength, and also low heat of hydration from the selection of materials and composition?

Low shrinkage is a very important performance for concrete materials because too much shrinkage will cause cracking, which will reduce the load-bearing capacity and working life. High strength and high durability are more special for multifunctional concrete. The multifunctional concrete developed by this invention has a strength of 150 MPa and an electric flux of less than 100 coulomb/6 h.

15.2 Technical standards for multifunctional concrete testing

15.2.1 Self-compactness

Through the U-shaped instrument test of self-compacting concrete, to detect the self-compacting function. Figure 15.1 is the U-shaped instrument for SCC to detect the self-compacting function.

In the U-shaped instrument test, if the concrete can raise ≥ 30 cm in height in 10-15 seconds, this concrete can meet the self-compacting function.

For the test of multifunctional concrete, it is usually also expressed by the slump, the flow diameter, and the flow time of concrete in the inverted slump cone. The flow time in the inverted slump cone is about 10-15 seconds. As shown in Figure 15.2.

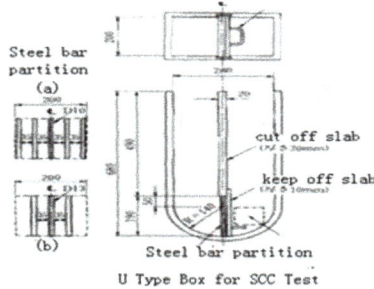

(a) The U-shaped instrument for testing SCC self-compacting function

(b) Measure the self-compacting function of MPC with a U-shaped instrument

Figure 15.1 Apparatus and application for testing MPC self-compacting function.

For the test of fresh concrete, if the slump ≥26 cm, the slump flow ≥680×680 mm, the flow time of the inverted concrete cone is 10-15 sec and the raise of concrete in U-shaped instrument is 30 cm within 10-15 sec, the concrete can meet the requirements of the non-vibrating self-compacting function.

(a) Slump

(b) Slump-flow

(c) Slump measurement

(d) Slump-flow measurement

(e) The flow time of inverted slump cone

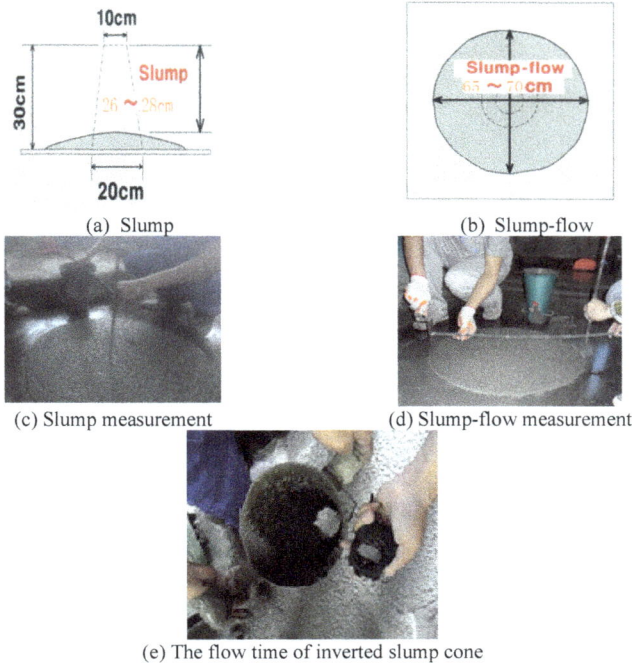

Figure 15.2 Fresh concrete test.

15.2.2 Self-curing performance

After the concrete is poured into a shape, the initial setting is completed, it must be watered for curing. The cement in the concrete continues to hydrate and absorbs the moisture in the internal pores and capillaries of the concrete. The curing can be done by covering with a plastic film and or to be watered for curing. This is to keep the surface of the concrete to be moist to avoid the moisture migrating outward and evaporating. This requires a lot of water resources and labor. Self-curing is to mix an organic or inorganic powder during the preparation of concrete. During the mixing process of concrete, it absorbs free moisture. After the initial setting of concrete, when water is required for cement hydration, it releases the absorbed water and supplies the water required for cement hydration to avoid auto-shrinkage and cracking of concrete due to dehydration of pores such as capillary tubes. As shown in Figure 15.3, after initial setting of MPC, the specimens were covered with plastic sheet, curing in curing room or outdoor.

(a) After demoulding, the MPC test cubes are sealed with plastic film

(b) The specimens were covered by sacks with watering for curing

Figure 15.3 MPC self-curing and water-curing.

The self-curing agent for multifunctional concrete in this study is a natural zeolite powder with a specific surface area of about 6000 cm2/g. The main rock-forming mineral of natural zeolite is zeolite of $\geq 60\%$. Natural zeolite powder is a porous crystalline powder, and in the process of concrete mixing, it absorbs free water and becomes a kind of zeolite water, which can be absorbed and released freely. Therefore, after the concrete hardens, it can supply the water required for cement hydration. The zeolite powder is evenly dispersed in the concrete, like many supply points for cement hydration needs. This will avoid auto-shrinkage and cracking of concrete due to the dehydration of capillary and other pores. Natural zeolite powder also contains high soluble silicon and aluminum, which can participate in the hydration reaction of cement and improve the strength of concrete. Porous natural zeolite powder can absorb free water in concrete, inhibit concrete segregation and settlement, and ensure the uniformity of concrete mixture, which is also an important performance of SCC.

15.2.3 Low heat hydration

On the premise of meeting the strength requirements, there are many ways to reduce the maximum temperature of concrete (≤ 78 °C) and the temperature difference between the center part and the surface of large-volume multifunctional concrete (≤ 25 °C), such as to reduce the amount of cement as much as possible, use low-heat cement, and use low-heat additives, such as microbeads or fly ash. This will help to avoid concrete thermal cracks.

When measuring the adiabatic temperature rise of the multifunctional concrete, sieve out the mortar from concrete, wrap it with plastic film, insert a thermometer, place it in a thermos bottle, expose the thermometer to facilitate the reading, and then seal it. As shown in Figure 15.5. Check the temperature of the mortar and the concrete. Generally, the temperature of the mortar is higher than the temperature of the concrete.

(a) Put the sample in the thermos   (b) Automatically record the temperature rise

Figure 15.5 Measure the temperature of multifunctional concrete and its mortar.

15.2.4 Low shrinkage

Low shrinkage includes autogenous contraction, early contraction, and long-term contraction. 3 d autogenous shrinkage ≤0.1‰, long-term shrinkage ≤0.6‰, so that the concrete structure will not crack. It is the key technology to inhibit the autogenous shrinkage and cracking of high-strength and ultra-high-strength MPC with blended NZ powder and a 1.5% sulphoaluminate expansion agent. The former provides an internal water supply to inhibit auto-shrinkage, while the latter expands slightly by hydration to compensate for auto-shrinkage, so that auto-shrinkage ≤0.1‰, thereby inhibiting auto-shrinkage and cracking. Figure 15.6 is a device that can measure autogenous contraction, early contraction, and long-term contraction.

1. Mold for autogenous shrinkage measurement: The test mold for measuring autogenous shrinkage of MPC is shown in Figure 15.6 (a). (1) The inner lining of the test mold is a rubber pad with a thickness of 3 mm (4 in the figure), and there are reserved holes on it, which correspond to the reserved holes in the end plate. The buried probe can extend out of the mold through the rubber gasket and the end plate; the gap between the reserved hole and the probe is sealed with Vaseline. the removable side panels (8 in the picture) are inserted long to the inside. The bottom is lined with a layer of Teflon backing plate (7 in the picture).

2. Fabrication of autogenous shrinking specimens: When the MPC is loaded with a thickness of 50 mm, the Cu50 temperature measuring thermal resistance is embedded in the center of the test mold, and the lead wire is drawn out. When the concrete is installed to about 5 mm from the upper edge of the test mold, cover the mold cover, tighten the bolts, and place the sealed test mold in the 20±1 °C constant temperature indoor curing room. At the same time as the autogenous shrinking specimen is formed, the initial setting time of MPC is measured. During the initial setting, the cover plate is removed, the side plates on both sides are pulled out, the test mold is resealed, and the weight is weighed (W0).

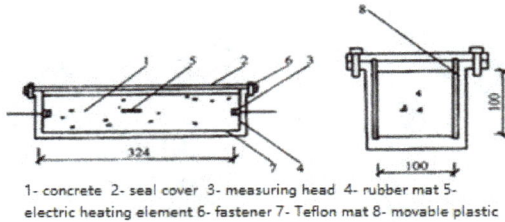

1- concrete 2- seal cover 3- measuring head 4- rubber mat 5-
electric heating element 6- fastener 7- Teflon mat 8- movable plastic

(a) Schematic diagram of test mold for autogenous shrinkage measurement

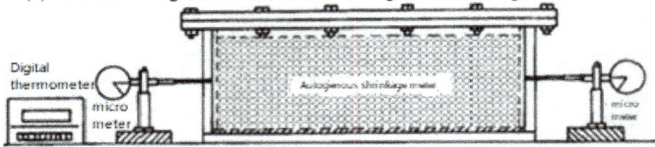

(b) Schematic diagram of the autogenous shrinking measurement device

(c) MPC self-shrinkage measuring device

Figure 15.6 Autogenous shrinkage measuring device and its principal analysis.

3. Measurement of autogenous shrinkage: In a constant temperature room of 20±1°C, place the test mold on the horizontal test bench; place the dial indicator brackets at both ends, so that the probe of the dial indicator and the probe embedded in the concrete are centered and contacted with each other. Read the initial reading value of the dial indicator. From the initial setting to 1d take the reading every 2 h. The mass loss during the whole measurement process shall not exceed 0.05%.

4. Examples

    (1) Low W/C and large content of cement per cubic meter concrete will increase the autogenous shrinkage of concrete. The concrete mix ratio is shown in Table 15.1 and the self-shrinkage measurement results are shown in Figure 15.7.

Table 15.1 Test concrete mix design (kg/m³)

NO.	W/B	Cement	Additives	Water	Superplasticizer	Sand	aggregate
AS-1	0.5	250	120	185	0.7%	840	980
AS-2	0.4	300	140	175	0.8%	800	1000

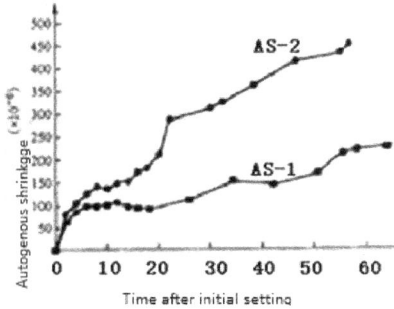

Figure 15.7 Influence of water-cement ratio and cement dosage on autogenous shrinkage.

It can be seen from Figure 15.7 that the water-cement ratio of AS-2 is decreased, the amount of cement is increased, and the autogenous shrinkage increases. W/C=0.4 of AS-2<W/C=0.5 of AS-1. The cement content of AS-1 is 250 kg, lower than that of AS-2(300 kg).

(2) The cement content per cubic meter concrete is low, although the W/B is low (the W/C is also low), the autogenous shrinkage of the concrete increases. The concrete mix is shown in Table 15.2 and the result of autogenous shrinkage is shown in Fig 15.8.

The W/B=0.45, (W/C=0.7) of AS-7, although the water binder ratio is lower than the W/B=0.49 (W/C=0.62) of AS-6, the cement content of concrete is lower than that of AS-6, so the autogenous shrinkage is low.

Table 15.2 Test concrete mix ratio (kg/m³).

NO.	W/B	Cement	Additives	Water	Superplasticizer	Sand	Aggregate
AS-6	0.49	297	78	185	0.5%	780	1040
AS-7	0.45	250	140	175	0.6%	780	1050

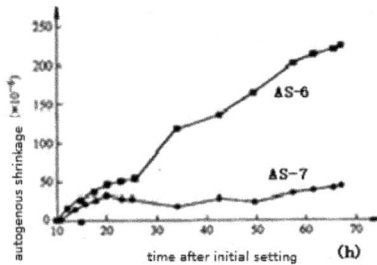

Figure 15.8 Influence of water-cement ratio and cement content on autogenous shrinkage.

(3) The natural zeolite powder self-curing agent has the effect of inhibiting autogenous shrinkage. The natural zeolite powder is free to absorb and release water. When the water-absorbing zeolite powder is mixed in MPC, after the concrete is initially set, the cement continues to hydrate, and the zeolite powder can release the absorbed water and supply for cement hydration. In this way autogenous shrinkage of MPC will be inhibited, as shown in Figure 15.9 and the concrete mix design is shown in Table 15.3.

Table 15.3. The ratio of natural zeolite powder self-curing agent to concrete (kg/m³)

NO	Cement	Zeolite	Water	Superplasticizer	Sand	aggregate
AS-3	495	55.0	160	1.5%	696	1044
AS-4	495	60.5	154,5	1.5%	696	1044
AS-5	495	60.5	160	1.5%	696	1044

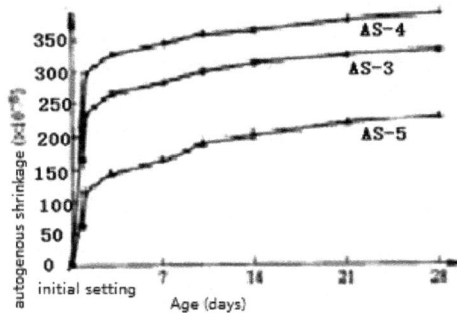

Figure 15.9 Inhibition effect of natural zeolite powder as self-care agent on autogenous shrinkage

In Table 15.3, AS-3 is mixed with ordinary zeolite powder, while for AS-4 and AS-5 zeolite powder is mixed with 10% water to form saturated zeolite powder. 5.5kg of zeolite mixing water was deducted for mix water for AS-4. AS-3 and AS-5 have the same amount of water for concrete, but AS-5 has lower autogenous shrinkage due to the self-curing effect of saturated zeolite powder. Although AS-4 also incorporated saturated zeolite powder, but due to the reduction of water content of concrete, its autogenous shrinkage is also high.

15.2.5 High-strength performance

Using commercially available concrete raw materials, the author has developed multifunctional concrete (MPC) with a strength of up to 150MPa. It has the above-mentioned multiple functions and can be pumped from the ground to a height of 521m for concrete construction. And when demolding, there is no autogenous shrinkage and cracking found. The size of the test specimens for testing compressive strength is 10x10x10cm.

Compared with ordinary strength or high strength MPC, ultra-high-strength multifunctional concrete (UHS-MPC) has many advantages and characteristics, but it has higher requirements for raw materials, more serious autogenous shrinkage, and cracking problems, and poor fire resistance and brittleness. The construction application is relatively difficult. I will discuss it in the relevant chapters later in this book.

15.2-6 High durability

The durability of concrete is related to its impermeability. But when the strength of concrete exceeds a certain range, the impermeability of concrete cannot be measured by the usual impermeability test method. HPC, UHPC, and MPC should be tested according to the American ASTM C1202 standard. The impermeability of concrete can be evaluated according to the electric flux. For example, the electric flux of UHS-MPC is generally 50-100 Coulomb/6h. It belongs to ultra-high durability concrete. MPC with strength grade $\geq$ C 100, 6h electric flux is also below 500 Coulomb with very high durability.

From the analysis of the causes of environmental degradation of concrete, whether it is single-factor or multi-factor degradation, water is the key factor. Neutralization requires water, chloride ion diffusion and penetration require water, and alkali-aggregate reaction requires water, freeze and thaw damage requires more water. Therefore, the key factor of concrete durability is to cut off the water source. Cutting off the contact between the water source can make the concrete structure very durable.

The author has tried to use emulsified stearic acid emulsion to replace 1/4 of the water consumption of concrete and mix the concrete to obtain water-repellent concrete, and its durability can be greatly improved.

Chapter 16

High Strength Multi-Purpose Concrete and Ultra High Strength Multi-Purpose Concrete

In China, Concrete with strength $\geq$ 100 MPa (equivalent to concrete with strength grade C80), is called ultra-high strength concrete. Therefore, in this chapter, the research and development and application of C80 MPC are introduced first.

16.1 Selected topic background

Cracks were found in the wall of the first floor of a super high-rise building after casting C80 ultra-high strength concrete. Exterior wall cracks are dense, mainly transverse and longitudinal cracks. The interior wall is dominated by oblique and vertical cracks. As shown in Figure 16.1 (a) (b).

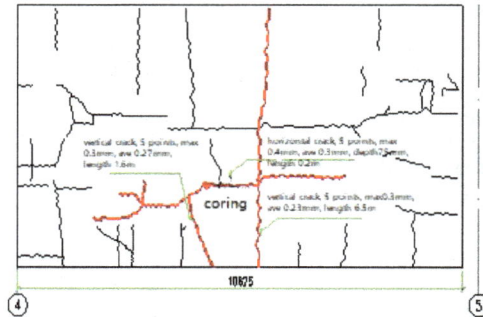

(a) Cracks condition on the South Wall No. 1 (external)

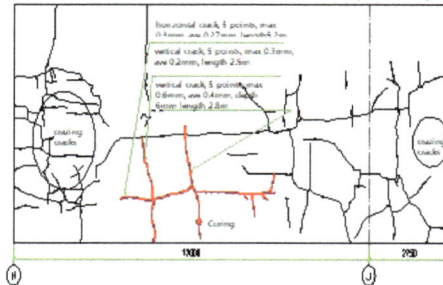

(b) Cracks Condition on the East Wall No. 1 (internal)

Figure 16.1 C80 Cracks of Ultra-high strength concrete shear walls.

Multifunctional Concrete Technology   Materials Research Forum LLC
Materials Research Foundations **127** (2022)   https://doi.org/10.21741/9781644901991

In Figure (a), there are about 25 cracks. Most Cracks are about 0.2 mm in width and less than 4 cm in depth. The wall is dominated by horizontal and vertical cracks and has a long length. One of the transverse cracks has the largest width and depth, with a width of 0.4 mm and a depth of 75 mm. In Figure (b), the cracks at the end of the wall are denser and mainly cracked. There are few cracks in the middle, mainly transverse and vertical cracks. The longest transverse cracks in the upper part is about 10 m. The deepest crack is shown in the thick red line, with a depth about 86 mm and a maximum crack width of 0.6 mm

The outer wall of the core tube is provided with double steel plate shear wall inside the concrete. The reinforcement of shear wall is very dense. As shown in Figure 16.2(a)(b).

(a) Steel plate shear wall is arranged inside reinforced concrete shear wall

(b) Reinforcement of shear wall dense, cause difficulty in pouring of concrete

Figure 16.2 Double shear wall and reinforcement.

During demoulding, cracks were found on the wall, thus the concrete was not affected with external conditions. It's due to self-contraction. During the initial set of concrete, cement continues hydration, absorb capillary moisture, make capillary generation from vacuum phenomenon, if the capillary tension of generation is greater than the concrete resist pull strength, the concrete produces craze. This is called autogenous shrinkage cracking.

Because of the high density of reinforcement, concrete is difficult to be compacted by vibration during pouring. Therefore, it is necessary to use self-compacting concrete, at the same time to solve the early shrinkage cracking, self-shrinkage cracking and other technical problems. The experiment and application of multifunctional high-performance concrete are carried out.

16.2 Development and application of C80 MPC

MPC has a variety of functions: 1. Vibration-free self-compacting. Use the self-compacting concrete U-Box to lift height inspection. MPC can still reach the requirement after 3 hours when it raised more than 30 cm after passing through the J Ring Test. Be it on site or casting yard, before pumping or during pumping MPC are able to meet the requirements. 2. Self-curing: within concrete itself, rely on moisture carrier for cement hydration, externally wrapped with film, prevent moisture evaporation; self-curing can restrain self-shrinkage cracking; Self-curing concrete strength ≥ water curing method. 3. Low heat of hydration: the key technology is the concrete mould temperature, maximum temperature, internal and external temperature difference to meet the requirements of the relevant specification. 4. Low shrinkage: including self-shrinkage, early shrinkage and long-term shrinkage; 3 d self-shrinkage ≤ 0.1%; long-term shrinkage ≤ 0.6%; concrete total shrinkage ≤ 0.7-0.8%. Suppressing auto contraction is key. 5. High retardation protection: performance of fresh mixed concrete: CFA retardation admixture, basically unchanged after the initial 3 hours. 6. High durability: MPC has high compactness and low porosity; salt resistance, sulphate corrosion resistance, etc. Select all or part of the functions according to the working environment of the concrete structure. To ensure the quality of concrete structure and service life.

16.2.1 Raw materials and MPC ratio selection of fresh mixed concrete performance

1. Raw material selection (1) Cement: Ordinary Portland Cement P2 52.5; (2) microbeads: ultrafine fly ash (MB), Shenzhen Tongcheng Company; (3) natural zeolite powder (NZ): curing agent, self-developed products; (4) fly ash (FA): sold in Guangzhou; (5) carrier fluidizing agent (CFA): a self-developed product with the function of reducing water and maintain workability; (6) Sulphoaluminate microexpansion agent (EHS): produced by a cement factory in Tianjin. (7) Polycarboxylic acid high-range water reducing agent: product of Bozhong Water reducing agent factory.

2. For the proportion and performance of MPC, see Table 16.1,2,3,4 and Figure 16.3,4,5,6,7. Compared No.5 and No.6, EHS and CFA are not the same. Hence, No. 6 works better.

Table 16.1 C80 MPC Mixing proportion (kg/m3).

NO.	CEM I	MB		NZ	FA	EHS	FA	CA	HRWRA	CFA	Water
5	320	60		15	170	---	800	900	1.9%	---	142
6	320	60		15	170	8.8	800	900	1.9%	1.5%	142

Table 16.2 Heat of Hydration and early shrinkage (NO.6).

Heat of Hydration(°C)			Early shrinkage (/100,00)			
Temperature rise start time	initial Hour of peak temperature	Initial temperature of peak temperature	24 h	48 h	72 h	8 d
25 h	35 h	25-75 °C	0.62	0.81	0.99	1.1

Table 16.3 Strength difference of Self-curing vs Water curing (MPa)(NO.6).

Age / Condition	3 d	7 d	28 d	56 d
Wet water curing	50.6	66.5	86.1	90
NZ self-curing	56.6	75.6	92.1	96

In the summer, when the outdoor temperature reaches 35°C, ice powder will be used to mix with concrete instead of using water according to the concrete ratio in Table 16.1. The performance of fresh concrete is shown in Table 16.4. Ice slag mixed concrete instead of water Figure 16.3

Table 16.4 Performance of fresh concrete (Ice slag instead of water NO.6).

Age	Slump (cm)	Flow spread (mm)	Discharge time from slump cone (S)	Height of U Box (cm)
Initial	27	670x700	4.4	32
3 hr.	27	700x730	4.0	32

When test concrete temperature changes. Ambient temperature of 35 °C, mortar mixing temperature of 11 °C, after added fine and coarse aggregates temperature of 17 °C. While concrete mixing was completed, temperature reached 20 °C. At room temperature, when laboratory raw material are tested, the temperature of fresh concrete is about 28 °C; in this test, ice slag was used instead of water, and the temperature of fresh concrete was about 20 °C. Other performance are unchanged, the effect is obvious. The temperature rise of hydration heat of No. 6 concrete mixed with 1.5% Sulphoaluminate microexpansion agent (EHS) was slightly increased from 68 °C to 74 °C. However, the self-shrinkage and early shrinkage of ordinary C80 concrete decreased from 192/10,000 to 99/10,000 which is a significant reduction.

(a) ice powder replacement (b) flow spread of ice slag mixed concrete

Figure 16.3 Ice slag mixed concrete.

16.2.2 C80 MPC Small scale simulation test

The purpose of L Shaped simulation test is to test the fluidity of concrete, through the performance of reinforcement, filling and surface smoothness, etc. in addition, the self-shrinkage performance and self-curing effect of the prepared C80 MPC were tested. As shown in Figure 16.4.

(a) reinforcement layout in the L-shaped mould (b) flowability of MPC in the L-shaped mould

Figure 16.4 L shaped Small Simulation test.

1. Mixing proportion of C80MPC for the Simulation test as shown in table 16.5

Table 16.5 MPC mixing proportion of simulation test (kg/m3).

Binder	W	F.A	C.A	Ag	CFA
320+80+170+15	142	800	900	1.9%	1.5%

Binder：Cement 320 kg; MB 80 kg; FA 170 kg; NZ15 kg; Sulfoaluminate Expansion Agent 8.8 kg.

2. Fresh MPC performance as Shown in table 16.6.

Table 16.6 Fresh performance of MPC.

Test	Slump (cm)	Flow spread (cm)	Discharge time from slump cone (s)	Height of U box (cm)
Fresh	26	68x71	6	32
3 h	25	68x68	5	32

It shows that concrete has good flowability and self-compacting performance. The workability and viscosity of freshly mixed concrete are very good and is able to fulfil the 3 hours construction operation.

3. Autogenous shrinkage and early shrinkage of concrete.

The autogenous shrinkage and early shrinkage of concrete were measured, results as shown in table 16.7.

Table 16.7 C80 MPC autogenous shrinkage and early shrinkage.

Age	24 h	48 h	72 h	10 d	20 d	25 d	29 d	35 d
Shrinkage ‰	0.103	0.106	0.099	0.177	0.313	0.389	0.407	0.417

The C80 MPC autogenous shrinkage and early shrinkage of concrete in 72 hours is 0.99/10,000, which is half as much as the original C80 concrete which autogenous shrinkage and early shrinkage was 1.9/10,000. According to the Japan Society of Civil Engineers, the shrinkage rate of concrete (including autogenous shrinkage and early shrinkage and long-term shrinkage) is 5/10,000 to 7/10,000. Therefore, the shrinkage rate of C80 self-compacting concrete in this study is relatively low. Especially autogenous shrinkage and early shrinkage. C80 MPC autogenous shrinkage is low, shrinkage deformation is small, the early cracking of concrete structure is low.

4. The initial temperature for heat of hydration and temperature rise was 28 °C, the temperature began to rise at → 14 hours, and reached the peak at 28 hours, and the maximum temperature was 74 °C. The temperature difference between the center part of the component and the surface layer is less than 25 °C, which will not cause thermal cracking.

5. Pressure bleeding test result. As shown in Table 16.8.

Table 16.8 Pressure bleeding test result.

Pressured time	10 seconds	45 seconds	140 seconds
Bleed volume(ml)	0	Start to bleed	3

It can be seen from the pressure bleeding test that MPC concrete in this study does not produce bleeding phenomenon in the pumping process and has good pumpability.

6. The strength of concrete. Placed outdoor under same ambient condition. The strength of self-curing and wet curing concrete will be shown in Table 16.8.

Table 16.8 C80MPC compressive strength (MPa).

Age	3 d	7 d	28 d
Self-curing	56.6	75.6	92.1
Wet-curing	50.6	66.5	86.1

Self-curing is when the concrete has been demoulded, covered with a plastic film and does not need watering maintenance, it occurs only by internal Zeolite Power as a curing agent from the internal water of the concrete. Wet curing is water curing every day, there is no self curing agent in the concrete.

(a) wet curing method, spray water twice daily. b) Self curing method, cubes were concealed in a plastic bag. No watering is needed.

Figure 16.5 Difference in the concrete cube curing condition.

As seem from Table 16.8, the strength of self-curing concrete is higher than wet curing concrete. In the small-scale simulation test, the strength of self-curing concrete core sample is also higher than the wet curing concrete.

7. Durability of Multi Propose Concrete.
   1) Electrical conductivity. C30 MPC concrete age of 28 hours has conductivity ≤ 800 coulomb/6H, C80 MPC age of 56 days has conductivity ≤ 500 coulomb/6H.
   2) Sulphate resistance. According to ASTM C1202, the expansion rate of 2.5x2.5x28.5 cm specimens soaked in sulphate solution for 14 weeks is $\leq 0.4\%$.
   3) Alkali-Aggregate Reaction: Cement is only 300- 320 kg/m3, while total amount of microbeads, fly ash, mineral power and natural zeolite power is ≥ 250 kg/m3, it is completely feasible to inhibit the alkali aggregates reaction (can effectively inhibit the expansion of AAR)

8. Crack resistance (fracture energy)

MPC in production and construction process, due to low heat of hydration, autogenous shrinkage and early shrinkage, reduce the early shrinkage cracking, for hardening concrete fracture toughness can increase 30%.

According to the test, the characteristic length of ordinary C80 concrete is 40.9 cm, while C80 MPC is 59.5 cm.

16.2.3   Simulated shear wall structure test

Shear wall simulation test is also to test the various functions of MPC applied in the structure. Observe the cracks on the surface when demoulding. Concrete according to the above C80 MPC ratio. First, 10 cm thick reinforced concrete wall panel are poured, and nails are arranged according to the actual structure. Every 20 cm, a steel bar is pierced into the newly poured 35 cm concrete wall. The constraint of steel plate shear wall on the concrete was simulated.

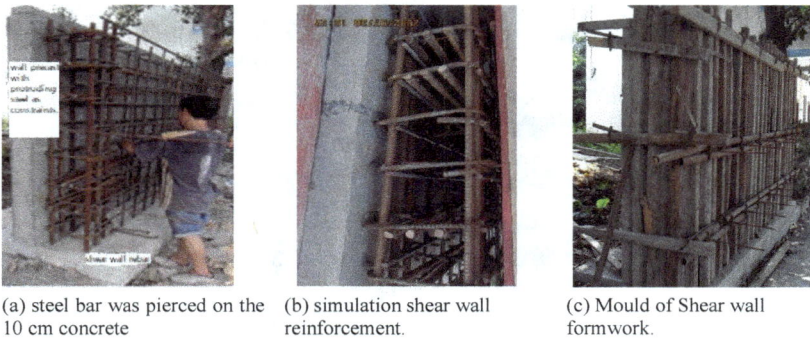

(a) steel bar was pierced on the 10 cm concrete

(b) simulation shear wall reinforcement.

(c) Mould of Shear wall formwork.

Figure 16.6 Simulation shear wall test mould assembly.

1. Simulation test of MPC composition, structure simulation test. C80 MPC composition as shown in Table 16.9.

Table 16.9. C80 MPC composition (kg/m3)

NO.	Cement	MB	NZ	FA	EHS	FA	CA	HRWR	CFA	Water
6	320	60	15	170	8.8	800	900	1.9%	1.5%	142

2. Properties of Fresh mixed MPC

The ratio of concrete is the same as smaller simulation test. But it was batched in the concrete production mixer. The properties of fresh MPC as follow. Shown in Figure 16.7.

(a) Slump ≥ 26 cm

(b) flow spread 680x700 mm

(c) Height of U-Box ≥32 cm

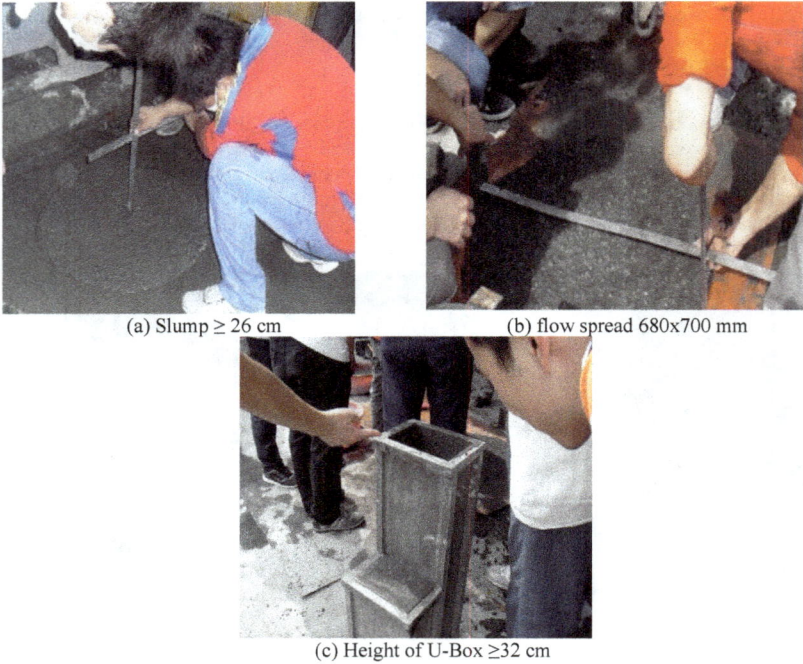

Figure 16.7 structural test performance of fresh mixed concrete.

Time taken to discharge concrete from slump cone is 4 seconds. Temperature of MPC when placing is 31 ℃.

3. Casting of MPC. The mixed concrete is sent to the test site by concrete mixer truck and pumped to the mould. As shown in Figure 16.8, after pouring into the mould, the concrete is formed by shock free self compacting and then demoulded after 2 days. The structural surface is shown in Figure 16.9.

4. Compressive strength of MPC, MPC was tested in the difference age and using different curing method. As shown in Table 16.10

Table 16.10 Compressive strength of simulation structure (MPa).

Age	7 d	14 d	28 d	60 d	90 d
Wet curing	67.3	82.2	96.1	101.1	101.2
Self curing	77.8	84.4	100.3	103.9	104.0
Standard curing	73.8	83.2	90	98.1	

(a) C80MPC from mixer truck to pump car

(b) MPC been pumped to the mould with shock free self compacting

Figure 16.8 MPC simulation structure and test site.

Figure 16.9 Demould 2 days after MPC casted.

After casting and demould 2 days later, no cracks were found on the surface of the test component.

16.2.4 Structural test

Through the simulation test, we have a comprehensive understanding of the performance of C80 MPC and meet the relevant technical requirements; The construction performance and corresponding technical effect of C80 MPC are further verified by shear wall structure simulation test. On the basics of these researches and tests, the shear wall solid structure test are further carried out. Combined with the shear wall of Guangzhou East Tower Project. The C80 MPC concrete test of real large structure is carried out.

1. Large structure shear wall: Wall thickness of 160 cm, inner thickness of 80 cm stell plate shear wall; height of 3.5 m and 4.0 m long. The shear wall is separated from the centre line. One part of which is poured with C80 MPC, and the other part is pour with C80 concrete which commonly used in the market. As comparison shown in Figure 16.10.

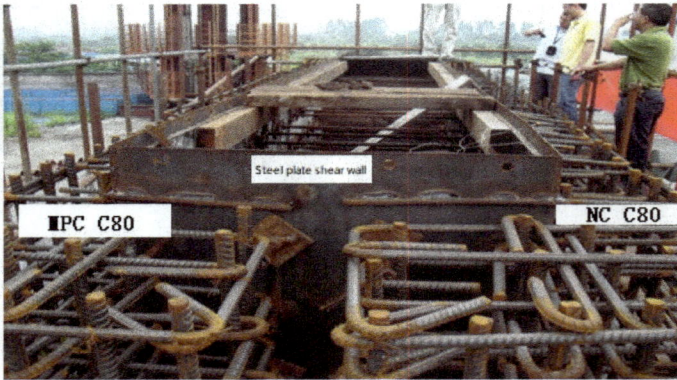

Figure 16.10 Large Structure shear wall reinforcement layout and internal steel shear wall layout.

2. C80 MPC vs conventional C80 concrete composition in Table 16.11.

Table 16.11 Concrete Composition.

Type	Binder	EHS	Water	FA	CA	HRWR	CFA
MPC	320+60+170+15	8.8	142	800	900	1.4%	1.5%
NC	320+60+190	--	142	700	1000	1.4%	1.5%

Note: in MPC mix, cement 320, MB 60, FA 170, NZ 15.
    In NC mix,   cement 320, MB 60, FA 190,

3. Fresh properties of concrete. Properties as shown in Table 16.12

Table 16.12. Fresh properties of concrete.

Type	Slump mm	Flow Spread mm	Discharge time from Cone sec	U box cm
MPC	230	660x660	10	32
NC	22.5	595x590	17	--

4.   Autogenous shrinkage and early shrinkage. As shown in Table 16.13.

Table 16.13 Autogenous shrinkage and early shrinkage.

Type/Age	24 h	48 h	72 h
MPC	0.062‰	0.081‰	0.094‰
NC	0.142‰	0.17‰	0.19‰

5. Heat of hydration. Temperature for the heat of hydration shown in Table 16.14

Table 16.14 heat of hydration in concrete.

Type	Initial age for Heat of hydration	Initial temperature to the peak temperature	Time to reach peak temperature
MPC	23.5 (h)	26~75 °C	11 (h)
NC	20 (h)	28-77 °C	35(h)

The casting temperature of two group of low-heat concrete is 26-28 °C. Maximum temperature is 75-77 °C. It also takes longer time for the temperature raise.

6. Compressive strength. As shown in Table 16.15.

Table 16.15 Compressive strength of the concrete (MPa).

Type/Age	3 d	7 d	28 d	56 d
MPC	58.9	68.1	88.9	92.5
NC	62.2	73.8	86.6	90.4

7. Casting of concrete. Casting structure was shown in Table 16.10 above.

Half of the wall is poured with C80 MPC, the other half is poured with conventional C80 concrete (NC). The steel plate shear wall in the concrete of the structure is filled with sand, so that the steel plate shear wall will not be deformed by casting of both sides. And it is also convenient for construction operation. Figure 16.11 C80 NC been pumped and compacted. As shown in Figure 16.12.

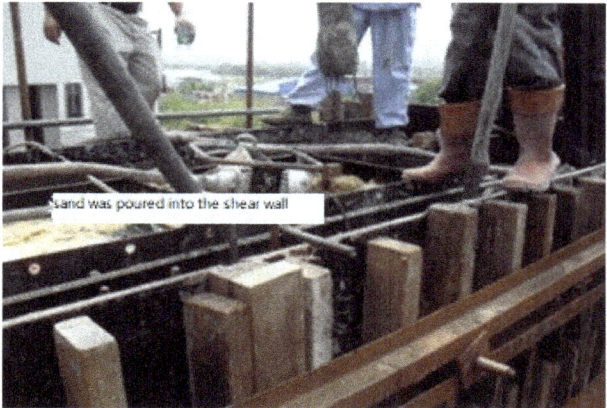

Figure 16.11 Sand is filled in the steel plate shear wall.

Materials Research Forum LLC
https://doi.org/10.21741/9781644901991

C80 MPC concrete is directly pumped into the mould and left to self-compacting. The concrete structural was demoulded the next day and, the C80 NC concrete wall was found to have micro cracks. The C80 MPC concrete wall has no cracks, only some pot hole on the bottom slab. This may be due to the high temperature in the summer and the long distance transportation of concrete, which thickens the concrete and makes it difficult to flow. As shown in Figure 16.13.

Figure 16.12 NC C80concrete is been pumped and compacted.

(a) C80 MPC Surface after demould   (b) C80 NC Surface after demould

Figure 16.13 surface difference between C80 MPC and C80 NC.

8. Experiment conclusion.

1)   Multi Purpose Concrete (MPC) can be prepared easily by controlling the quantity and quality of cementitious material, organic and inorganic additives, and reasonable mix ratio of concrete.

2) MPC contains water carrier natural Zeolite Powder (NZ); the current use of moisture carriers are: organic, such as SAP resin used in concrete; inorganic , such as pottery power; in this study, NZ powder was used to absorb water, release water and enhance strength. In MPC, it has various functions of self-curing, thickening and strengthening.

3) In this study, a small amount of sulfoaluminate expansion agent is used to compensate the early shrinkage and autogenous shrinkage of concrete in the early stage (age 12-72 hours). At the same time, the MPC with low shrinkage can be obtained because the self-curing further reduces the autogenous shrinkage.

4) by reducing the amount of cement, the use of low calorific value of admixtures, MB and fly ash; The peak curing temperature rise of concrete can be reduced to ≤ 78 °C.

5) By reducing the water to binder ratio, filling of cementitious material particles, and self-curing measure, the MPC can obtain high strength.

6) in the process of research and development of this topic, developed a compound efficient water reducing agent, thickening agent, self-curing agent and retardation agent; it is the key material to complete the performance of this topic.

Chapter 17

Research and Application of C120 MPC

In 2009, under the organization and guidance of Professor Feng Naiqian from Tsinghua University, China Construction Fourth Engineering Group launched the development and application of C120 MPC (Super High Strength Multifunctional Concrete) in conjunction with the Jingji Building Project. Shenzhen Municipal Bureau of Housing and Urban-rural Development has invested hundreds of thousands to support the research of the project. Shenzhen Zhengqiang Investment Co., Ltd., Jinzhong Concrete Company, Guangzhou Tianyu Concrete Company, and China State Construction Concrete Company, etc., also invested in this research. Before the topic was opened, Chinese and Japanese experts held a seminar on the subject. After three years of arduous test and research, the project research team used domestic existing materials, equipment, and production facility to formulate and produce C120 strength grade MPC and pump it to a structure with a height of 146 meters. Every step from production, transportation to the site, and pumping to the super-high structure is very smooth, paving a new way for the development and application of UHPC in China.

17.1 Preparation technology of C120 MPC

Raw materials for testing: Microbeads are ultra-fine spherical fly ash with an average particle size of 1.0-1.2 μm. Coarse aggregate is crushed andesite stone with particle size ≤16 mm, and admixture is self-developed and produced sulfamic acid-based high-efficiency water-reducing agent, combined with naphthalene-based high-efficiency water-reducing agent. This admixture with a solid content of 35% provides a high water reduction rate and a good function of controlling slump loss. The mix design of MPC is as shown in Table-17.1.

Table 17.1 The mix design of concrete (kg/m^3).

Water/binder	Cement	Silica fume	Mineral additive	Fine aggregate	Coarse aggregate		admixture
	PII 52.5	Elken	Microbeads	Washed sea sand	Small	Big	Depending on workability
0.15-0.18	500	75	175	700	300	700	About 3.0%

According to the mix design of the above-mentioned constituent materials, 50 liters of concrete were mixed to prepare 13 groups of 100 mm × 100 mm × 100 mm cube specimens, and the cubes were cured under standard conditions for 28 days. The compressive strength test was carried out and the standard deviation was calculated, as shown in Table 17.2. At the same time, part of the test specimens was sent to the Shenzhen Municipal Bureau of Housing and Urban-rural Development for testing. The results are shown in Table 17.3.

Table 17.2 The testing results of compressive strength.

Results (MPa)	136	136	137	136.33	132	139	143	138.00	
	140	134	144	139.33	133	138	139	136.67	
	131	135	145	137.00	133	135	131	133.00	
	130	123	142	131.67	137	128	133	132.67	
	135	118	134	129.00	137	136	139	137.33	
	131	125	137	131.00	139	139	136	138.00	
	114	137	117	117.00					
Statistic	Average (MPa)		135.62		Standard deviation (MPa)	5.93			

Table 17.3 The test results of the Shenzhen Quality Inspection Center.

Strength grade	C120	C120	C120
Curing	Standard	Standard	Standard
Cube size	100 (mm)	100 (mm)	100 (mm)
Compressive Strength (MPa)	123.4	122.9	126.4
	125.1	130.1	126.5
	129.0	118.9	124.6
Average strength	125.8 (MPa)	124.0 (MPa)	126 (MPa)
Relative (%)	105	103	105

As the current code does not specify the size effect coefficient of the small concrete specimens for C120 strength grade, the testing center still uses a coefficient of 0.95 for calculation.

From the strength test results in Table 17.2, the average concrete strength reached 135 MPa, the mean square error was 5.93 MPa, and the strength fluctuation was small.

17.2 Test of plastic retention performance for C120 MPC

C120 MPC (Super High Strength Multifunctional Concrete) is a research and development project combined with the Jingji Building project, but the experimental research and development site is at Baoan Zhengqiang Concrete Mixing Plant. It takes about 1.5 hours to transport the concrete from the production site to the construction site, and sometimes it takes longer in traffic jams. Therefore, the C120 concrete developed needs slump retention for up to 3 hours to ensure the pumping construction requirements.

Whether concrete can maintain slump retention performance, the key is the function of the superplasticizer. We have developed a superplasticizer that uses Lignosulfonate grafting technology and is copolymerized with sulfamate. Another high-efficiency naphthalene-based superplasticizer from BASF was also used. In our experimental research, the sulfamate superplasticizer and naphthalene-based water-reducing agent are compounded in a certain proportion, and then a certain amount of zeolite ultrafine powder is mixed to obtain a three-component, solid and liquid composite superplasticizer (amino liquid + naphthalene liquid + NZ superfine powder). This water reducer has good dispersibility to cement and good plastic retention performance! The concrete composition of the experimental study is shown in Table 17.4.

Table 17.4 Mix design of concrete.

No	W/B	Cement	Microbead	Silica fume	Washed sea sand	Sone	Water	superplasticizer
801	0.18	500	170	80	700	1000	130	4.0%

40 liters of concrete was mixed to measure the slump, slump flow, and flow time of the concrete over time under indoor and outdoor sunlight, as shown in Table 17.5.

Table 17.5 Comparison of workability of fresh concrete under indoor and outdoor in summer.

Indoor	Slump (mm)	Slump flow (mm)	Flow time (sec)	Outdoor	Slump (mm)	Slump flow (mm)	Flow time (sec)
Initial	255	590*600	4	initial	255	590*600	4
1 h	250	570*545	4.5	1 h	240	580*570	5
2 h	240	580*570	6	2 h	245	550*540	6
3 h	215	480*450	11.5	3 h	240	480*490	6.5

Under outdoor conditions (30 °C), the initial slump was 255 mm, and it still maintained 240 mm after 3 h. The initial flow time was 4 seconds, and it could still be maintained for 6.5 seconds after 3 h. It shows that the plasticity is excellent, and it can fully guarantee long-distance transportation and ultra-high pumping requirements.

The blending of microbeads and silica fume was used in the composition material to make the cementitious material get a dense filling, reducing the water consumption of unilateral concrete. This will reduce the water-binder ratio, and make the concrete obtain high strength. But the key technology for flowability slump retention and dispersion of concrete is water reducing agent. This is the function of this special water reducing agent which has three components composed of solid and liquid phases. The details will be discussed later.

17.3 Testing of autogenous shrinkage, early shrinkage, and long-term shrinkage of concrete

17.3.1 Autogenous shrinkage

Under the condition that there is no medium exchange between the concrete and the environment, due to the hydration of the cement, the water in the capillary is absorbed, so that the capillary is in a self-vacuum state, resulting in capillary tension. The shrinkage of the concrete at this time is called auto-shrinkage. The testing of autogenous shrinkage of C120 concrete is shown in Figure 17.1. The concrete mix design is shown in Table 17.6.

Figure 17.1 Autogenous shrinkage test.

The two-end contact measurement method is adopted. After the test specimens are cast, the specimens are covered with a plastic film and sealed. After the initial setting, the two long side mold plates and two end mold plates were removed. The test specimen can slide in the remaining mold. The autogenous shrinkage test can be started. The results are shown in Table 17.7.

Table 17.6 Mix design for autogenous tests (composite admixture, PII 52.5 OPC).

NO	Type	W/B	C	SF+FMB	S	G(small)	G2(big)	AG
2-1	reference	0.18	500	250	700	300	700	4.0%
2-2	1 kg/m³ Grace fibre	0.18	500	250	700	300	700	4.0%
2-3	2 kg/m³ Grace fibre	0.18	500	250	700	300	700	6.0%
2-4	10% Natural anhydrous gypsum	0.18	450	250	700	300	700	4.0%
2-5	10% Sulphoaluminate cemen	0.18	450	250	700	300	700	4.0%
2-6	10% Ground sulphoaluminate cement	0.18	450	250	700	300	700	4.0%

The water consumption of concrete is 130 kg/m3, and AG is a three-component composite of naphthalene series, sulfamate and NZ powder;

For No. 2-4, there is 50 kg anhydrous gypsum, and for 2-5, 2-6 50 kg sulphoaluminate cement powder was not included in the water content.

It can be seen that the autogenous shrinkage of the reference concrete is the largest, and the autogenous shrinkage of the concrete has been significantly reduced after the use of Grace fiber. The sample with the natural anhydrous gypsum powder content has a self-shrinkage of 99×10^{-6} mm/m in 3 days, which is less than 1/10,000 and also less than 1/3 of reference concrete. The autogenous shrinkage of concrete mixed with 1 kg/m³. Grace fiber is also low, which is about 1/2 of the autogenous shrinkage of the reference concrete.

Table 17.7 The autogenous shrinkage of C120 MPC.

No	Autogenous shrinkage at 3 days（$\times 10^{-6}$ mm/m）
2-1	329.9
2-2	181.4
2-3	253.6
2-4	99.0
2-5	/

17.3.2 Early shrinkage

After the concrete is cast and formed, the test starts after 2 hours. The upper surface of the specimen is opened to the air, which means the moisture in the concrete can be discharged into the air. Therefore, early shrinkage also includes a part of autogenous shrinkage. As shown in Figure 17.2. The test results are shown in Table 17.8.

The test results of early shrinkage show that the reference concrete, concrete containing 1kg/m^3. Grace fiber, concrete containing 2 kg/m^3. Grace fiber, and concrete containing natural anhydrous gypsum powder all have the same level of early shrinkage in 3 days. And the early shrinkage is greater than the autogenous shrinkage. After mixing with 10% sulphoaluminate cement, the concrete not only does not shrink but slightly expands in the early stage.

Figure 17.2   Early shrinkage test.

Table 17.8 Early shrinkage of C120 MPC.

NO	3 days early shrinkage（$\times 10^{-6}$ mm/m）
2-1	340.2
2-2	338.1
2-3	482.5
2-4	303.1
2-5	expansion 296.9
2-6	Expansion 59.8

17.3.3 Long-term shrinkage

The shrinkage caused by drying when concrete is exposed to the air for a long time, of course, also includes autogenous shrinkage. This study measured the long-term shrinkage of two types of specimens: (1) After the early shrinkage measurement, the long-term shrinkage test was carried out: 1-1, 1-2, 2-1, 2-2, 2-3, 2-4. The results are shown in Table 17.9 and Figure 17.3. (2) After the auto shrinkage test, perform the long-term shrinkage test: 1-3, 1-4, 1-5, 1-6. The results are shown in Table 17.10 and Figure 17.4.

It can be seen from Table 17.8, Table 17.9, Figure 17.3, Figure 17.4 that no matter whether the test specimens was used after the auto-shrink test, or after the early shrink test for the measurement of the long-term shrinkage, there was little change. In general, most of the specimens had a stable shrinkage after 14 days of age.

Table 17.9 Long-term shrinkage after early shrinkage measurement.

NO	3d	7d	14d	28d	45d	60d
1-1	746.4	746.4	800.0	861.9	888.	896.9
1-2	558.8	604.1	670.1	709.3	655.7	721.6
2-1	340.2	441.2	562.9	630.9	614.4	637.1
2-2	338.1	406.2	492.8	554.6	552.6	567.0
2-3	482.5	562.9	703.1	812.4	866.0	-
2-4	303.1	422.7	443.3	476.3	459.8	476.3

Table 17.10 Long-term shrinkage after autogenous shrinkage measurement.

No	3d	7d	14d	28d	45d	60d
1-3	142.3	237.1	317.5	364.9	356.7	389.7
1-4	327.8	367.0	461.9	449.5	459.8	457.7
1-5	263.9	356.7	470.1	478.4	445.4	389.7
1-6	185.6	233.0	299.0	311.3	321.6	313.4

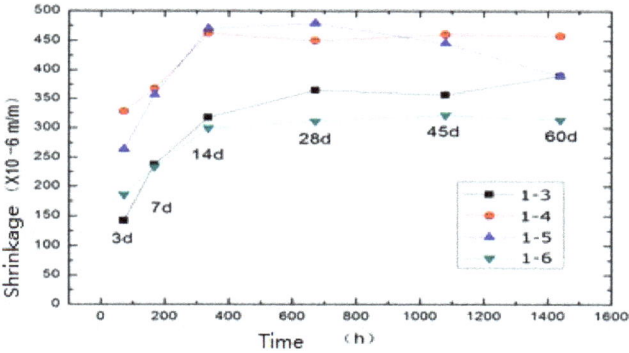

Figure 17.3 Shrinkage at 60 days.

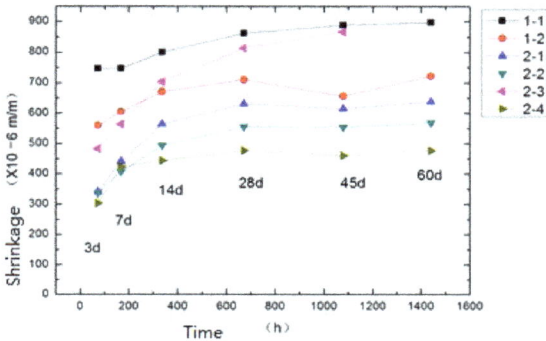

Figure 17.4 Shrinkage at 60 days after early shrinkage measurement.

17.3.4   The shrinkage of the specimens with sulfoaluminate expansion agent

It is different from other groups of specimens, shrinkage and expansion are listed in Table 17.11. The effect of sulfoaluminate expansion agent with different fineness is shown in Figure 17. 5.

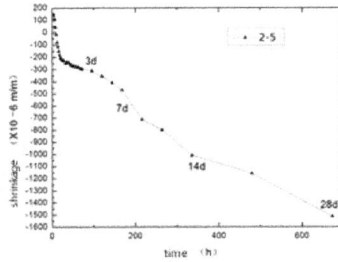

(1) Coarse original sulfoaluminate expansion agent

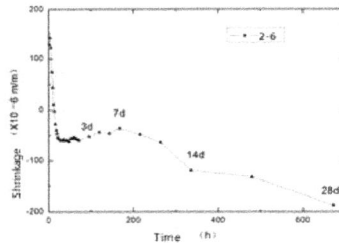

(2) Ground sulfoaluminate expansion agent

Figure 17.5 Effect of sulphoaluminate expansion agent with different fineness.

Table 17.11 Long-term shrinkage of ground and original expansion agent (specimen number 2-5, 2-6).

Time	2-5	2-6	Time	2-5	2-6
2 h	142.3	127.8	72 h	-296.9	-59.8
4 h	146.4	142.3	4 d	-311.3	-51.5
6 h	113.4	121.6	5 d	-354.6	-43.3
8 h	43.3	74.2	6 d	-408.2	-45.4
10 h	-16.5	43.3	7 d	-468.0	-35.1
12 h	-80.4	10.3	9 d	-711.3	-47.4
14 h	-113.4	-2.1	11 d	-800.0	-63.9
16 h	-150.5	-26.8	14 d	-1010.3	-119.6
18 h	-181.4	-39.2	20 d	-1156.7	-132.0
20 h	-202.1	-45.4	28 d	-1513.4	-187.6

From Table 17.11 and Figure 17.5, it can be found that the specimens mixed with sulphoaluminate expansion agent are still in an expansion process in 28 days, but the expansion value of the ground expansion agent is much lower than the original expanding agent (coarse).

17.3.5 Conclusions

1) C120 MPC super-high-strength concrete with a water-binder ratio of $0.18 \sim 0.2$ has a reduced autogenous shrinkage of $(190 \sim 330 \times 10^{-6} mm/m)$ and does not crack. This is due to the use of a three-component water-reducing agent, which contains natural zeolite powder saturated with the water-reducing agent, which can release high-efficiency water-reducing agent and make cement particles better dispersed.

2) Incorporating polypropylene fiber can obviously inhibit the shrinkage of concrete. This test showed that the autogenous shrinkage value of MPC ultra-high-strength concrete can be reduced by about 40%.

3) Natural anhydrous gypsum mixed into MPC ultra-high-strength concrete can control its autogenous shrinkage in a lower range, which is beneficial to the combination of reinforced concrete components. But the concrete strength is affected.

4) In MPC ultra-high-strength concrete, adding a certain fineness of sulphoaluminate expansive cement can play a micro-expansion effect and control the early expansion value of MPC ultra-high-strength concrete to a very low level.

5) In the test, most of the specimens gradually stabilized in shrinkage deformation under the age of 14 to 28 days. Compared with the early shrinkage and autogenous shrinkage, the shrinkage deformation in the later period was smaller.

17.4 Toughness of C120 MPC Ultra High Strength Concrete

It is generally believed that as the strength of concrete increases, the toughness decreases, and the brittleness increases. For this project only the toughness of C120 MPC has been tested but the further study also was performed on the methods to improve the toughness. The toughness of MPC ultra-high-strength concrete was improved by adding polypropylene fiber. The concrete mix design is shown in Table 17.12. By blending polypropylene fiber, the change in fracture toughness was measured. The fiber blending amounts were: Grace fiber, 1 kg/m3; Grace fiber, 2 kg/m3; Shenzhen short fiber, 2 kg/m3.

Table 17.12 Mix design for toughness test.

No	W/B	C	SF+MB	S	G1	G2	Polycarboxylic admixture	fiber
1#	0.18	500	250	750	285	665	2.5%	Reference, no fiber
2#	0.18	500	250	750	285	665	2.5%	Grace fiber，1 kg/m3
3#	0.18	500	250	750	285	665	2.5%	Grace fiber，2 kg/m3
4#	0.18	500	250	750	285	665	2.5%	Shenzhen Short fiber，2 kg/m3

Three 100 mm × 100 mm × 400 mm prism specimens with crack notch were prepared for each group of concrete. The fracture toughness test was performed to determine the flexural strength and crack opening displacement at a certain loading rate or constant displacement rate.

Figure 17.7 Specimen with crack notch for measuring crack opening displacement.

The determination of fracture parameters is shown in Figure 17.8, which is tested by the Institute of Building Materials, Tsinghua University. The results are shown in Table 17.13. From the results in Table 17.13 it can be seen:

With the increase of fiber content, the fracture energy of concrete increases. At the same dosage, the type of fiber will cause the fracture energy of concrete to be different. In the test, the order of the fracture energy is: reference specimen <1 kg Grace fiber specimen <2 kg Shenzhen fiber specimen <2 kg Grace fiber Specimen. Due to the addition of fiber, fracture energy is greatly improved. And the brittleness parameter is significantly increased, indicating that the toughness of ultra-high-performance concrete has been greatly improved.

Figure 17.8 Determination of fracture parameters of C120 MPC.

Table 17.13 Summary of fracture parameters of concrete.

Mix	No	E (GPa)	P_{fc} (N)	P_{max} (N)	σ_c (MPa)	σ_{fc} (MPa)	σ_t (MPa)	σ_f (MPa)	G_f (J/m²)	l_{ch} (cm)
Reference concrete	1	51.9	5000	14000	127.0	6.82	6.86	9.07	213.8	23.5
	2		4600	14500	121.0	6.27	-	9.39	-	-
	3		4700	13800	124.0	6.41	-	9.59	-	-
	Ave		4767	14100	124.0	6.50	6.86	9.35	213.8	23.5
1 kg Grace fiber	1	52.2	4800	13800	129.0	6.55	6.67	8.94	291.8	33.8
	2		5000	15500	126.0	6.82	7.08	10.04	302.9	27.9
	3		5000	15300	119.2	6.82	-	9.91	-	-
	Ave		4933	14867	124.7	6.73	6.88	9.63	297.35	30.9
2 kg Grace fiber	1	52.2	4800	12000	140.8	6.55	6.56	7.78	322.8	39.2
	2		5000	15000	123.6	6.82	-	9.73	-	-
	3		5200	14000	141.4	7.09	7.22	9.08	322.1	33.4
	Ave		5000	13667	135.3	6.82	6.89	8.86	322.5	36.3
2 Kg Shenzhen fiber	1	52.0	5100	14000	120.0	6.95	6.97	9.07	300.3	30.0
	2		5000	14000	115.0	6.82	6.84	9.07	321.8	30.7
	3		5200	14000	-	7.09	-	9.07	-	-
	Ave		5100	14000	118.0	6.95	6.91	9.07	311.1	30.4

(In the table, P_{fc} is the cracking load, P_{max} is the maximum load, E is the modulus of elasticity, σ_c is the compressive strength, σ_{fc} cracking strength, σ_t is the tensile strength, σ_f is the flexural strength, G_f is the fracture energy; l_{ch} is the brittleness parameter, The smaller the l_{ch}, the greater the brittleness of the concrete.)

The l_{ch} of the reference concrete is 23.5, and the l_{ch} of the specimen of 1 kg Grace fiber is 30.9. The l_{ch} of concrete with the 2 kg Grace fiber specimens is 36.3, and the l_{ch} of the concrete specimens of 2 kg Shenzhen fiber is 30.4. It can be seen that the addition of organic fibers has a very significant effect on improving the toughness of concrete, especially Grace fiber has the best effect.

17.5 Fire resistance of C120 MPC

As the strength of the concrete increases, the density increases. Under the action of high temperature, the water inside the concrete evaporates, and the steam pressure is larger, but the water vapor is not easy to be discharged outside the concrete. As the heating temperature increases, the steam pressure further increases, when the steam pressure exceeds the tensile strength of the concrete, the concrete will crack, causing the structure to lose its load-bearing capacity.

The fire resistance test adopts the method of in-situ measurement. That means loading, heating, and test observation are performed at the same time. This is also one of the innovations of this research. The test concrete samples were composed of the reference concrete, concrete with kg/m³ polypropylene fiber, and concrete with 2 kg/m3 polypropylene fiber. Three types of specimens were used: prism 100 mm × 100 mm × 300 mm, cylinder Φ100×150 mm, Φ100×300 mm. As shown in Figure 17.9. There are two heating jackets for heating the specimen, one is prism, the other is cylinder, as shown in Figure 17.10.

| (1) 100x100x30 0mm | (2) φ100x150 mm | (3) φ100x300 mm |

Figure 17.9 Test specimens for fire resistance.

In-situ measurement: (1) Loading: a constant load of 30% of the ultimate strength of concrete was maintained on the specimen. (2) Heating: the changes of the test specimen at three temperatures of 200 °C, 300 °C, 400 °C were observed. The heating was done through electric wiring. (3) Constant temperature: observe the condition of the test specimen at the specified temperature for 30-40 minutes as shown in Figure 17.11. The specimens with different fiber dosages are shown in Figure 17.12. The reinforced concrete specimens are shown in Figure 17.13.

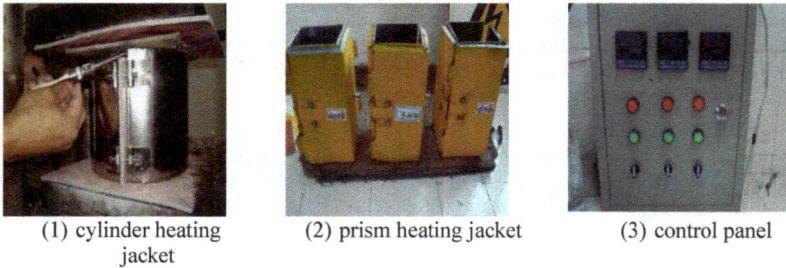

| (1) cylinder heating jacket | (2) prism heating jacket | (3) control panel |

Figure 17.10 Heating jacket and control unit.

(1) reference concrete @200°C
Constant temperature and
constant load

(2) reference concrete @300°C
Constant temperature and
constant load

(3) reference concrete @300°C
Constant temperature and constant
load

(4) reference concrete @400°C
Constant temperature and constant
load

Figure 17.11 Reference concrete under constant loading, different temperature.

(1) 200°C   (2) 400°C   (3) 400°C   (4) 400~
                                                    500°C

(a) The change the specimen mixed with 1 kg/m³ polypropylene fiber
under constant load and different temperature

(1) 200°C   (2) 400°C   (3) 400°C   (4) 400~500°C

(b) The change the specimen mixed with 2 kg/m3 polypropylene fiber
under constant load and different temperature

Figure 17.12 Fire resistance of concrete with different dosage of fiber.

(1) 300°C constant temperature and load
Steel concrete, φ100x150mm

(2) 300°C constant temperature and t load
Steel concrete, φ100x300mm

Figure 17.13 Changes of reinforced concrete at different temperatures under constant load.

Summary on improving the fire resistance of ultra-high strength MPC:

1) The higher the heating temperature of the ultra-high-strength MPC, the more bursts and the more serious it will be. The width of the cracks on the surface of the specimen increases, and the number of cracks increases, and eventually an explosion occurs, causing the specimen to peel off in a large area, and it cannot continue to bear the load.

2) Under the same conditions, the damage of the steel-reinforced high-strength high-performance concrete specimens is more serious than reference concrete specimens. This is mainly caused by the thermal expansion of the steel bar itself. Because the reference concrete specimen is only affected by the stress generated by its own thermal expansion under high temperature conditions, the phenomenon of burst occurs. While the reinforced concrete specimen is not only affected by its own thermal expansion but also bears the stress generated by the thermal expansion of the steel bar. Therefore, the Bursts and damage are more frequent and more serious.

3) Incorporating a certain amount of polypropylene fiber can significantly improve the fire resistance of concrete. In the test specimens mixed with polypropylene fiber, during the entire heating and constant temperature process, the polypropylene fiber will gradually melt after being heated to a certain temperature, resulting in an increase in the number of capillary pores in the concrete, and providing the channels of moisture and steam escape in the test specimens. The channel also provides a certain space for the expansion stress and heat conduction of the concrete itself, which reduces the possibility of concrete bursting, and finally ensures that the concrete specimens are still relatively complete after being subjected to high temperature and fire burning.

4) The vapor pressure mechanism means that the moisture in the high-performance concrete body evaporates into water vapor under high temperatures (fire). Water vapor creates vapor pressure inside the concrete, and when this vapor pressure reaches a certain value, it triggers a burst. The fibers mixed in the concrete will melt under high temperatures (fire) and become the drainage pipe of moisture and heat inside the concrete, reducing the occurrence of concrete bursting.

5) Under the action of 30% of the ultimate load of concrete, when the specimen is heated to 200 °C, 300 °C, 400 °C and 500 °C, and maintained for 30-40 min, the reference specimen has been cracked and damaged, and its bearing capacity is lost, and the building will collapse, causing personal casualties and property damage. But the specimens mixed with polypropylene fiber are still intact, no cracks are found. The people in the building still have time to escape. This is the most important function of fiber in ultra-high-strength ultra-high-performance concrete mixed with polypropylene. It is also the main purpose of improving fire resistance.

17.6 Conclusion of this chapter

1) The C120MPC in this study is made of raw materials used in the production of ordinary concrete such as OPC 52.5 cement, sub-nano microbeads and silica fume, crushed granite stone with a particle size of 5-20 mm, graded river sand or washed sea sand with fineness modulus 2.6-2.8, and combined admixture with naphthalene based, ammonia Sulfonate based superplasticizer and ultrafine zeolite powder.

2) Among the raw materials, there are two unique new materials: (1) Microbeads: emissions from coal-fired power plants. The average particle size of the dust recovered from the smoke is less than or equal to $1\mu m$, which is a kind of spherical glass beads with water-reducing effect, enhancement effect and durability effect. It is a unique industrial waste in China. (2) A three-component water reducing agent composed of naphthalene series, sulfamate series and ultrafine zeolite powder, with high water reduction rate, dispersibility to cement particles, good plasticity retention, and strengthening effect.

3) The average strength of C120 concrete can be 135 MPa, and the mean square error is 5.5 MPa, which shows that the studied concrete has good homogeneity, high strength guarantee rate.

4) The concrete has a slump of 240-250 mm, a slump flow of 600-680 mm, a flow time of 5-6.5 seconds, and slump retention for more than 3 hours. And there is no bleeding and retardation, indicating that this kind of concrete has excellent workability and pumpability.

5) Incorporating polypropylene fiber can improve the toughness of ultra-high-strength and ultra-high-performance concrete (Table 17.14)

Table 17.14 The comparison of the toughness of concrete with polypropylene fiber.

Specimen	G_f (fracture energy)	l_{ch} (brittleness index)
2-1(reference)	213.8	23.5
2-2(1 kg fiber)	297.35	30.9
2-3(2 kg fiber)	322.5	36.5

It can be seen that the addition of polypropylene fiber can effectively improve the toughness of super-strength MPC.

6) Improve the fire resistance of ultra-high-strength and ultra-high-performance concrete

Reinforced ultra-high-strength ultra-high-performance concrete is not fire-resistant. Under a constant load, the concrete will burst and lose its bearing capacity. But the test specimens mixed with 1-kg/m3 polypropylene fiber had no cracks that occurred under constant load and constant temperature for 30 mins under different temperatures. In addition to effectively inhibiting early shrinkage and auto-shrinkage, the blending of polypropylene fiber also improves the toughness of ultra-high-strength and ultra-high-performance concrete, and also effectively improves the fire resistance. Therefore, to make ultra-high-strength and ultra-high-performance concrete with special properties such as low shrinkage, high toughness, and fire resistance it is necessary to add polypropylene fiber and sulphoaluminate cement expansion agent, in addition to the preparation of commonly used materials.

7) Incorporating polypropylene fiber can reduce the autogenous shrinkage, early shrinkage and long-term shrinkage of ultra-high-strength MPC.

Chapter 18

Research and Performance Testing of 150 MPa Strength MPC

18.1 Selection of raw materials

Cement: Jinyang P2 52.5; Silica fume (SF) containing $SiO_2 \geq 95\%$; Microbeads (MB): Shenzhen Tongcheng product; Mineral powder (BFS): S95 mineral powder; Combined water reducing agent (NS): self-developed; coarse Aggregate (G), particle size 5-10 mm, impact crushed limestone stone; fine aggregate (S), river sand, medium to coarse; plastic preserving agent (CFA), self-developed; thickening agent (NZ) with self-curing function, (self-developed); Sulphoaluminate expansion agent (EHS-), supplied from Tianjin.

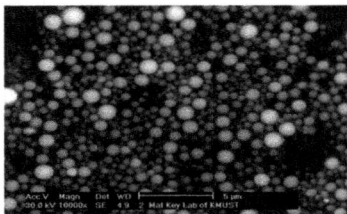

Figure 18.1 SEM of Microbeads.

18.2 Preparation and performance testing of MPC with strength of 150 MPa

The mix design of MPC is shown in Table 18.1, and the properties of fresh concrete are shown in Table 18.2. The strength of self-curing at different ages is shown in Table 18.3.

Table 18.1 Mix design of MPC with a strength of 150MPa (kg/m³).

NO.	Cement	Microbeads	SF	NZ powder		Sand	Stone	Fiber	Water	admixture	CFA
1	600	190	90	20		720	880	1-2	130	2.0%	2.0%
2	650	190	90	20		720	880	1-2	130	2.2%	2.0%

Table 18.2 Properties of fresh concrete.

NO.	Height in U Channel (cm)		Slump (cm)		Slump flow (mm)		Flow time (s)	
	Initial	2 h	Initial	2 h	Initial	2 h	Initial	2 h
1	31	31	27	27	740	680	7	8.5
2	31	31	27	28	680	740	6	7.7

W/B=15-16%; The U-Channel test of fresh concrete showed that the initial rise height is 31 cm, and after 2 hours, it could still rise by 31 cm, indicating that the plasticity retention and self-compactness of the concrete were very good.

Table 18.3 Concrete strength at different age under self-curing.

Age	3 d	7 d	28 d	56 d
NO1	125.3	136	147	151
NO2	127.2	136.5	148.5	152

18.3 Application of powder effect and carrier effect

In the mix design, the micro-filling effect of the three-component cementing material and the effects of microbeads and silica fume spherical particles are used to improve the fluidity of fresh concrete; as shown in Table 18.4. In Table 18.4, C+MB+SF (cement+microbeads+silica fume) three-component particles are filled with each other, the microporosity is the lowest among the three groups of powders, especially the microbeads with spherical glass bodies, under the same fluidity, requires lower water content. Therefore, among the three groups of cementitious materials, cement+microbeads+silica fume has high fluidity and low viscosity.

Table 18.4 The flowability of concrete with different powder combination

NO.	Binder	Slump (mm)	Slump flow (mm)	Flow time (s)
1	C+BFS	265	650x630	18
2	C+BFS+SF	275	660x670	9
3	C+MB+SF	285	680x690	5

In Figure 18.2, the electric double layer effect was created after cement particles absorbed the superplasticizer. The electrostatic repulsion was also produced by ultrafine powder filled in the pores of the cementitious material. This is like chopping wood with an axe, which plays the role of cleavage and separates the powder particles. The electric double layer effect of cement particles and the cleavage effect of powder particles are superimposed, the dispersion effect is more significant, and the flowability is improved. The natural zeolite ultrafine powder in this study was used as the carrier of water reducing agent. There were two kinds of water-reducing agents, solid and liquid as shown in Figure 18.3.

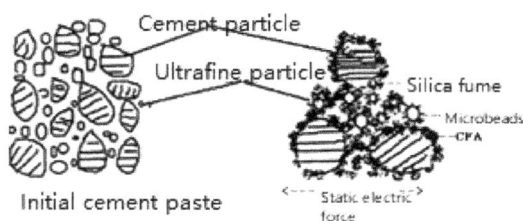

Figure 18.2 Binding material of cement, microbeads and silica fume, mixed with water reducing agent and CFA.

(a) Solid CFA

(b) Liquid CFA

Figure 18.3 Carrier superplasticizer.

The carrier fluidizing agent (CFA) slowly releases the adsorbed water-reducing agent, maintaining the adsorption capacity of the cement particles to the water-reducing agent. This will maintain the electric double layer effect on the surface unchanged so that the MPC can maintain plastic. The MPC can be easily pumped and self-compacting.

18.4 The choice of coarse aggregate

18.4.1 Choice based on the fracture stress of coarse aggregate

Compare the fracture stress of the two aggregates with particle size D1=10 mm and particle size D2=20 mm:

The fracture stress relationship is as follows: σ2=0.7σ1.

σ2-D=20 mm fracture stress; σ1-D=10 mm fracture stress.

That means using coarse aggregate with D=10 mm to prepare concrete, the skeleton effect of aggregate is better than that of aggregate with bigger particle size. For UHPC, as the strength increases, the maximum particle size D max of the aggregate becomes smaller, and super high Strength MPC should choose coarse aggregate with a particle size of 5-10 mm.

18.4.2 Choice based on testing results

In Figure 18.3, three types of coarse aggregates are shown. (1) Sanya crushed stone, with more needles and flakes, and larger particle size; (2) Haikou crushed stone, with a particle size of about 10 mm, and less needles and flakes; (3) Sanya crushed stone produced by impact crusher, with a particle size of about 10 mm and less needle and flakes. Concrete was prepared by using these three aggregates under the same conditions, the results are shown in Table 18.5.

Figure 18.3 Three types of aggregate.

Table 18.5 Comparison of strength and flowability of concrete with different aggregate.

NO.	Slump (cm)	Compressive Strength (MPa)			
		1 d	3 d	7 d	28 d
1	3.5	52.3	72.1	77.7	86.5
2	6.7	77	86.8	92.2	107
3	6.7	78	87.1	87.2	107

The test results show that the workability and strength of the concrete prepared with Sanya (1), are poor, and the concrete with crushed stone 2 or 3 has better performance.

Through theoretical analysis and experimental verification, with the increase of strength of MPC the particle size of coarse aggregate should be D≤10 mm. Products should be produced by impact crusher. Tests have proved that the amount of coarse aggregate in MPC is ≤400 L/m^3.

18.5 Inhibition of autogenous shrinkage and cracking of ultra-high strength MPC

According to the discussion in the relevant chapters above, the main technical ways to suppress the autogenous shrinkage cracking and early shrinkage cracking of UHPC are as follows: 1) Reduce the cement content in concrete, use low-heat cement or medium-heat cement with low C_3A and C_3S content; 2) Use natural Zeolite ultra-fine powder as self-curing agent, which evenly distributes the water required for cement hydration from the inside of the concrete. With a sulfoaluminate micro-expansion agent at 1% of cement to compensate for early shrinkage and self-shrinkage; 3) Incorporated in a small amount of Organic fiber (1-2 kg/m^3) can also effectively reduce the autogenous shrinkage cracking and early shrinkage cracking of UHPC. Table 18.6 shows an example.

Table 18.6, Mix design of concrete for inhibition of autogenous shrinkage and early cracking

C	MB	SF	NZ	W	S	G	EHX	WRA	CFA
600	190	90	20	135	720	880	1.5%	2.0%	2.0%

The concrete prepared according to Table 18.6 has a strength of ≥150 MPa at 56 d age, and the concrete has good flowability and plasticity retention effects. It was used in shear walls and other structures, and their ass no auto-shrinkage cracking and early shrinkage cracking.

18.6 Properties of 150MPC

18.6.1 Fresh performance and mechanical properties

According to the mix design in Table 18.6, multifunctional concrete was produced and tested. The properties of fresh concrete are shown in Table 18.7, and the mechanical properties of concrete are shown in Table 18.8.

Table 18.7 Properties of fresh concrete.

Slump (mm)	Slump flow (mm)	Flow time (s)	U Channel (cm)
275	760x720	2.72	330/30(s)

Table 18-8 Mechanical properties of concrete

Age	Testing item	Age	Testing item
28 d	Compressive strength 147	28 d	Tensile split 5.10
28 d	Flexural strength 9.9	28 d	E-Modulus 5.3x10 4
28 d	Cylinder strength 118.7	28 d	

18.6.2 Chloride diffusion and RCPT

Chloride ion diffusion coefficient: 0.96907×10^{-8} cm^2/s. Electric flux 23.338 Coulomb/6 h.

18.6.3 Microstructure

1) SEM analysis as shown in Figure 18.4.

(a) Interface between cement and aggregate x110

(b) CSH gel x18000

(c) Calcium hydroxide x 10000

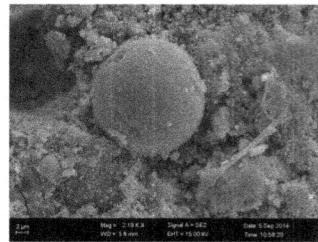

(d) Micro beads

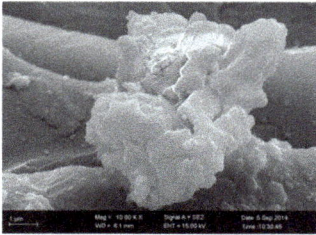

(e) CSH gel x10000   (f) Ca(OH)$_2$ and CSH gel x 8000

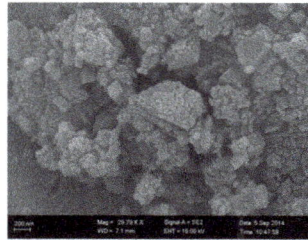

Figure 18.4 SEM of hydrates.

2) The XRD spectrum is shown in Figure 18.5. There are characteristic peaks such as calcium silicate hydrate and calcium hydroxide.

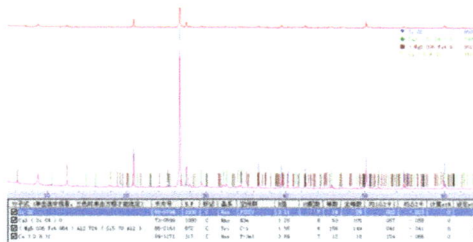

Figure 18.5 XRD spectrum of cement stone in Ultra-high strength MPC.

3) Measurement of pore structure. The measurement was done by a laboratory in Tsinghua University. The results are shown in Table 18.9.

Table 18.9 Pore structure of ultra-high strength MPC.

Specimen NO	Result (%)	Average (%)
1	3.82	
2	3.23	3.26
3	2.74	

The total porosity of ultra-high-strength MPC is only 3.26%, indicating that the density is very high. It has also high impermeability and durability.

In the United Kingdom, a cement stone with porosity as low as 2% has been developed, and its compressive strength has reached 665 MPa. As shown in Figure 18.6.

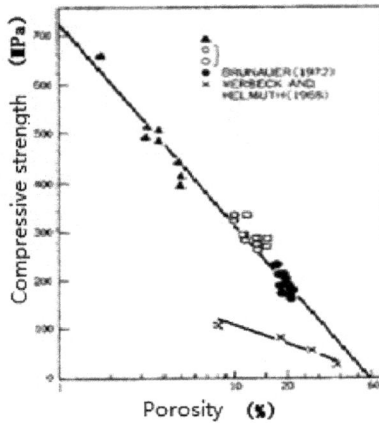

Figure 18.6 The relationship between cement stone strength and porosity (Teamura, Sakai).

18.7 Research and application of UHPC in USA

In 1988, at 225 West Wacker Drive, in Chicago, USA, 97.3 MPa HSC was used for the bottom column, which greatly reduced the size of the column.

Before 2008, the United States had produced and applied UHPC with a strength of 250 MPa (36000 Psi). It is also believed that when the compressive strength of UHPC exceeds 153 MPa (22000 Psi), it should be reinforced with fibres to ensure that brittle fracture does not occur. The amount of cementing material is high, and special aggregates are used. In the United States, traditional high-strength concrete (strength <139 Mpa) is not fiber-reinforced.

18.8 Research and application of UHPC in Japan

Ultra high-strength concrete with a strength of more than 150 Map as produced in a ready-mixed concrete factory. Based on experiments, it was used in super high-rise building structures. The cementing materials used in this concrete are ordinary Portland cement, slag, gypsum and silica fume. Tests have proved that the slump, slump flow value, air content and compressive strength are all in line with the technical requirements of ultra-high strength concrete. Production and construction applications were carried out. It was used in the building shown in Figure 18.7. The UHPC used in Japan is shown in Table 18.10, and the strength is shown in Figure 18.9. The figure shows the concrete strength of 3 d, 7 d, 28 d and 56 d. UHPC with W /B=0.15 56 d has achieved the strength of ≥150 MPa.

(a) The building with UHPC

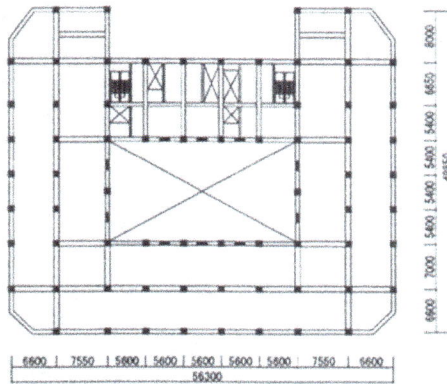

(b) Floor plan

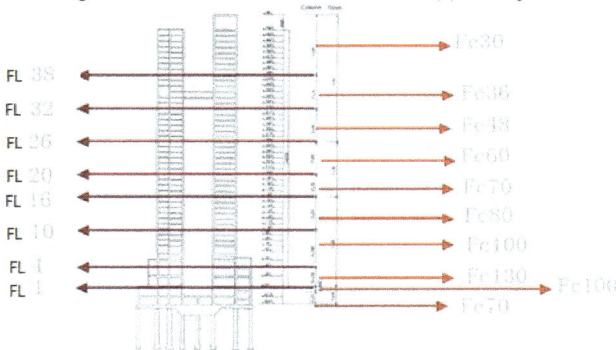

(c) Reinforced concrete columns at different structural positions, strength of HS/HPC

Figure 18.8 HS/HPC applied in Reinforced concrete columns in different parts of super high-rise buildings.

Table 18.10 The mix design and workability of UHPC in Japan.

W/B	Target		Usage per cubic meter (kg/m3)				
	Air content	Slump flow	Binder	Water	Fine agg	Coarse agg	admixture
0.15	1.0%	700 mm	1000	150	501	835	26.8
0.17	1.0%	700 mm	883	150	604	835	15.9
0.22	2.0%	600 mm	682	150	755	835	9.4

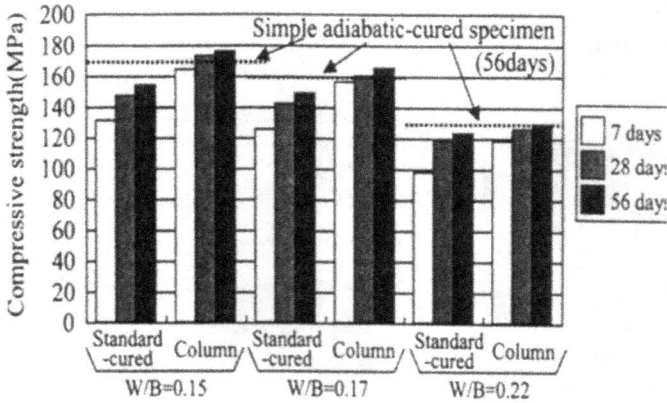

Figure 18.9 Compressive strength of UHPC at different age under different curing conditions.

18.9 The opportunity and risk of HPC and UHPC

FH Wittmann proposed the compression-strain curves of different concretes, as shown in Figure 18.10.

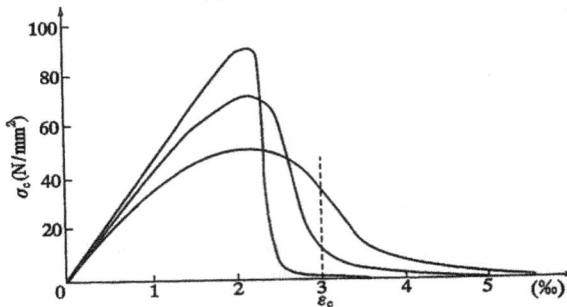

Figure 18.10 the compressive stress-strain curve of concrete.

It can be seen from the figure that concrete members are often in a state of strain rather than stress, such as strains due to humidity and temperature gradients. When the strain of ordinary concrete reaches 3‰, its bearing capacity can still maintain more than half. But if the same strain value is added to UHPC, the actual bearing capacity has approached zero. This means that we only observe crack formation in UHPC. Therefore, by adding fibers, the toughness loss of UHPC can be compensated. But in corrosive environments, steel fiber is not suitable.

Chapter 19

Development and Application of Papermaking White Clay for C30, C60 Strength Grade MPC

19.1 Introduction

White clay, also called papermaking white clay, is the waste slag treated by alkali recovery technology from papermaking black liquor in large and medium-sized pulp mills. One ton of lime can recover one ton of soda from papermaking black liquor, and the economic benefits are very significant. But the problem caused is that for every ton of caustic soda recovered from papermaking black liquor, one ton of waste residue-white clay is produced. Some companies in China use white clay to produce premixed mortar and have achieved good technical and economic effects. Nanning Sugar Co., Ltd. used vacuum washing machines for slag and pre-hanging vacuum filters to treat white clay, reducing the moisture content of papermaking white clay from 60% to 40%, and then dried in the factory to replace part of the limestone for the production of cement in the dry-process vertical kiln. It reduced environmental pollution and the cost of cement production. However, due to the high energy consumption of cement production in the vertical kiln, heavy pollution, and poor cement, the process of producing cement in the vertical kiln is abolished, and the corresponding white clay treatment method also stopped!

Guangzhou Tianda Concrete Company and Hainan Lubang Tiancheng Company used the original white clay from the paper mill to develop and produce C30 and C60 multifunctional concrete, and C30, C35 and C40 ordinary concrete, and achieved excellent technical and economic effects. It opened up a new way for the application of papermaking white clay!

19.2 Microscopic analysis and physical and chemical properties of papermaking white clay

19.2.1 The physical and chemical properties of white clay

The white clay is off-white slurry, with a pH value of 11, and the main component is $CaCO_3$. As shown in Table 19.1.

Table 19.1. The main chemical composition of white clay (mass%)

CaO	MgO	SO3	K2O	Na2O	LOI%
53.7	0.58	0.27	minor	1.25	41.5

The specific gravity of white clay is 2.35, the loose dry bulk density is 6 20 ~ 6 30 kg/m³, the natural bulk density is 14 0 0 ~ 1 500 kg/m³, the consistency is 6 ~ 8 cm, and the residue on 75 mesh sieve is 0. The residue on the 180 mesh sieve is 8.4%.

19.2.2 X-ray (XRD) analysis of white clay

1) Pure cement and white clay are mixed in a ratio of 7:3, and 30% water is added, and the XRD pattern of the sample is shown in Figure 19.1.

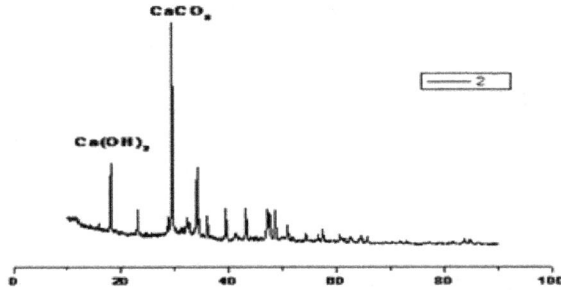

Figure 19.1 XRD pattern of cement + white clay.

It can be seen from the XRD graph that the peak value of $CaCO_3$ mixed with white clay is much higher than that of $Ca(OH)_2$.

2) The XRD pattern of the sample of pure cement + 30% water is shown in Figure 19.2.

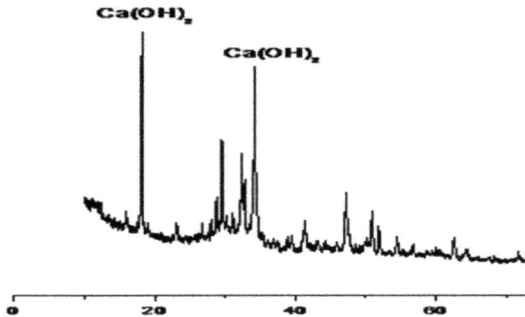

Figure 19.2 XRD pattern of cement specimen.

It can be seen from the XRD diagram that the XRD diagram of pure cement is all $Ca(OH)_2$ and $CaCO_3$ is basically not visible.

19.2.3 SEM analysis of specimens with and without white clay

1) SEM of the specimen with pure cement and 30% water

X5000

X8000

X10000

Figure 19.3 SEM of the specimen with pure cement and 30% water.

2) SEM of the specimen with cement + white clay=7:3 with 30% water

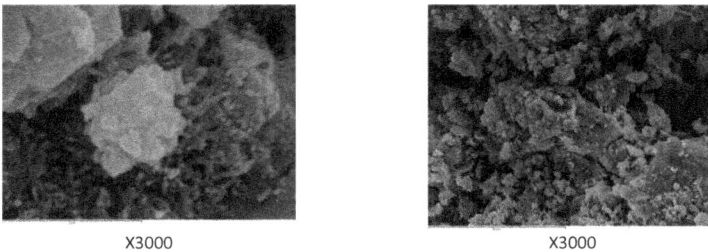

X3000

X3000

X8000

Figure 19.4 SEM of the specimen with cement + white clay=7:3 with 30% water.

The main morphology is cement gel and $CaCO_3$ crystals, as well as $Ca(OH)_2$ crystals. It can be seen from the above that the white clay-containing specimens mainly contain $CaCO_3$; while the specimens without white clay are all $Ca(OH)_2$. $CaCO_3$ is basically not visible. Therefore, the main component of white clay is $CaCO_3$.

19.2.4 Radioactivity detection of white clay

The white clay from the Yangpu Paper Mill was tested by the China Construction Fourth Engineering Institute for radioactivity. The results showed after a 6-hour radioactivity test on the white clay the indoor and outdoor radioactivity coefficients were all within the national regulations. Hence, the building materials products using white clay as raw material are not radioactive harmful.

19.3 Papermaking white clay C30 multifunctional concrete

19.3.1 Raw materials

The raw materials used for this study are shown in Table 19.2.

Table 19.2 Raw Materials.

Materials	Supplier	Materials	Supplier
Cement	Pingnan Huarun PII42.5R	Fly ash	Power plant in Nanning, Guanxi
Microbeads	Power Plant from Wuhan	White Clay	Hainan Yangpu Paper Mill
Slag	Shaogang S95 slag	Admixture	Guangzhou Bozhong
Crushed stone	Boluo Shunda Stone Quarry	Sand	Guangdong Xijiang River Sand

19.3.2 Mix design in laboratory

The mix design of C30MPC is shown as Table 19.3.

Table 19.3 Mix design of C30MPC.

NO.	Cement	Fly ash	White clay	Water	Sand	Aggregate	Admixture
1	250	120	70	150	745	955	1.85%
2	250	140	50	150	745	955	1.75%

19.3.3 Properties of concrete

The properties of fresh concrete are shown in Table 19.4, the strength of concrete at different ages is shown in Table 19.5, and the autogenous shrinkage and early shrinkage of concrete are shown in Table 19.6. The heat of hydration of concrete is shown in Table 19.7.

The MPC containing papermaking white clay has a 72-hour auto-shrinkage of 0.92-0.95/10,000, less than 1/10,000 and no auto-shrinkage cracking.

Table 19.4 Properties of fresh concrete.

No	Time	Flow time (s)	Slump (mm)	Slump flow (mm)	U channel
NO.1	initial	3.22	265	760*680	340 mm/10.0"
	2 h	6.09	260	660*640	330 mm/23.0"
NO.2	Initial	2.25	265	660*670	330 mm/23.0"
	2 h	4.41	270	680x680	330 mm/9.81"

Table 19.5 Strength of concrete.

No	Strength (MPa)		
	3 d	7 d	28 d
NO.1	23.0	28.8	40.3
NO.2	25.2	31.1	41.5

Table 19.6 Autogenous shrinkage (/10,000).

NO.	1 d	2 d	3 d	4 d	5 d	6 d	7 d	14 d	28 d
1	0.41	0.72	0.92	1.03	1.18	1.32	1.46	2.05	3.11
2	0,43	0.81	0.95	1.06	1.22	1.45	1.52	2.22	3.14

Table 19.7 Heat of hydration of concrete.

No	Heat of hydration (°C)				
	initial	Starting	Max temperature	Start dropping	Room temperature
NO.1	24.5	5 h-25.5	22 h to 45.5	24 h to 45.4	42 h to 24.9
NO.2	24.5	5 h-25.6	22 h to 45.3	24 hto 45.7	43 h to 24.9

From the above test data, it can be seen that the mechanical properties of the two proportions of concrete can meet the requirements of C30MPC. Other properties also meet the requirements.

Taking into account the performance of the workability and viscosity of the concrete, the mix design of NO.2 is better.

19.4 C60 MPC with papermaking white clay

19.4.1 Raw materials

The raw materials are shown as Table 19.8.

Table 19.8 The raw materials of C60 MPC.

Materials	Supplier	Materials	Supplier
Cement	Pingnan Huarun PII42.5R	Fly ash	Power plant in Nanning, Guanxi
Microbeads	Power Plant from Wuhan	White Clay	Hainan Yangpu Paper Mill
Slag	Shaogang S95 slag	Admixture	Guangzhou Bozhong
Crushed stone	Boluo Shunda Stone Quarry impact crusher	Sand	Guangdong Xijiang River Sand

19.4.2 Mix design and properties of C60 MPC

1) The mix design of concrete is shown in Table 19.9.

Table 19.9 The mix design of C60 MPC.

NO.	Cement	Microbeads	Fly ash	White clay	Water	Sand	Crushed stone	admixture
1	350	30	100	70	150	745	955	2.0%
2	350	30	70	100	150	745	955	2.0%

2) The properties of fresh concrete are shown in Table 19.10.

Table 19.10 Properties of fresh concrete.

NO.	Slump (mm)	Slump flow (mm)	Flow time (s)	U channel (cm/s)
1	275/260	710/670	3.07/5.03	32cm, 5.66/15.13
2	265/250	710/630	3.67/6.02	32cm.7.66/20.13

Note: The numerator value is the initial value, and the denominator value is the measured value after 2 h.

As the amount of white clay increases, the viscosity of the concrete increases. The amount of white clay should not exceed 100 kg/m^3.

3) Heat of hydration of concrete is shown in Table 19.11.

Table 19.11 Heat of hydration of concrete.

No	Heat of hydration (℃)				
	initial	Starting	Max temperature	Start dropping	Room temperature
NO.1	24.5℃	3 h-25.5℃	36 h to 51.5℃	38.5 h to 51.4℃	65 h to 24.6℃
NO.2	24.5℃	3 h-25.6℃	35 h to 52.1℃	39.5 h to 52.1℃	65 h to 24.6℃

4) The early shrinkage and shrinkage at 28 days are shown in Table 19.12.

Table 19.12 Early shrinkage and shrinkage at 28 days.

Age No	Age of concrete (d)								
	1 d	2 d	3 d	4 d	5 d	6 d	7 d	14 d	28 d
NO.1	-0.23	-0.19	-0.02	0.03	0.17	0.352	0.49	1.04	2.16
NO.2	-0,31	-0.22	-0.05	0.02	0.22	0.48	0.56	1.32	2.16

The early shrinkage and 28 d shrinkage of C60 MPC are lower than C30MPC. It may be that C60MPC contains more papermaking white clay.

5) Compressive Strength of concrete is shown in Table 19.13.

Table 19.13 Compressive strength of concrete.

NO	Compressive strength at different age (MPa)		
	3 d	7 d	28 d
NO.1	43.4	54.4	67.7
NO.2	42.5	53.6	65.9

6) RCPT of concrete

The RCPT at 28 d age is 800 Coulomb/6h, and at 56 d age is 700 Coulomb/6h. It has high resistance to chloride ion penetration.

The above C30 and C60MPC have been tested many times and have excellent performance, which can be used in production.

19.5 Production and application of papermaking white clay C30 and C60 MPC

After many repeated tests, the performance of the papermaking white clay multifunctional concrete has been proven to be reliable. With the support of the Vanke Project Department of China Construction Fourth Engineering Bureau, on December 26, 2013, the papermaking white clay multifunctional concrete was produced and applied. All floor slabs, stairs, and balconies of a certain floor of the project were cast with C30 MPC, while beams and columns used C60 MPC. The quantity of the experimental production and application of concrete was about 300 m^3, which was produced and supplied by Guangzhou Tianda Concrete Company. The construction was completed in one afternoon. All concrete is vibration-free and self-compacting. After pouring, it is covered with a plastic film. All concrete is free of watering and self-curing.

19.5.1 The mix design and properties of C30, C60 MPC

The raw materials used in the production of C30 and C60 MPC are shown in Table 19.2. and Table 19.8. The mix design of concrete used in production and application is shown in Table 19.14.

Table 19.14 Mix design of C30 and C60 MPC for production.

Strength	Cement	Microbeads	Fly ash	White clay	Water	Sand	Crushed stone	admixture
C30	250	0	140	50	150	760	940	7.7
C60	350	30	100	70	150	760	940	9.08

1) Properties of fresh concrete

When the concrete leaves the factory, samples are taken from the unloading hopper of the mixer to determine the slump and other properties, as shown in Table 19.15.

Table 19.15 Properties of fresh concrete.

Strength	Slump (mm)		Slump flow (cm)		Flow time (s)		U Channel (cm/s)	
	initial	2 h	initial	2 h	initial	2 h	initial	2 h
C30	260	260	69x68	65x65	2.13	4.05	32/5.6	32/11.4
C60	275	270	72x71	68x68	3.09	5.12	32/5.4	32/14.2

2) Heat of hydration of concrete

The heat of hydration of C60 concrete with papermaking white clay is shown in Table 19.16.

Table 19.16 Heat of hydration.

Initial	Start	Max	Dropping	Till room temp
14.5°C	After 3 h 17.6°C	after32 h to 48.6°C	After 32 h, 48.6°C	After 114 h to 14.6°C

3) Shrinkage of concrete

The shrinkage of concrete with white clay as shown in table 19.17.

Table 19.17 Shrinkage of C60 concrete (/10,000).

1 d	2 d	3 d	4 d	5 d	6 d	7 d	8 d	9 d
-0.14	0.37	0.72	0.74	0.78	0.83	1.03	1.09	1.13

4) Strength of concrete

The compressive strength of C30 and C60 concrete with white clay as shown in Table 19.18.

Table 19.18, Compressive strength of C30 and C60

	3 d	7 d	2 8d
C30	26.2	36.2	41.5
C60	39.7	52.4	71.2

19.5.2 Concrete casting and construction quality tracking

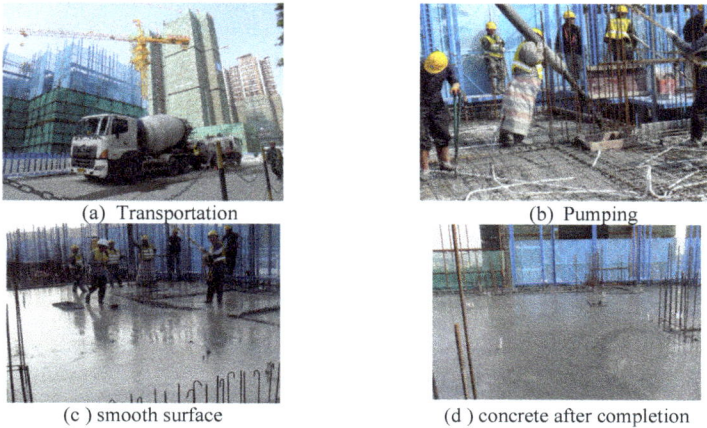

(a) Transportation

(b) Pumping

(c) smooth surface

(d) concrete after completion

Figure 19.5 Concrete construction of white clay C30 and C60 MPC.

The white clay C30MPC and C60MPC were produced and transported from the mixing plant to the construction site within about 1 hour, which required high slump retention and self-compacting performance. After the concrete is cast, the engineering and technical personnel tracked the quality. After demoulding, the surface of the concrete component was smooth and there was no appearance of honeycomb or pitted surface, cracks, and the quality was high. Refer to Figure 19.5(a)-(d) for the construction process. After demoulding, the appearance of the concrete is shown in Figure 19.6(e)(f).

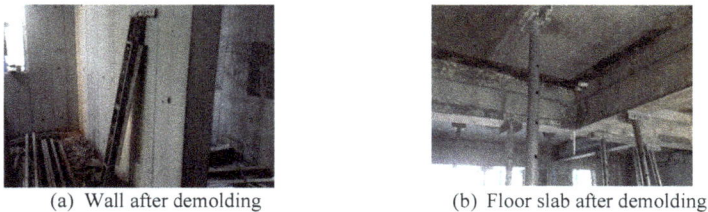

(a) Wall after demolding

(b) Floor slab after demolding

Figure 19.6 MPC construction process and appearance quality.

19.6 Development and application of C35 MPC with papermaking white clay

The China Mobile E-commerce Center in Haizhu District, Guangzhou was also a construction project under China Construction Fourth Engineering Bureau. On December 5, 2014, the construction and casting of C35MPC with papermaking white clay was conducted to the outdoor wall of the basement on the northwest side of the second floor of the project. The total volume of concrete is about 100 m³.

19.6.1 Raw materials and mix design of C35 MPC with white clay

The raw materials were the same as those used in C30 MPC, and the mix design of concrete is shown in Table 19.29.

Table 19.19 Mix design of C35 MPC (kg/m³).

Class	Cement	Microbeads	Fly ash	White clay	Water	Sand	Crushed stone	admixture
C30	250	0	140	60	150	760	940	8.0

19.6.2 Properties of C35 MPC

The workability of fresh concrete is shown in Table 19.20; the autogenous shrinkage and early shrinkage of concrete are shown in Table 19.21. The heat of hydration is shown in Table 19.22.

Table 19.20 Properties of fresh concrete.

Time	Flow time (s)	Slump (mm)	Slump flow (mm)	U Channel
0 h	1.52	265	680*690	340mm/9.23"
2 h	2.83	265	680*680	330mm/18.0"

Table 19.21 Autogenous shrinkage and early shrinkage.

Class	12 h	24 h	48 h	72 h	7 d	28 d
C35	-0.82/10000	-0.47/10000	-0.08/10000	0.02/10000	0.39/10000	1.94/10000

Table 19.22 Heat of hydration.

Time to start（h）	Time to Max（h）	Initial-Max temp°C
13	35	27~52

19.6.3 On-site MPC performance

After the concrete arrived at the construction site, the concrete was tested for its workability before pumping. The interval between exiting the mixer and the test was about 2 to 3 hours. The following test results are averaged.

Table 19.23 Properties of fresh concrete.

Flow time/S	Slump flow/mm	slump/mm	U Channel(mm/S)
3.11	670×680	260	320/18.48

Table 19.24 Heat of hydration

Class	Heat of hydration（°C）				
	Initial	Time to start	Max temp	Start to drop	Room temp
C35	27.0°C	After 13 h 27.5°C	After 35 h to 52.0°C	After 35 h to 52.0°C	After 92 h to 27.0°C

19.6.4 Concrete pouring and quality tracking after construction

Figure 19.7 The surface of C35MPC after casting and demoulding.

After the concrete was cast, the technicians of Tianda Concrete Company tracked the quality of the concrete construction. There were no cracks in the whole concrete, the surface was smooth, there was no honeycomb or pockmarked surface, and the forming quality was good. The strength of the remaining samples for construction is shown in Table 19.25. It can be seen from the table:

1) The strength of self-curing concrete is higher than that of standard-cured concrete.

2) The strength of self-curing concrete is 48.5 MPa at 28 days, and the strength level is above C40; the strengthening effect is obvious.

Table 19.25 The compressive strength of C35 MPC

Self-curing			Standard curing		
3 d	7 d	28 d	3 d	7 d	28 d
26.7	33.2	48.5	23.4	26.5	42.3

19.7 Conclusions

1) Using papermaking white clay to prepare C30, C35, and C60 and other strengths of MPC, and the test application in the project, showed that the use of papermaking waste as a useful resource for concrete is feasible. Papermaking white clay is a very useful featured concrete resource.

Materials Research Forum LLC

https://doi.org/10.21741/9781644901991

2) The effect of thickener and self-curing admixture for self-compacting concrete was achieved by replacing the natural zeolite powder in multifunctional concrete with paper-making white clay. This is because the calcium carbonate particles in the white mud are very fine and adsorb a large amount of free water. After the initial setting of the concrete, when the cement continues to hydrate, the calcium carbonate particles release the adsorbed free water to supply the water required for hydration of the cement and inhibit the autogenous shrinkage and early shrinkage of the concrete. Because it can absorb the free water, it thickens the concrete and inhibits the segregation and bleeding of concrete.

3) In the process of preparing multifunctional concrete, papermaking white clay also has shortcomings. If the amount is too much, it will greatly increase the viscosity of MPC, and it is prone to scratching at the bottom, which has a great impact on flowability and filling performance, to a certain extent. The dosage of white clay in concrete has to be limited.

Chapter 20

Ultra-high Pumping Technology of Ultra-high Strength MPC

20. Introduction

The author has supervised the research team to develop C100 ultra-high performance concrete, C100 self-compacting concrete, C120 ultra-high performance concrete, and C130 multifunctional concrete in combination with the West Tower Project, Jingji Building and East Tower Project. With the teamwork between ready-mixed concrete plants, construction contractors, and pump manufacturers the concrete has been successfully pumped to heights of 416 m, 420 m, and 510 m, respectively. Through the practice of these projects, the research team has gained experience. This chapter will discuss the lessons learned on the control of super-high pumping parameters and performance evaluation.

C100 self-compacting concrete and C130 multifunctional concrete were chosen as examples of ultra-high pumping for the study, as these two kinds of concrete must meet the requirements of self-compacting when leaving the factory and should also meet the requirements of self-compacting when delivered to the construction site. It is important that the concrete discharged from the pump after ultra-high pumping must meet the requirements of self-compaction to achieve vibration-free self-compaction. That means the rising height of the concrete measured by the U-Channel must be greater than 30 cm, which is the most basic requirement. In addition, UHP-SCC has a high viscosity and high viscous resistance between the pump pipe, and the flow in the pump pipe is more difficult than other concretes. If super high pumping for this kind of concrete is successful, the super high pumping for other concretes will be easily solved.

20.1 The transport characteristics of different types of concrete in pump pipes

Ultra-high-performance self-compacting concrete (UHP-SCC), ultra-high-performance concrete (UHPC), multifunctional concrete (MPC) and ordinary concrete (NC) are composed of different quantities and quality of raw materials. And the viscosity and flowability of fresh concrete are also different. Therefore, the pumping parameters are different. As shown in Figure 20.1, high-performance ultra-high-performance concrete (including HPC, UHPC, UHP-SCC and high-strength MPC) has high structural viscosity and low shear strength, and ordinary concrete (NC) has low structural viscosity but high shear strength while the plastic viscosity and shear strength of high-flow high-performance concrete is roughly the same. Therefore, high-flow concrete is easier to pump than ultra-high-performance and ultra-high-strength SCC.

Materials Research Forum LLC

https://doi.org/10.21741/9781644901991

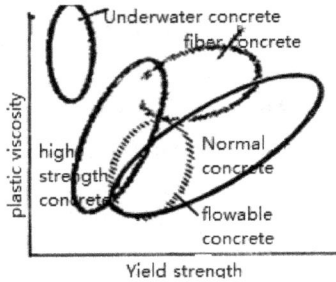

Figure 20.1 Plastic viscosity and shear strength of different concrete.

Refer to Figure 20.2 for the pressure loss of normal concrete (NC) and high-performance ultra-high-performance concrete during the pumping process.

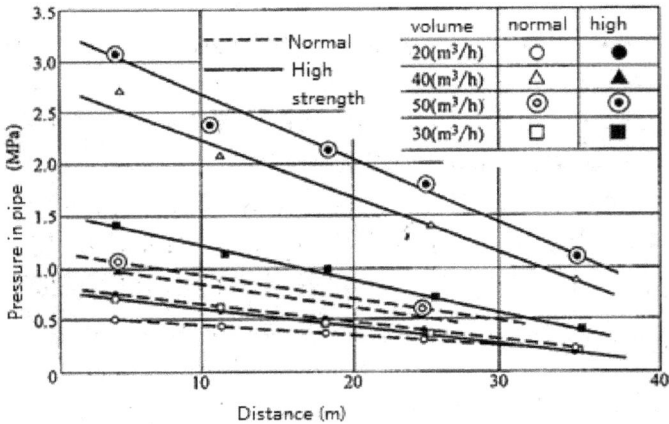

Figure 20.2 The pressure loss during pumping for NC and UHPC.

In Figure 20.2, the dashed line shows ordinary concrete with a design strength of 50 MPa. The solid line shows high-performance ultra-high performance concrete with a design strength of 100 MPa. High-strength concrete uses a large amount of cementitious materials and low water cement ratio, thus the viscosity is high, and pressure loss during pumping is also high. In the figure, when the pumping volume is 30 m^3/h, the pressure loss is about 0.2 MPa, and when the pumping volume is 50 m^3/h, the pressure loss is about 0.4 MPa. But when the UHPC pumping volume is 30 m^3/h, the pressure loss is about 0.4 MPa, and when the pumping volume is 50 m^3/h, the pressure loss is about 0.6 Map, which is about 50% more than the pressure loss of ordinary concrete.

20.2 The relationship between friction resistance and pumping speed during pumping

The flow rate of concrete in pumping pipe: $\mho = v/\pi r^2 l$ (1)

Where: r-radio of pumping pipe; l- length of pipe , V-volume of concrete pumped;

The friction resistance of concrete in the pipe: $f=F/2\pi r \, l$ (2)

$f=k1+k2 \, \mho$; k1- viscous factor; k2- speed factor;

For example, the lower the W/B of the concrete, the greater the viscosity, and the greater the k1. The higher the pressure required for the concrete to flow in the pipe, the greater the viscous resistance during pumping, and the greater the pumping volume, and the k2-speed coefficient will be greater, and it will be more difficult to increase the concrete pumping volume. k1 and k2 should be as small as possible. It is related to the composition and proportion of concrete as well as to the pumping volume per unit time during construction. According to the previous construction experience of ordinary concrete, K1 and K2 are only related to the slump of concrete. Therefore, the following two formulas was proposed by some researchers:

$K_1 = (3.00 - 0.10S) \times 10^{-3} \text{ kg f/cm}^2$ (1)

$K_2 = (4.00 - 0.10S) \times 10^{-3} \text{ kg f/cm}^2/ \text{ m/sec}$ (2)

In the above two formulas, K_1 is the viscosity coefficient, which reflects the viscosity of concrete, and is related to the constituent materials and proportions of the concrete. The value of the concrete slump (S in the formula) directly reflects the value of K_1. If the slump is high, viscosity will be low and K_1 value will be low too. The concrete will be easy to pump. The fundamental purpose of slump retention is to maintain the parameters of fresh concrete such as slump and flow time, so that the K_1 value remains unchanged, and the viscosity resistance between the concrete and the pipe wall remains the same. Concrete will be easy to pump to a super high level.

K_2 is the speed coefficient, reflecting the size of the pumping volume per unit time. But the speed coefficient is also related to the slump. The slump is high, the K_2 value is small, the movement resistance of concrete is in the pipeline will be small. The pumping volume can be increased.

Therefore, after the concrete slump (S) is measured, K_1, K_2 can be calculated. Thereby its pumpability can be determined.

When concrete is pumped, the relationship between the frictional resistance of the pipe wall and the pumping speed is shown in Figure 20-3.

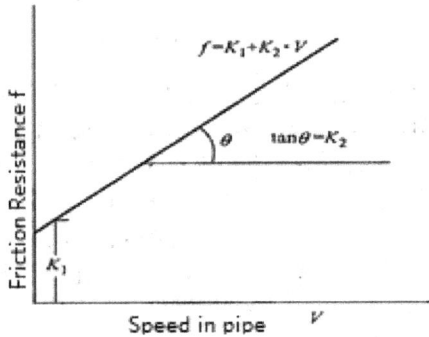

Figure 20.3 The relationship between pipe wall frictional resistance and pumping speed during concrete pumping.

It can be seen from Figure 20.3 that if K_1 is relatively high, the pressure required for the concrete to start flowing in the static state of the pipe will be relatively high. If K_2 is relatively high, it will be difficult to increase the pumping volume of concrete. Therefore, we hope K_1, K_2 as small as possible so that the small pressure is required for pumping. The pumping of ultra-high performance concrete requires smaller K_1 and K_2 and these two values are related to the mix design of concrete.

20.3 Determination of the friction resistance between concrete and pipe wall

When concrete is pumped, the pressure required for transporting concrete must be able to overcome the frictional resistance between the walls of the concrete pipes before the concrete can be pumped. So how can this frictional resistance be determined? The friction resistance between concrete and pipe wall can be measured by the device shown in Figure 20.4.

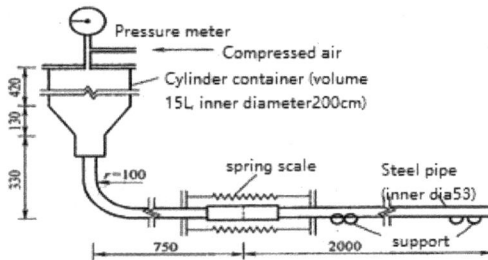

Figure 20.4 Measuring device for friction resistance during concrete pumping.

The concrete is placed into the cylindrical container, and the container is then sealed. The pressure gauge is installed, and the compressed air is then connected. The compressed air will apply pressure on the concrete in the cylindrical container. Due to the frictional resistance between the concrete and the pipe wall, the conveying pipe will move in the direction of pressure delivery. The connection between the container and the conveying pipe is a flexible connection. The conveying pipe is supported on rollers. As the restraining force does not work, the force in the conveying pipe can be measured.

Suppose the volume of concrete discharged per unit time is V, and the force generated is F (deduct part of the pressure loss from the reading of the pressure gauge, or read it from the spring scale). Both V and F can be measured by experiments. The friction force of concrete in the pipe f per unit area can be calculated according to the above formula (2).

20.4 Selection of pumps for ordinary concrete pumping

Figure 20.5 shows the layout of concrete pumping pipes for a certain project. The pumped concrete is C40 strength grade, slump 20-22 cm, pumping volume 30 m^3/h. The calculation of the pressure of the piping, selection of the pump, and estimation of the pumping limit are discussed below.

20.4.1 Calculation of the pressure of pumping

After K_1 and K_2 are determined, the pumping pressure can be calculated at any place in the pipe. The concrete pumping pipeline of a certain project is shown in Figure 20.5. The concrete slump was 20 cm, and the pumping volume was 30 m^3/h.

$f=k_1+k_2 \mho$; (pumping pressure)

$\mho =v/\pi r^2 l$; (pumping speed of concrete) (l-length of pipe)

$K_1=(3.0-0.1\ S)\times 10‰$ MPa

$K_2=(4.0-0.1\ S)\times 10‰$ MPa ;

V- volume of pumping 30 m3/h; S- slump 20 cm .

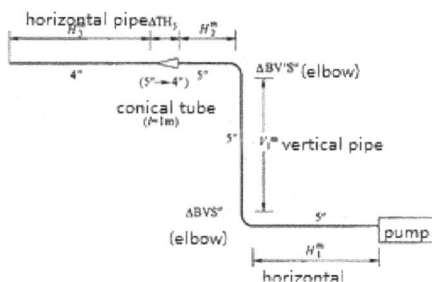

Figure 20.5 Illustrated example of concrete pumping piping.

1) Calculation of pressure loss of horizontal pipe (see table):

According to the pump type (piston pump), pipe diameter (5" pipe), slump (20 cm), pumping capacity (30 m3/h), the horizontal pipe pressure loss can be found to be 0.08 MPa/m.

2) Vertical pipe conversion factor:

It is expressed as the ratio of the pressure loss of a horizontal pipe per meter to the pressure loss of a vertical pipe of the same pipe diameter: 3.5 x 0.08=2.4 MPa/m (equivalent to the pressure loss of a horizontal pipe). It can be seen that the conversion factor for the vertical pipe, the vertical bending pipe, horizontal elbows, tapered pipes, straight pipes, etc., can be found and converted to horizontal pipes, then the corresponding horizontal pipe length can be obtained.

The vertical pipe conversion factor is 3.5

The conversion factor for vertical elbows is 6.1

The conversion factor for horizontal elbows is 3.0

The conversion factor for cone pipe is 1.8

The conversion factor for straight pipe from 5"→4" pipe is 1.8

The converted length L to horizontal pipe (5" tube) can be calculated:

L= 30 +6.1 + 3.5X30+ 3.0 +10 + 1.8 +1.8X20=192 m

The basic pressure of the piping = 0.08x192=1.53 MPa. That means the piston pump with an outlet pressure $\geq$ 1.53 MPa should be selected to meet the requirements of concrete pumping.

20.4.2 Pumping limit estimation

If the pump outlet pressure is 2.5 MPa, the pumping limit horizontal length is 310 m:

If the outlet pressure of the pump is 3.0 MPa, the pumping limit length is 370 m;

According to the pumping conditions, the horizontal pipe length is converted to 192 m; and the pumping limit is 310 m horizontal, therefore, the pumping capacity is sufficient.

20.5 Pumping of high-strength and high-performance concrete

High-strength high-performance concrete has low W/B, high viscosity, and large frictional resistance between the concrete and the pipe wall, which has a great impact on the pumping performance.

20.5.1 The relationship between pumping volume and pumping pressure loss

The relationship between the pumping volume of high-strength and high-performance concrete and the pumping pressure loss is shown in Figure 20.6. The figure shows the results of the pumping test of three types of concrete.

1) W/B= 28-30%, high-strength high-performance concrete with superplasticizer, when the pumping discharge volume is about 30-35 m³/h, the pumping pressure loss can reach 0.05-0.06 MPa/m. It may be due to its high viscosity, the friction resistance between the concrete and the pipe wall is high during the pumping process.

2) W/B=37% high-strength high-performance concrete is also mixed with superplasticizer. The test proves that the adhesion of this concrete is 0.01 MPa, while the high-strength high-

performance concrete with W/B=0.28 is 0.04 MPa. It is 4 times higher. In other words, as the W/B decreases, the greater the adhesion of the concrete, the more difficult it is to pump.

3) For ordinary concrete with W/B=45-57%, the slump is 15-25 cm, even if the pumping speed changes, the pressure loss is within the range of 0.01-0.02 MPa/m.

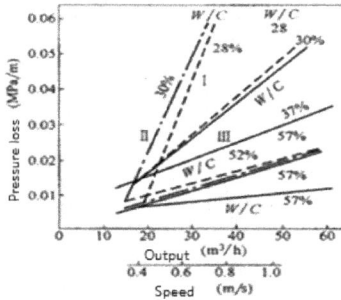

Figure 20.6 Pressure loss and pumping volume.

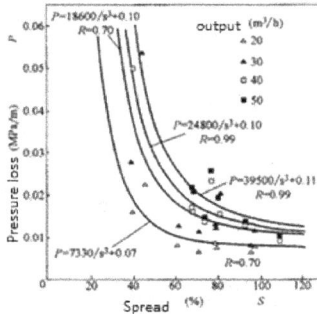

Figure 20.7 Pressure loss and viscosity.

20.5.2   Pressure loss and viscosity

The relationship between pressure loss and slump flow (ASTMC 124) is shown in Figure 20.7. The curve in the figure is a regression curve obtained using the actual pumping volume as a parameter. The greater the pumping volume, the greater the pressure loss. The greater the slump flow, the pressure loss is reduced. The pumping volume is different, the pumping pressure loss is also different.

When the pumping discharge volume is:

At 20 m^3/h, pressure loss p=7330/s^3 + 0.07 correlation coefficient R=0.7.

At 30 m^3/h, pressure loss p=18600/$s^{3.}$ + 0.10 Correlation coefficient R=0.7.

At 40 m^3/h, pressure loss p=24800/s^3 + 0.10 correlation coefficient R=0.99.

At 50 m^3/h, pressure loss p=39500/s^3 + 0.11 Correlation coefficient R=0.99.

It can be seen that when the pumping discharge volume is the same, increasing the slump flow S during the slump test of the fresh concrete can reduce the pumping pressure loss. For the concrete with the same slump flow, the greater the pumping discharge volume, the greater the pressure loss.

When pumping high-strength and high-performance concrete, the appropriate pump should be selected according to the slump flow S and the pumping outlet volume (m^3/h) (to determine the pumping pressure required by the pump). If the performance of the pump is already fixed, it is necessary to adjust the slump flow S during the slump test, or the pumping discharge volume, to meet the requirements of the pump.

20.6 Ultra high pumping for UHP-SCC

UHP-SCC (C100) was developed in conjunction with the West Tower project. The systematic test and study for UHP-SCC were carried out and completed according to the technical requirements of the project. Several trials on ultra-high pumping were performed in accordance with the progress of the West Tower project. The trial with a pumping height of 411 m on December 6, 2008 will be discussed as below.

In addition to the relevant units of the China State Construction System, the research team involved in the development of UHP-SCC included Yuexiu Group, Guangzhou Construction Group, Zoomlion and Tsinghua University.

20.6.1 Technical requirements for UHP-SCC for super-high pumping

Through the ultra-high pumping of UHP-SCC, the comprehensive capabilities of the pump, concrete materials and construction technology are tested. In the West Tower project, Zoomlion provided a high-pressure pump with an outlet pressure of 40 Map. According to the outlet pressure and the performance of UHP-SCC, the pumping height can be calculated.

Figure 20.8 West Tower project under construction.

1) Mix design of UHP-SCC

After nearly a year of test and research, after the mix design of C100 UHP-SCC and raw materials were confirmed, the mixing plant has mixed about 2 cubic meters of C100UHP-SCC, loaded it into the truck, and the truck moved around the plant. The performance of fresh concrete was tested at 1 hour, 2 hours, and 3 hours after mixing. The mix design of C100UHP-SCC was adjusted and confirmed after the test met the performance requirements. As shown in Table 20.1, W/B =20% is C100 UHPC; W/B=22% is C100 UHP-SCC; CFA-control slump loss additive, NZP- is natural zeolite powder.

Table 20.1 Mix design of UHP-SCC (kg/m³).

W/B	C	BFS	SF	CFA	NZP	S	G	W	HWRA
20%	500	190	60	/	/	750	900	150	18
22%	450	190	60	14	28	750	900	154	15.4

Table 20.2 Main technical requirements of UHP-SCC.

Property time	permeability (bleeding)	slump (mm)	Slump flow (mm)	Flow time (Sec)	U Channel (mm)
initial		265	720	4.0	310
1 h		270	720	4.0	310
3 h		275	730	3.5	318
site		275	750	3.3	310
After pumping		280	760	3.2	310

2) Main technical requirements of UHP-SCC

The main technical requirements for ultra-high pumping UHP-SCC are shown in Table 20.2. The pressure bleeding test is shown in Figure 20.9. Under the pressure of 30 MPa, the bleeding test results are shown in Table 20.3.

Figure 20.9 Bleeding test under 30 MPa pressure.

Table 20.3 Pressure bleeding test results under 30 MPa.

	NO	V_{10}(ml)	V_{140}(ml)	S_{10}(%)
UHP-SCC	3	0	3	0

"The Technical Specification for Concrete Pumping Construction" (JGJ/T10) stipulates that the pumpability of concrete can be controlled by pressure bleeding test combined with construction experience. Generally, the relative pressure bleeding rate S10 at 10s should not exceed 40%. The S10 (%) of UHP-SCC for the West Tower super high pump is 0 which is in line with the technical requirements of super high pumping. Before and after pumping. The concrete raising height is greater than 300 mm in U-Channel, so it's self-compacting performance is guaranteed.

20.6.2 The estimation of pumping height for construction

Based on the pressure loss per meter length = 0.01 MPa, the vertical pipes, elbows, and tapered pipes need to be converted into horizontal pipes and the total length of the horizontal pipe can be calculated. The total pressure required for pumping can be calculated by the total length of the horizontal pile multiplied by the pressure loss per meter length of the horizontal pipe = 0.01 MPa.

For this project:

1) The length of horizontal pipe 800 m: 800x0.01=8.0 MPa;

2) The length of vertical pipe is 500 m, converted to horizontal pipe length = 500x4 = 2,000 m, 2,000x0.01 = 20 MPa;

3) The number of elbows and cone-shaped pipes:

   elbows: 90°, 20 pcs, pressure loss 0.1 Map/pc;

   45°, 2 pcs; pressure loss 0.05 Map/pc

   Cone tube 1 pc, pressure loss 0.1 Map/pc.

   Shut-off valve: 2 pcs, pressure loss 0.05 Map / pc.

   Distribution valve: 1 pc, pressure loss 0.2 Map /pc.

   The total pressure loss of elbow and tapered pipe p3 = 21x0.1+4x0.05+1x0.2 = 2.5 MPa.

Therefore, the total pressure loss of horizontal pipes, vertical pipes, elbows, and tapered pipes is: 8.0 + 20 + 2.5 = 30.5 MPa. In addition, there is a 30% pressure reserve, so when choosing a pump, the outlet pressure should be considered ≥30.5 +30%×30.5= 39.65 Map. The outlet pressure of the existing pump is 40 MPa, and the pumping capacity is sufficient. During the first 416 m ultra-high pumping test of the C100 UHPC and C100UHP-SCC of the West Tower project, the outlet pressure of the pump was 28 MPa. During the pumping process, due to a problem with the distribution pipe, it must be shut down for maintenance. After 30 minutes, the on-site commander was afraid that the parking time was too long and the problem would occur and ordered to start the pump to pump out the concrete from the pump pipe. At that time, the outlet pressure of the pump reached more than 34 Map. The reason could be that the concrete moves from static to moving, the adhesion coefficient increases, the resistance increases, and the pumping pressure increases. The calculation of the pumping height in engineering is much more practical than the actual theoretical calculation, and it is much simplified, and it is also in line with the actual operation.

20.6.3   The properties of hardened concrete

1) Compressive strength at different ages. As shown in Table 20.4.

Table 20.4 Compressive strength at different ages (10x10x10 cm),

Testing No	Compressive strength (MPa)			
	3 d	7 d	28 d	56 d
1	87.0	108.7	130.8	130.0
2	99.0	96.3	115.0	115.9
3	84.9	88.7	94.9	107.9
4	91.2	106.5	117.3	118.0

Note: The size factor of 100 mm×100 mm×100 mm specimen converted into 150 mm×150 mm×150 mm specimen is 0.93

2) Flexural strength at different ages. As shown in Table 20.5

Table 20.5 Flexural strength of concrete at different ages (MPa)

NO	3 d	7 d	28 d	56 d
	100×100×400	100×100×400	100×100×400	100×100×400
1	8.4	9.6	10.0	12.4
2	9.0	11.1	12.5	12.6
3	8.5	11.8	11.8	12.2
4	8.3	9.9	11.3	13.1

3) Tensile split strength at different ages. As shown in Table 20.6.

Table 20.6 Tensile split strength of concrete at different ages (MPa).

No	3 d	7 d	28 d	56 d
	100×100×100	100×100×100	100×100×100	100×100×100
1	6.05	7.15	7.58	8.31
2	7.09	7.13	7.42	8.11
3	5.87	7.17	8.09	8.29
4	6.36	7.09	7.92	8.44

4) Cylinder compressive strength and E-Modulus at different ages. As shown in Table 20.7.

Table 20.7 Cylinder compressive strength and E-Modulus.

NO	Size (mm)	Cylinder strength	E-Modulus	W/B
1	150×150×300	101.7	47900	20%
2	150×150×300	114.7	59500	20%
3	150×150×300	119.5	63300	20%
4	150×150×300	113.2	48400	22%

5) Compressive strength before and after pumping. As shown in Table 20.8.

Table 20.8 Compressive strength before and after pumping.

Compaction method	Sampling	3 d	7 d	28 d	56 d
vibration	Before pumping	85.7	104.1	121.4	120.8
No vibration		89.5	102.3	122.5	118.4
vibration	After pumping	84.7	100.5	104.9	119.8
No vibration		77.7	103.5	120.1	114.1

6) Testing of shrinkage and cracking of concrete.

   a.   The autogenous shrinkage is shown in Table 20.10.

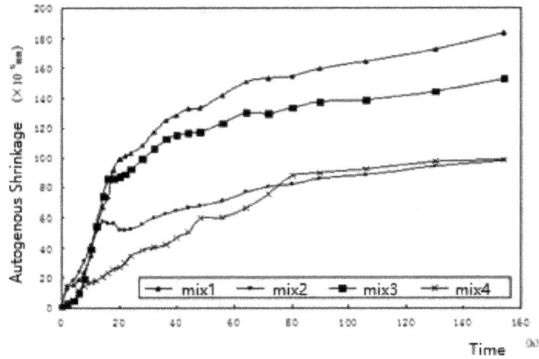

Figure 20.10 Autogenous shrinkage curves of UHPC and UHP-SCC with different mix designs. Mix 1, 2, 3 are UHPC, and mix 4 is UHP-SCC.

   b.   Plate cracking test. In this test, four testing slabs with different mix designs are shown in Figure 23.11. After casting, they were all exposed to the sun, and a high-power fan was used to blow air to the surface of the specimen at the same time. The test data was recorded at the age of 24 h. Table 20.9 is the test results on the slab cracking test. The "average cracked area per crack" indicates the size of the concrete cracks, and the "number of cracks per unit area" indicates the number of concrete cracks. The product of the two data "Total cracked area" represents the overall situation of concrete cracks.

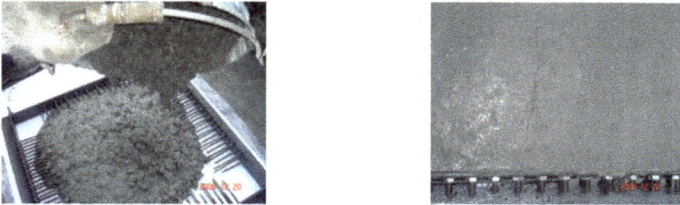

Figure 20.11 Slab cracking testing of concrete.

From the data analysis in Table 20.9, the following conclusions can be drawn:

Under the conditions of construction and application, the ability of concrete to resist cracking from strong to weak is as follows: Mix 1>4>2>3. The anti-cracking ability of concrete with mix 1 and mix 4 is relatively close and far better than mix 2 and mix 3. The difference of the "total cracked area per unit area" of the concrete is nearly ten times.

Table 20.9 Testing data of slab cracking test.

No	W/B	Crack (No)	Average crack area (mm^2)	Total crack (No/m^2)	Total crack area (mm^2/m^2)
1		7	10.11	19	196.53
2	0.20	65	11.54	181	2083.4
3		30	44.22	83	3685.35
4	0.22	8	12.66	22	281.33

20.6.4   Conclusion of the trial for 411 m ultra high pumping of UHPC and UHP-SCC concrete

1) The viscosity modifier is not necessary for UHP-SCC prepared with natural zeolite ultra-fine powder. The concrete has high fluidity, good cohesiveness, no bleeding, no segregation, no delamination. During the U-channel flow test, it can flow evenly through the steel bar, rising 320 mm, and after 2 hours, the height is still 300 mm.

2) In the composition of UHP-SCC, carrier fluidizing agent (CFA) is also used, which can make the water reducing agent slowly precipitate into the concrete and maintain the plasticity of the concrete for 4 hours without bleeding, segregation and retardation. The mixture will not bleed even under a pressure of 30 MPa. After the comparison of sampling before and after the pump, the original working performance is still maintained, which meets the requirements of ultra-high pumping construction.

3) The compressive strengths of UHP-SCC test specimens with or without vibration are almost the same, indicating that its self-compactness is good, which is conducive to the construction

requirements of the high-congest of steel bars. UHP-SCC 28-day strength ≥108 MPa (tested in laboratory 28-day strength ≥117 MPa) meets the design requirements of mix design.

4) Compared with UHPC with the same mix design, UHP-SCC has relatively lower autogenous shrinkage, early shrinkage, and long-term shrinkage. The slab crack test data also proves that the UHP-SCC in this study has better crack resistance. This is because the natural zeolite ultrafine powder is used, and the concrete can be self-curing.

It can be seen that the ultra-high pumping performance of our UHPC and UHP-SCC are all due to the reasonable application of this special functional powder—natural zeolite ultrafine powder.

20.7 Ultra-high pumping and application of MPC with compressive strength 150 MPa

Around October 2015, the East Tower project entered the topping stage. According to the suggestion of Mr. Ye Haowen, Director of the China Construction Fourth Engineering Bureau and East Tower Project Manager, the MPC with a compressive strength of 150 MPa was tested for ultra-high pumping.

20.7.1 The mix design of MPC is as shown in table 20.10

Table 20.10 Mix design of MPC for ultra-high pumping.

C	MB	SF	NZ	W	S	G	EA	WRA	CFA
600	190	90	20	136	720	880	10	6.5	9.0

20.7.2 The properties of fresh concrete is shown in Table 20.11

Table 20.11 The properties of fresh concrete.

Testing location	Slump (mm)	Slump flow (mm)	Flow time (s)	U-Channel (cm/s)
Plant	275	760x720	3.0	330/30
Site	275	750x720	3.0	330/30
After pumping	270	720x720	5.0	310/32

20.7.3 The strength of MPC at different ages is shown in Table 20.12

Table 20.12 Strength of MPC at different ages.

Age	1 d	3 d	7 d	28 d	56 d
Strength	104	125.3	136	147	150

(a)East Tower under
construction

(b)Construction site for super high pumping

Figure 20.12 Super high pumping of MPC for East Tower project.

20.7.4 Pumping construction site

The pumping construction site is shown in Figure 20.12. Using the "Sany Heavy Industry" concrete pump with an outlet pressure of 40 MPa, the multifunctional concrete with a strength of 150 MPa at 56 d was pumped from the ground to a height of 521 m, and components such as shear walls and floor slabs were poured. Refer to 20.6.2 section for pumping pressure calculations. MPC bulk density is 2595 kg/m^3, strength at 7 d is 136 MPa and at 28 d is 147 MPa, at 56 d is 150 MPa. After the shear wall was demolded, there were no cracks on the surface of the wall, as shown in Figure 20.13.

(a) The surface of floor slab

(b) The surface of shear wall

Figure 20.13 The surface of components cast with MPC with a strength of 150 MPa.

20.8 Summary of concrete construction with super-high pumping

Concrete construction with pumping, vibration-free and self-compacting, self-curing instead of watering curing, concrete can inhibit autogenous shrinkage cracking and early shrinkage cracking, which is a breakthrough in concrete materials technology and construction technology.

Chapter 21

Lightweight Aggregate Multifunctional Concrete and its Application

21.1 Introduction

The lightweight aggregate multifunctional concrete developed in this project used various industrial waste and domestic waste as raw materials, achieving low consumption of natural resources and low energy consumption in the production process of building materials and products. Ultra-light concrete is used as wall and roof insulation materials to reduce the energy consumption of the maintenance and management of the buildings.

The internal structure of ultra-light concrete often contains air bubbles. That means all ultra-light concrete belongs to porous concrete. For example, in the 1950s, foam concrete insulation materials and foam concrete blocks were produced and applied in Chian. It is made by mixing foaming agent into cement fly ash slurry, stirring, forming, and steaming (or autoclaved). At that time, this kind of concrete was used as roof insulation material throughout the country. And autoclaved porous concrete blocks were often used as insulation walls. For example, in Tianjin carpet scraps were used as a foaming agent to produce autoclaved porous concrete blocks.

In the 1960s, aerated concrete emerged, which used the chemical reaction between aluminum powder and other components in the slurry to generate gas to form a porous structure. The dry bulk density is 400~800 kg/m^3, the strength is about 1.5~10 Mpa, the thermal conductivity is 0.08~0.25, and it is made into wallboard, roof panel or block, etc.

In the 1970s, ceramisite aerated concrete was developed in China. Ceramisite was mixed into the porous concrete made of aluminum powder. It makes the concrete form a double porous structure-porous ceramisite and porous slurry. Although it is a steam-cured product, the shrinkage of the concrete is reduced due to the presence of lightweight aggregate and at the same time, if there are cracks, the lightweight aggregate will break the cracks and turn continuous cracks into discontinuous cracks. The dry bulk density of this concrete is 900~950 kg/m^3, and the compressive strength is 7.5~8.0 MPa. It was used as exterior wall panels. Later, natural zeolite powder was used as carrier gas instead of aluminum powder to introduce gas into lightweight aggregate concrete to obtain gas-carrying porous concrete. This kind of steam-cured concrete has a dry bulk density of 800-1250 kg/m^3 and compressive strength of 6-13.5 MPa. In the 1980s, in Zhangjiakou, China, natural zeolite powder was used as the carrier gas to produce porous concrete wall panels for internal partition walls of buildings, as shown in Figure 21.1.

(a) casting of natural zeolite powder panels

(b) Building with zeolite powder panels

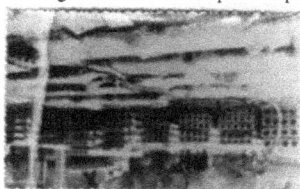

(c) Curing of panels

(d) Finish products

Figure 21.1 Production and application of natural zeolite powder panels.

Our research works in the Faculty of Engineering of Meiji University and Yamato Housing Company, Japan has developed a new type of lightweight concrete, as shown in Figure 21.2. The bulk density of the lightweight concrete with natural zeolite as gas-carrying agent is 700-900 kg/m^3, and the corresponding strength is only 5-7MPa. While the bulk density of newly developed lightweight concrete is 900-1000 kg/m^3, and the strength is 12-14 MPa.

(a) lightweight concrete with natural zeolite

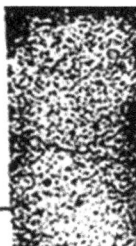

(b) specimen without and with ceramsite

Figure 21.2 Lightweight concrete with natural zeolite and lightweight aggregate concrete.

In the 1990s, Japan invented 3L concrete. Japan's Kubota Corporation introduced a foaming agent from Germany. This foaming agent was mixed with lightweight aggregate concrete to form a porous structure of foaming agent and a porous structure of artificial lightweight aggregate in the concrete. Through steam curing, the dry bulk density of concrete was about 1400 kg/m3, and the compressive strength reached 12~14 Mpa. Kubota used this concrete to make exterior wall panels for houses.

At the end of the 20th century, Guangzhou University developed lightweight concrete made of organic microbeads And thermal insulation materials were made of this organic microbead, foaming agents, and cement paste.

Ultra-lightweight concrete can be summarized as the following illustration: The key is pore-forming material and ultra-lightweight aggregate.

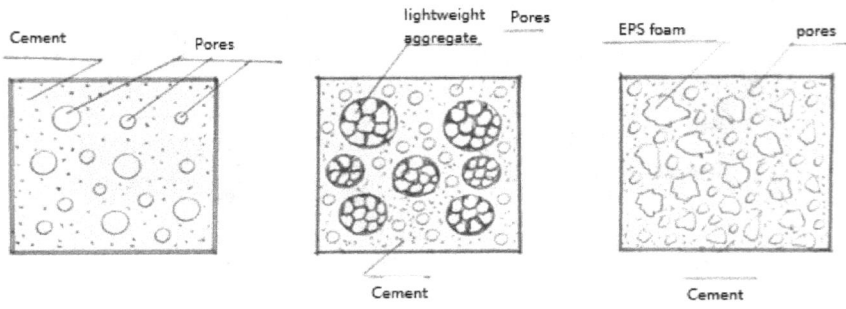

(a) porous concrete   (b) lightweight porous concrete   (c) foam plastic porous concrete
Pores - the pore-forming material can be aluminum powder, foaming agent, carrier gas
Ultra-light aggregate - can be artificial ultra-light aggregate and polystyrene foam beads

Figure 21.3 Technical approach to ultra-light concrete.

This chapter introduces the lightweight (ultra-light) aggregate multifunctional concrete made of ultra-light ceramsite, vitrified microbeads, polystyrene foam pellets and air-entraining additives, etc. It has self-compacting, self-curing, low shrinkage, and low heat of hydration, high plastic retention, and high durability. Its dry bulk density is less than or equal to 1000 kg/m^3, compressive strength is more than or equal to 12 MPa, and the thermal conductivity is about 0.22. Through the small-scale test of the panel in the laboratory and the use of the production equipment of the existing precast plant, the pilot production has produced more than 100 internal partition panels, which provided engineering applications.

21.2 Raw materials of experiment

1) Cement: PII42.5R Portland cement. The properties are shown in Table 21.1.

Table 21.1 The testing results of cement.

80 μm residue	Specific surfaces area m²/kg	Setting time（min）		Flexural strength (MPa)		Compressive Strength (MPa)		soundness	L.O.I
		Initial	Final	3 d	28 d	3 d	28 d		
3.5%	420	198	233	5.4	8.4	34.3	58.5	pass	1.78%

2) Fly ash: Class II fly ash and the properties are shown in Table 21.2.

Table 21.2 The testing results of class II fly ash.

fineness （residue）	Water demand（%）	L.O.I（%）	SO₃(%)
18.9	99	2.84	0.54

3) River sand: The properties of river sand are shown in table 21.3.

Table 21.3 The testing results of river sand

Fineness Modulus	density (kg/m³)	Bulk density (kg/m³)	Compacted density (kg/m³)	Clay Lump （%）	Clay content （%）	Chloride content （%）
2.6	2630	1450	1570	2.0	0.6	0.001

4) Polystyrene foam plastic pellets: unit weight 10 kg/m3~14 kg/m3. The particle size is about 0.1~1 mm. Waste materials for recycled application.

5) Vitrified beads: Expanded perlite treated by glazing, and the bulk density is about 100 kg/m³.

6) Microbeads: ultra-fine product of fly ash collected by air separation, technical properties indicators are shown in Table 21.4.

Table 21.4 Technical properties of microbeads.

Density (kg/m³)	Water demand（%）	L.O.I（%）	Bulk density (kg/m³)
2640	99	1.2	970

7) Silica fume: the technical properties of silica fume are shown in Table 21.5.

Table 21.5 The technical properties of Silica fume.

SiO₂ content（%）	Residue（%）	L.O.I（%）	Bulk density (kg/m³)
93.8	0.4	1.9	336

Slag powder: Class S95 slag powder, the technical properties are shown in Table 21.6.

Table 21.6 The technical properties of slag powder.

Density （g/cm³）	Flowability （%）	Reactivity index（%）	Specific surface area (m²/kg)

8) Expanded clay: The properties of expanded clay are shown in Table 21.7.

Table 21.7 The properties of expanded clay.

Strength in tube （MPa）	density (kg/m³)	Bulk density (kg/m³)	porosity （%）	1 h water absorption （%）	grading
1.2	510	235	42	20.5	graded

21.3 Experiment program

21.3.1 The technical requirements of the experiment

The requirements of the experiment for lightweight aggregate multifunctional concrete are shown in Table 21.8.

Table 21.8 The technical requirements of Lightweight multifunctional concrete.

	Dry density（kg/m³）	≤1000
	Compressive strength（MPa）	≥12
	Water absorption（%）	<15
Basic properties	Water permeability	≥P8
	U Channel rising（mm）	≥300
	slump（mm）	≥250
	Slump flow（mm）	≥650
	Engineering performance	Can be nailed and sawed

21.3.2 The mix design of lightweight multifunctional concrete

The mix proportion for 35 liters of lightweight multifunctional concrete is shown in Table 21.9, and the slump, expansion, U-Channel rising height, wet bulk density, dry bulk density, and compressive strength of 7 days and 28 days of self-curing were tested .

Table 21.9 Mix proportion for 35 litres of lightweight concrete.

No.	Cement	Slag	Lightweight aggregate	Water	Admixture (g)
1	8.75	10.5	16.9	4.9	327.3
2	10.5	9.9	12.3	5.08	139.2
3	10.5	9.9	12.3	5.08	139.2
4	10.5	9.9	12.3	5.08	139.2
5	10.5	7.1	11.9	5.25	270
6	10.5	7.1	11.2	5.25	378

No.1 used polystyrene foam pellets.

No. 2, 3, and 4 use vitrified microbeads;

others are air-entraining agents;

No.1: ordinary mixing process;

No. 2, 3, and 5: cement slurry mixed first, and then putting ceramsite;

NO.4: cement slurry mixed first, and air-entraining agent and water are mixed to produce bubbles, and then poured into the slurry;

NO.5, 6: aggregate was replaced with vitrified beads.

The test of fresh concrete is shown in Figure 21.4 and the test results are shown in Table 21.10.

Table 21.10 The testing results of lightweight concrete.

No.	Slump (mm)	Slump flow (mm)	U- Channel (mm)	Strength (MPa)		Dry density (kg/m³)
				7天	28天	
1	230	550	/	2.6	4.3	760
2	270	700	340	7.9	10.4	998
3	265	680	340	3.8	6.2	950
4	270	670	340	/	/	955
5	275	700	340	13.4	17.6	1000
6	270	690	340	10.3	13.1	860

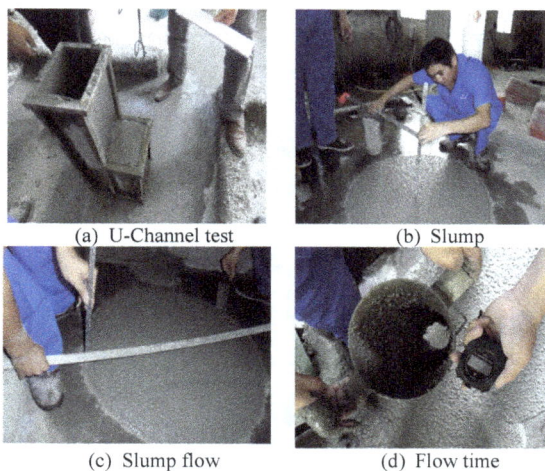

(a) U-Channel test

(b) Slump

(c) Slump flow

(d) Flow time

Figure 21.4 The testing of fresh lightweight concrete.

It can be seen from the test results in Table 21.10:

1) No.1 has a small bulk density but low strength, and fresh concrete cannot meet the performance requirements of multifunctional concrete.

2) No. 2 and No. 3, fresh concrete can meet the performance requirements of multi-functional concrete. The strength of No.2 is also higher, but the strength of No.3 at 28 d drops too much.

3) No. 4, using different mixing techniques, the performance of the mixed lightweight (ultra-light) aggregate multifunctional concrete is consistent with the bulk density, but the strength is lower. The concrete specimen shrinks greatly.

4) No. 5 and 6 use part of vitrified microbeads instead of part of light aggregate to achieve dry bulk density ≤1000 kg/m3, 28-day strength ≥12 MPa, and can achieve self-compacting performance.

During the test, it was found that the No.5 and 6 lightweight aggregate multifunctional concrete had the phenomenon of ceramsite floating when the specimen was formed. Part of the river sand was added to replace the fine aggregate. At the same time, the air-entraining agent was added to give a certain amount of bubbles in the slurry. Air bubbles ensured the stability and uniformity of lightweight aggregate multifunctional concrete.

On the basis of the previous tests, further tests are shown in Table 21.11.

Table 21.11 Mix design of lightweight multifunctional concrete.

No.	Cement	Addition	Lightweight aggregate	River sand	Water	admixture
1	300	300	300	/	145	9.0
2	300	300	280	/	130	8.4
3	300	300	280	/	135	9.0
4	300	300	240	100	145	7.8
5	300	300	280	100	150	7.8
6	300	300	250	120	130	7.8
7	300	270	260	120	135	7.41
8	300	270	280	120	135	7.41

The test was carried out according to the proportion of lightweight aggregate multifunctional concrete in Table 21.11, and the results are shown in Table 21.12.

Table 21.12 Testing results of lightweight multifunctional concrete

No.	Slump (mm)	Slump flow (mm)	U-Channel (mm)	Strength (MPa)		Wet density (kg/m^3)	Dry density (kg/m^3)
				7天	28天		
1	255	680	330	11.5	13.8	1082	970
2	255	660	330	10.7	12.8	1025	940
3	255	660	330	9.2	12.4	1020	945
4	260	680	340	10.3	12.7	1113	990
5	260	680	340	10.0	13.0	1048	975
6	255	680	340	10.3	13.5	1133	986
7	260	690	340	11.5	14.2	1052	965
8	260	690	340	12.1	15.8	1036	960

It can be seen from Table 21.12 that for the No.1-8 concrete mix the properties of fresh concrete can meet the self-compacting requirements. The dry bulk density is below 1000 kg/m^3, and the 28 d strength is above 12 MPa. In the case of mixes with partial replacement of river sand, the most suitable dosage of vitrified beads is 80 kg.

21.4 Testing of mechanical properties, durability, and fire resistance of lightweight aggregate multifunctional concrete

According to the mix design of No.8 in Table 21.11, the performance of strength, autogenous shrinkage, early shrinkage, split tensile strength, flexural strength, axial compressive strength, etc. were tested as in Table 21.13 for details.

Table 21.13 The properties and strength of light (ultralight) multifunctional concrete.

Slump (mm)	Slump flow (mm)	U-Channel (mm)	Compressive strength (MPa)			Wet density (kg/m^3)	Dry density (kg/m^3)
			7 day	28 day	56 day		
260	690*690	340	12.1	15.8	16.6	1036	960

Table 21.14 The autogenous shrinkage and early shrinkage.

Time	24 h	72 h	7 d	14 d	28 d
Early shrinkage	-0.8/10K	0.56/10K	1.39/10K	/	/
Autogenous shrinkage	-0.32/10K	-0.76/10K	0.19/10K	1.19/10K	1.81/10K

Table 21.15 The tensile-split, flexural and compressive strength and E-Modulus.

28 d	Tensile split	Flexural	Compressive	E-Modulus
Result (MPa)	1.2	2.92	13.6	

Note: In the experiment, the cross section of the test block for measuring the axial compressive strength is 100×100, and the final strength value should be multiplied by a coefficient of 0.95. Therefore, the representative value of the final axial compressive strength is 12.9 MPa.

Table 21.16 Fire resistance of lightweight concrete.

Time and temperature	Standard compressive strength	1.5 hours after 400°C	1.5hours after 600°C
Change in strength (MPa)	18.3	8.3	5.6

Table 21.17 Freeze and thaw resistance of lightweight concrete.

No.	Size (mm)	Initial mas (kg)	Mass after freeze (kg)	Mass loss	Initial E-Modulus (MPa)	Modulus after freeze (MPa)	Loss of Modulus
1	100×100 ×400	5.05	5.09	0.79 (%)	13.45	12.35	8.4 (%)
2		5.09	5.13		14.02	12.81	
3		5.06	5.10		13.51	12.60	

Note: Test results of 25 freeze-thaw cycles (4 h each time).

Table 21.18 Impermeability Test of Lightweight Aggregate Multifunctional Concrete.

No	Pressure when leaking （MPa）	Impermeability class
CH1	1.3	
CH2	1.3	
CH3	1.3	P12
CH4	1.3	
CH5	1.3	
CH6	1.3	

Table 21.19 RCPT of lightweight aggregate multifunctional concrete (Coulomb/6h).

No	CH1	CH2	CH3	CH4	CH5	CH6	Average
Coulomb	373	323	288	251	286	330	309

The charge passed was only 309 Coulomb/6h, less than 500 Coulomb, indicating excellent impermeability.

Summary of durability:

1) The tested lightweight aggregate multifunctional concrete has a charge passed of ≤500 coulomb/6h for 56 days. The impermeability is excellent.

2) For the lightweight aggregate multifunctional concrete, there is only 300-320 kg/m^3 of cement, while the total amount of microbeads, Fly ash, mineral powder, silica fume and natural zeolite powder, etc. is ≥250 kg/m^3. This will give the concrete excellent performance in inhibiting alkali-aggregate reaction;

3) Due to the addition of appropriate amount of fiber, the tensile splitting strength is high, and the autogenous shrinkage and early shrinkage are small, which reduces the early shrinkage cracking and concrete hardening fracture;

4) The results of the fire resistance test showed that the lightweight aggregate multifunctional concrete has higher fire resistance performance.

21.5 Lightweight aggregate multifunctional concrete partition wall test

In order to apply the results of this research to the project, under laboratory conditions, several partition wall panel tests were carried out. And then batch production were carried out at Wanyou Precast Factory. More than 100 panels were trial produced.

1) Partition wall panel test at laboratory

Under the conditions of the laboratory, a set of 2 moulds were used to form a set of lightweight aggregate multifunctional panels. And the solar curing shed was used for curing. After the test was successful, trial production at Wanyou Concrete Products Factory was performed. The conditions of the test panels in the laboratory are as shown in Figure 21.5.

(a) Lightweight concrete test

(b) Curing in solar shed

c) The tested lightweight aggregate
multifunctional partition wall panel,
28 d compressive strength≥12.5 MPa, dry bulk
density is about 1000 kg/m3.

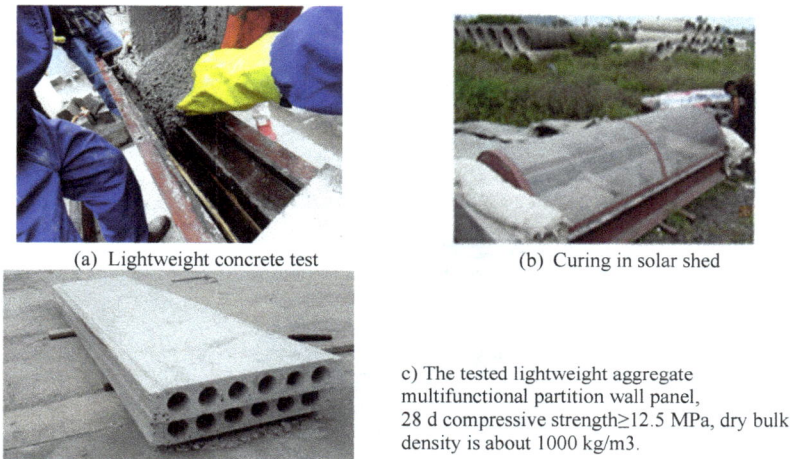

Figure 21.5 Partition wall test under laboratory condition.

2) Trial production in a concrete product factory

In Wanyou Concrete Products Factory, a group of 5 panels had been cast for trial production. The design size of the mold was: thickness 100 mm, length 2400 mm, and width 600 mm. The mix design of the lightweight aggregate multifunctional concrete used in the trial production was the same as that in laboratory, as shown in Table 21.20.

Table 21.20 The mix design of lightweight concrete (kg/m³).

Cement	Additives	Lightweight aggregate	Microbeads	River sand	Water	admixture
300	270	280	20	120	135	7.5

The production mixer was used to mix concrete in the prefabricated component factory. The concrete performance and wet bulk density are shown in Table 21.21. The trial production process of panels is shown in Figure 21.6.

Table 21.21 The performance of lightweight concrete at discharging from mixer.

Slump (mm)	Slump flow (mm)	Flow time (s)	U Channel (mm)	Wet Density (kg/m³)
260	680×680	3.20	340	1043

(a) Pouring of lightweight concrete into mould

(a) Levelling of panels

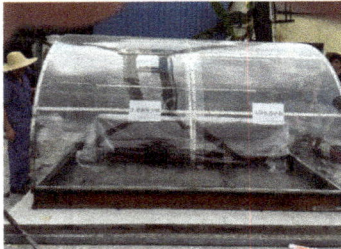

(c) Curing in solar shed

(d) the panels after curing in solar shed

Figure 21.6 Production process of lightweight aggregate multifunctional concrete panels.

The 28 d and 56 d strengths of lightweight aggregate multifunctional concrete were 15.3 MPa and 16.9 MPa, respectively. Dry bulk density was ≤1000 kg/m³.

3) Summary

From the trial production of panels in the factory the following conclusions can be summarised:

a. Using commercially available materials and ordinary concrete mixing equipment, lightweight aggregate multifunctional concrete can be produced and constructed. This kind of concrete has: low bulk density, self-compacting, self-curing, low shrinkage, low heat of hydration, high plastic retention and high durability, etc.

b. In multifunctional concrete, ceramsite and vitrified microbeads can reduce the bulk density of concrete, improve the fluidity and strength of mortar, reduce shrinkage and improve fire resistance.

c. In multifunctional concrete, natural zeolite powder is a thickening agent, self-curing agent and can reduce autogenous shrinkage.

d. Air-entraining water reducing agent not only introduces bubbles, but also has a high water reducing effect. There is a certain amount of bubbles in the slurry, which ensures the stability and uniformity of the mixture.

e. The effect of self-curing is good. This is because the self-curing agent and ceramsite uniformly supply water for the cement hydration inside the concrete, and the surface of the concrete panels are wrapped with plastic sheets or covered by the solar shed, so the water in the concrete is not easy to escape.

21.6 Concluding remarks

From the selection, preparation and performance tests of raw materials, wall panels production test, and trial production of lightweight aggregate multifunctional concrete, the preliminary conclusions are as follows:

1) Expanded clay produced from waste clay, vitrified microbeads, industrial Waste residue—microbeads, fly ash, slag powder, silica fume, and some cement are used as raw materials to prepare lightweight aggregate concrete with a dry bulk density of 850~1000 kg/m³, compressive strength ≥12 MPa as well as self-compacting properties. This kind of concrete has the functions of labor-saving, energy-saving, and resource-saving and compatibility with the environment.

2) The thermal conductivity of this kind of concrete is about one-third of that of clay bricks, and it has good fire resistance. The thermal effect with only 100 mm~120 mm thickness of this kind of concrete can achieve the same as that of 370 mm thick clay brick wall. Therefore, using this lightweight aggregate multifunctional concrete as non-load-bearing wall and roof insulation can greatly reduce the internal energy consumption of the building during operation and give people a comfortable living environment.

Chapter 22

Electronic Protection Technology for Multifunctional Concrete Structure

22.1 Rebar corrosion in concrete structure

Rebar corrosion is the primary cause of concrete deterioration and damage. The corrosion product rust expands in volume, leading to concrete cracking and gradual loss of load capacity (Fig. 22.1).

(a) pH-neutralization   (b) chloride

Fig. 22.1 Corrosion damage of reinforced concrete

According to standards in Norway and UK, when Cl⁻ content at rebar surface reaches 0.4 *w.t.*% of cement, corrosion is considered to start. In Japanese standard the corresponding value is 0.3 kg/m³. A more general agreement is that when $Cl^-/OH^- \geq 0.6$ corrosion is likely to happen. The alkalinity of the concrete layer protects rebar from corrosion. In case of a pH-neutralization, the passive layer on the rebar surface is destabilized, leading to electrochemical corrosion and rusting. In Fig. 22.1 illustrates concrete corrosion damage caused by (a) pH-neutralization of the protection layer and (b) excessive chloride ingress.

22.1.1 Mechanism of rebar corrosion in concrete

The rebar is protected from corrosion by a passive film on its surface, which is table in the alkaline condition in a normal concrete. This film is depicted as a 20-60 Å – thick oxides containing water. In a corrosive environment, the passive film cannot be dynamically repaired. The rebar surface then enters an active state of electrochemical corrosion.

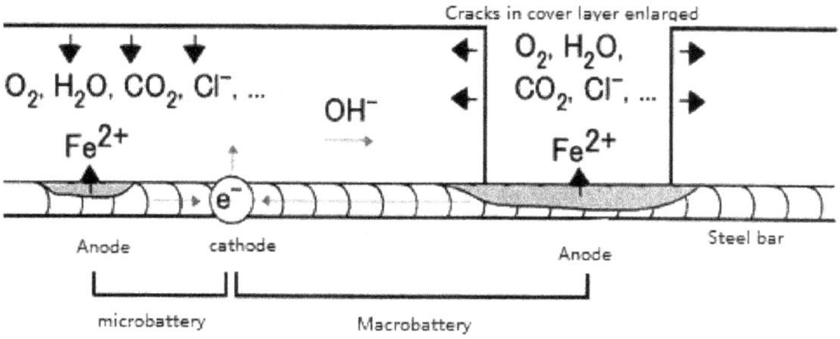

Fig. 22.2 Electrochemical corrosion of concrete rebar.

Micro and macro electrochemical corrosion may simultaneously take place on the same rebar. The reaction formula is as follows.

Anode: $2Fe \rightarrow 2Fe^{2+} + 4e^-$

$\implies Fe(OH)_2 \implies Fe(OH)_3$ rust

Cathode: $O_2 + 2H_2O + 4e^- \rightarrow 4OH^-$

On the left part of Fig. 22.2, the micro defects near the rebar surface facilitates the oxidization of Fe to Fe^{2+}. The released electrons travel to the cathode and form OH^-, which migrate to the anode (through concrete pore structure) and form $Fe(OH)_3$ (rust) with Fe^{2+}.

22.1.2 Measurement of metal (rebar) corrosion

In a corroded metal the anode is negatively charged and its natural potential becomes more negative (i.e. chemically more active). When external current is introduced to the metal, the anodic potential of the metal will reach balance with the external current. The corrosion of the metal can then be assessed by the balanced potential. This approach is called natural potential method.

The ASTM C876 Non-Destructive Testing Method of Rebar Corrosion was established in 1997. This method has been globally adopted, as it allows identifying the location of rebar corrosion and estimating the percentage of corrosion damage. The Japanese Concrete Institute later modifies the method and in 2000 released the JSCE-E 601:2000 'Determination of Natural Potential in Concrete Structure'. Equipment to measure natural potential are now available in Chinese market. Researchers in Tsinghua University has measured the corrosion of rebar in mortar samples made with sea sand (Fig. 22.3).

(a) Rebar embedded in mortar containing various contents of Cl⁻

(b) Measuring the natural potential of rebar

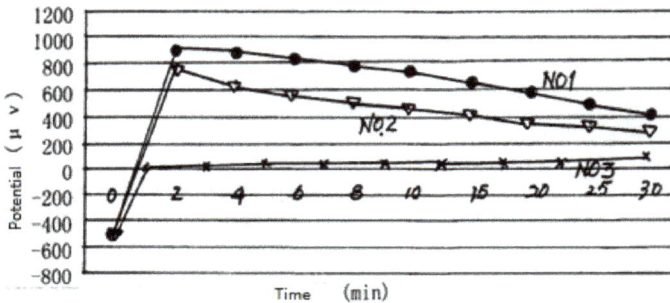

(c) Rebar potential as a function of time

Fig. 22.3 Rebar potential as a function of time after dry-wet cycling.

After 6 months of dry-wet cycles in 3% NaCl solution, the natural potential of rebar in three mortar specimens were measured (Fig. 22.3). The potential of rebar in normal mortar made of river sand and sea sand decreased with time. In contrast, the specimen with 10% inter-mixed Cl-absorber did not decrease within 30 min of measurement, suggesting that the rebar is not corroded.

22.2 Mechanism of electrochemical protection technology

In active corrosion, the corrosion current flows from anode to cathode, as shown in Fig. 22.4. When introducing direct current to the rebar, the potential of both anode and cathode decreases, along with the potential difference between them (Fig. 22.4b). As long as such difference exists, corrosion is still active. If the external current is large enough to eliminate the potential difference between cathode and anode, the corrosion will stop. This approach is called electrochemical protection.

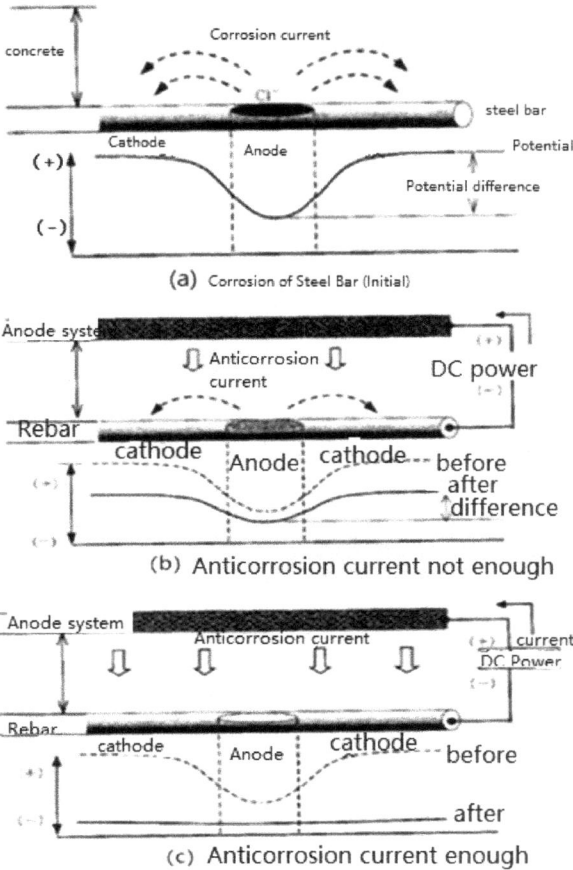

Fig. 22.4 Mechanism of electrochemical corrosion protection.

22.3 Electrochemical protection methods

22.3.1 External current

In the external current protection, an anodic system is installed on the surface of corroded concrete. The anode of an external current is connected to this system, while the cathode of the external current is connected to the rebar. When the current is switched on, it circulates from the anodic system, through concrete protection layer to the cathode (rebar). When there is sufficient current in the circuit, the entire rebar becomes cathode which is unlikely to corrode. This method is called external current protection.

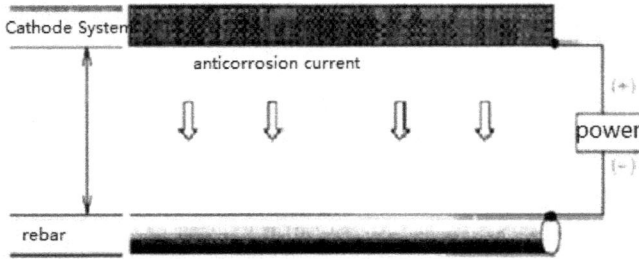

Fig. 22.5 Electrochemical protection using external current.

22.3.2 Internal current (sacrificial anode)

Metals with lower potential than rebar, such as zinc, aluminum and magnesium, can be placed near concrete and connected to rebar through conductive wires. These metals will work as anodes and supply electrons to rebar to prevent rear corrosion. This approach does not require external electronic power supply (Fig. 22.6).

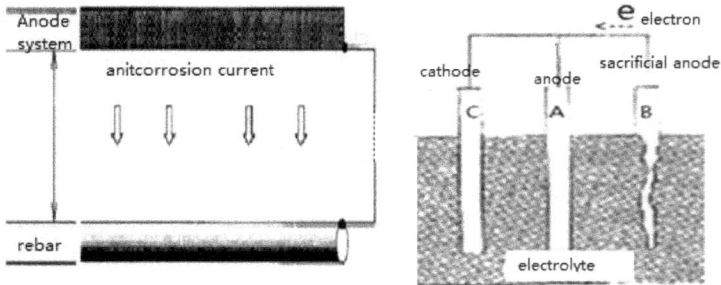

Fig. 22.6 Sacrificial anode protection method.

A suitable sacrificial anodic material should satisfy the following requirements. 1) Its electric potential has to be sufficiently lower than that of rebar. 2) The quantity of electricity from per unit consumption of the materials needs to be large. 3) The corrosion of the anode materials needs to be minimal. 4) The material needs to be cheap and mechanically strong. The most used sacrificial anode materials are zinc, aluminum, magnesium and their alloy. They can be applied in existing structure or newly constructed structures.

22.3.3 A comparison between the two approaches

The difference in the circuit between the two methods are shown in Fig. 22.7. In external current method, a temporary anode is used where the external current flows to the rebar to achieve a

Materials Research Forum LLC
https://doi.org/10.21741/9781644901991

cathodic polarization. The potential difference between the original anodic and cathodic regions (on the rebar) is reduced to as low as zero, so that corrosion current is eliminated. In internal current method the cathodic polarization is achieved through the relatively large potential difference between the sacrificial anode materials and the rebar.

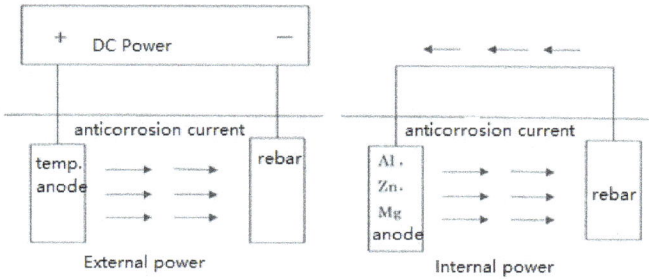

Fig. 22.7 A comparison between external and internal current methods.

1) Different types of external current protection methods

   a. Metal mesh used as panel anodic system (Fig. 22.8).

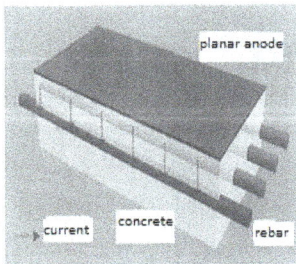

(a) The flow of protection current from anodic mesh to rebar

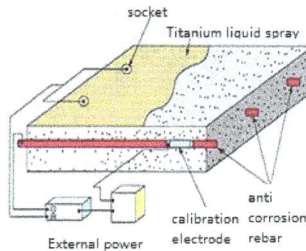

(b) The combination of anodic mesh and other devices.

Fig. 22.8. Panel anodic system for external current protection method. (a).

   b. Titanium rod used as anodic system. It is implemented on the surface or inside the concrete, where current flows from titanium to rebar (Fig. 22.9).

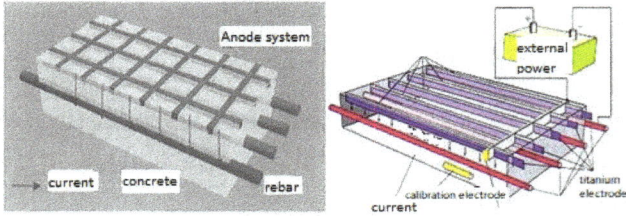

Fig. 22.9 Titanium rod used as anodic system.

c. Point-type anodic system. On top of the rebar, point-type anodes are installed to supply current. It requires holes (~12 mm diameter) to be drilled on concrete where titanium (or zinc) rods are installed and connected to external power supply through wires, as shown in Fig. 22.10.

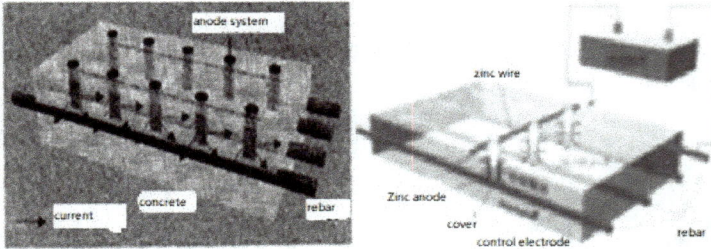

Fig. 22.10 Point-type electrochemical protection anode system.

2) Sacrificial anode protection methods

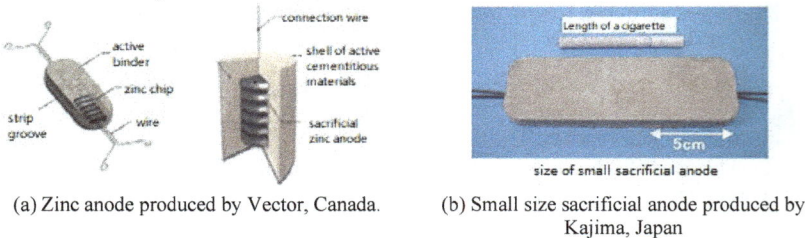

(a) Zinc anode produced by Vector, Canada.   (b) Small size sacrificial anode produced by Kajima, Japan

Fig. 22.11 Appearance of sacrificial anode products.

a. There are multiple appearances of the sacrificial anode. Two products from Canada and Japan are shown in Fig. 22.11.

b. The protection effect of sacrificial anode is illustrated in Fig. 22.12.

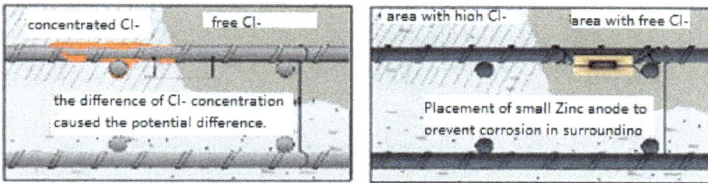

(a) without the protection of sacrificial anode.   (b) with the protection of sacrificial anode.

Fig. 22.12 Corrosion condition of rebar (a) without and (b).

c. The installation of small-size anode in new reinforced concrete.

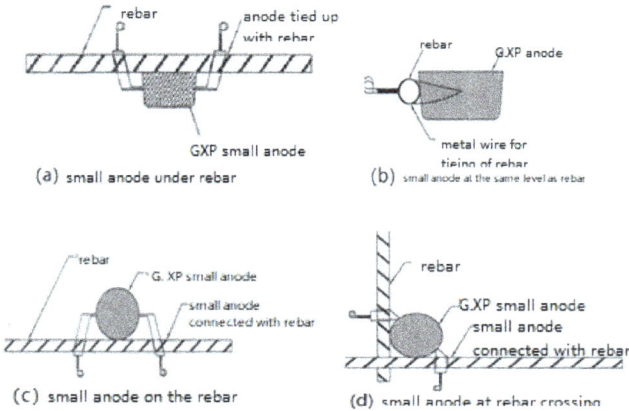

(a) small anode under rebar

(b) small anode at the same level as rebar

(c) small anode on the rebar

(d) small anode at rebar crossing

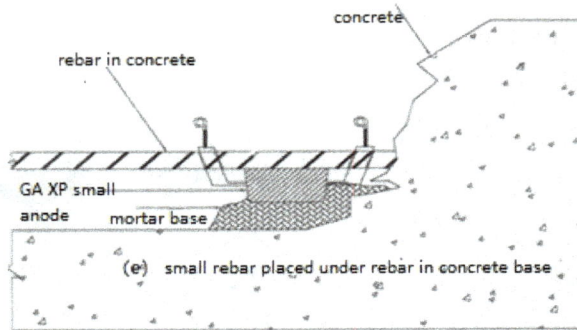

Fig. 22.13 The installation of small-size sacrificial anode.

d. The usage of sacrificial zinc anode in repair of concrete structure is shown in Fig. 22.14.

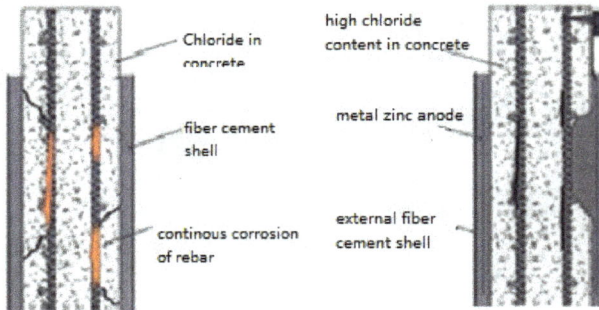

(a) corrosion continued without small anode (b) Fiber cement shell wrapped with zinc metal anode

Fig. 22.14 Application of sacrificial zinc anode in concrete repair.

e. Materials used in sacrificial anodes

The cathodic protection method of sacrificial anodes utilizes a large potential difference between the sacrificial anodes and the protected steel bars to generate current to protect the steel bars in concrete. Therefore, to be used as a sacrificial anode material, the following conditions must be met. ① Compared with the protected steel bars, it has a sufficiently low potential; ② The electricity generated per unit consumption is larger; ③ It can resist corrosion and has high current efficiency; ④ It has better mechanical strength and low cost. The sacrificial anode materials commonly used in engineering now include aluminum, magnesium, zinc and their combinations.

Table 22.1 Cases suitable for electrochemical protection.

Cases	Causes of degradation		
		Chloride	pH-neutralization
Land condition		○	○
Marine condition	Atmospheric zone	○	○
	Splash zone	○	○
	Tidal zone	▽	▽
	Submerged zone	▽	---
Structural elements	RC , PC	○	○
Existing structures		○	○
New structures		○	○

Note: '○', '▽' and '---' means fully suitable, generally suitable and not suitable, respectively.

22.4 Cases suitable for electrochemical protection

Electrochemical protection is generally suitable for concrete corrosion due to chloride ingress and pH-neutralization. With the installation of anode system, direct current flow from the anode, though concrete layer towards the rebar, leading to cathodic polarization of the rebar and the mitigation of corrosion. The common application cases are listed in Table 22.1.

22.5 Two major parameters for cathodic protection

1. Minimum current density. A minimum current density needs to be applied to eliminate the anodic current so that the rebar is fully protected. Assuming the total surface area of the rebar is A, the current density is $I = p/A$, (unit is A/m^2). The required minimum current density depends on the type of protected metal, the characteristics of the electrolyte, temperature, the polarization rate of electrode. In moisture condition, the minimum current density needed to protect concrete rebar is 0.055-0.27 (A/m^2).

2. Minimum protective potential. During cathodic protection of rebar, a minimum electric potential is needed to completely stop the rebar corrosion. This potential needs to be sufficient to reduce the rebar potential to an equal value as the corrosion potential E^0. Experiment data suggest that the minimum potential to protect iron in natural water or soil is -0.53 V *vs.* standard hydrogen electrode, or -0.77 V *vs.* saturated calomel electrode. The rebar will be in complete protection if the applied potential is more negative than these values.

References

[1]   (Japan) Yoshio Kasai, The Concrete by Technical College, 1982.

[2]   (USA) P. Mehta The structure, properties and materials of concrete. Translated by Zhu Yongnian, Shen Wei and Chen Zhiyuan. Shanghai: Tongji University Press, 1991.

[3]   Edited by Feng Naiqian, High Performance Concrete. Beijing: China Construction Industry Press, 1996.

[4]   Editor-in-chief Feng Naiqian, Practical Concrete Encyclopedia. Beijing: Science Press. 2001.

[5]   Adam Neville & Pierre-Clande Aitcin, High Performance Concrete –An overview Materials and Structures Vol.31, 1998.3. https://doi.org/10.1007/BF02486473

[6]   P.K. Mehta and P.C. Aitcin, Prineiples underlying Production of High Performance Concrete ASTM. C.C.A, Vol. 31, 1998.3.

[7]   N. Q. Feng, G.Z. Li and X.W. Zang, High-strength and Flowing Concrete with a Zeolitic mineral admixture ASTM. C.C.A, Vol.12, No.2, Winter 1990. https://doi.org/10.1520/CCA10273J

[8]   N. Q. Feng, Y. X. Shi and T. Y. Hao, Influence of ultrafine powder on the fluidity and strength cement paste Advances in Cement Research,Vol.12, No.3 July 2000. https://doi.org/10.1680/adcr.2000.12.3.89

[9]   R. C. Joshi and R.P. Lohtia, Type and Properties of Fly Ash , Mineral Admixtures in Cement and Concrete V.4 . abi New Delhi 1993 First Edition.

[10]   S. Popovics, What Do We know About the Contributions of Fly Ash to the Strength of Concrete, Second ICFSS, Madrid, ACI SP -114,1,1986.

[11]   M. Kokubu, Mass Concrete Practices in Japan, Symp. on Mass CONCRETE, ACI Publication Sp-6, 1963.

[12]   P.K. Mehta et al, Properties of Portland Cement containing Fly Ash and Condensed Silica Fume, Cement and Concrete Research (12), pp583,1982. https://doi.org/10.1016/0008-8846(82)90019-9

[13]   F.H. Wittmann Summary Report on Research Activities, Lausanne, November 1986.

[14]   ACI Committee 363 State-of-the Art Report on High-Strength Concrete, Chapter 5 – Properties of High-Strength Concrete, ACI Journal, July August, pp.383-389, 1984.

[15]   N.Q. Feng, F. Xing and F.G. Leng, Zeolite Ceramsite Cellular Concrete (J) Magazine of Concrete Research, 2000(4). https://doi.org/10.1680/macr.2000.52.2.117

[16]   N.Q. Feng, X.X. Feng, T.Y. Hao, F. Xing, Effect of Ultrafine mineral powder on the charge passed of the concrete (J) Cement and Concrete Research, 2002. https://doi.org/10.1016/S0008-8846(01)00731-1

[17]   J. Moksnes, High-Performance Concrete –A Proven Materials with Unfulfilled Potential. Seventh International Symposium on Utilization of HS/HPC. ACI, 2005.

[18]   M. Schmidt, E. Fehling, Ultra HPC, Research, Development and Application in Europe, Seventh International Symposium on Utilization of HS/HPC. ACI, 2005.

[19]   H. Okamura, K. Maekawa, T. Mishima, Performance Based Design for Self-Compacting Structural High Strength Concrete, Seventh International Symposium on Utilization of HS/HPC. ACI, 2005.

[20]   N.Q. Feng, J.H. Yan, G.F. Peng, Research Development and Application of HS/HPC in China, Seventh International Symposium on Utilization of HS/HPC, ACI, 2005.

[21]   N.Q. Feng, H.W. Ye, and Z.L. Lin, Research and Application of Multi-Properties Concrete, 10th International Symposium on HPC, Beijing, China ,2014

[22]   Feng Naiqian, High Strength Concrete Technology (Mandarin) Beijing: China Building Materials Industry Press,1992.

[23]   Feng Naiqian, Xing Feng, High Performance Concrete Technology (Mandarin) Beijing: China Atomic Energy Press, 2000.

[24]   Feng Naiqian, Fluid Concrete (Mandarin) Beijing, China Railway Press, 1986.

[25]   Feng Naiqian Natural Zeolite Concrete Application Technology (Mandarin) Beijing: China Railway Press, 1996.

[26]   Feng Naiqian, Xing Feng. Durability of Concrete and Concrete Structures (Mandarin) Beijing: Machinery Industry Press 2006.

[27]   Gu Jiexiang, Zeolite (Mandarin) Beijing, China Construction Industry Press. 1980.

[28]   Xi Zhige, Li Junwen, Wang Fuyu, Nano-Applied Technology for Environmental Sanitation (Mandarin) Beijing, Chemical Industry Press, 2004.

[29]   Hong Dinghai, Corrosion and protection of steel bars in concrete (Mandarin) Beijing: China Railway Press, 1998.

[30]   Standard of China Engineering Construction Standardization Association. Technical Specification for High Strength Concrete Application CECS207-2006 Beijing: China Planning Press 2006.

[31]   Institute of Building Technology. Proceedings of the 2003 Cross-Strait Symposium on Built Environment and Sustainable Management. Taipei: China Institute of Technology, 2003.

[32]   Feng Naiqian, High Performance and Ultra High Performance Concrete Technology (Mandarin) Beijing, China Construction Industry Press. 2015.

[33]   Kobayashi Ichisuke, Concrete Practical Handbook (Mandarin) Translated by Wang Xiaoyun, Deng Li, Beijing: China Electric Power Publishing House 2010.

[34]   Feng Naiqian, Modern Building Materials Handbook (Mandarin) Beijing: Machinery Industry Press 2021.

[35]   Peng Kai Fei, Civil Engineering Materials (Mandarin) Wuhan: Huazhong University of Science and Technology Press 2013.

[36]   Kasai Yoshio, Sakai Etsuro, Hybrid material for new Cement Concrete (Japanese), Japan: Technical College, 2006.

[37]   Zhuang Qingfeng, Macroscopic, mesoscopic and microscopic fracture analysis and hydration computer simulation of high performance concrete (Mandarin) Beijing: Tsinghua University 1997.

[38] Ma Xiaoxuan, Experimental study on the durability of silicate materials in the ground, (Journal) Concrete and cement products,1995(8).

[39] S. Nishibayashi, Concrete (shrinkage cracking) for performance evaluation (academic exchange report), Wuhan,2009 (10).

[40] Yoshio Kasai, Characteristics of Japanese high-strength and ultra-high-strength concrete (academic exchange report), Wuhan,2011 (4).

[41] Feng Naiqian, Development and application of high performance and ultra-high performance concrete (Journal) Construction technology, 2009(4).

[42] Feng Naiqian, Mukai Takeshi, Ehara Kyouji, Research on the Effective Application of Natural Zeolite as Concrete Reinforcement Accelerator (Journal) Proceedings of the Department of Architecture, Architectural Society of Japan 1988(6).

[43] Jiang He et al., Effects of composite admixtures on fluidity and long-term strength of cement mortar (Journal) Concrete and Cement Products 2021 (1).

[44] Wang Guan et al., Effects of different factors on workability and mechanical properties of ultra-high performance concrete (Journal) Concrete and Cement Products 2021 (12).

[45] Liu Zhongyuan et al., The effect of internal curing agent on the properties of self-compacting concrete (Journal) Concrete 2021 (6).

[46] Zhang Xinhua. Microstructure of High Performance Concrete (Mandarin) Beijing: Atomic Energy Press 2000 (6).

About the Authors

Professor Feng Nai Qian is a Professor at Tsinghua University China, guest Professor of Meiji University Japan, Visiting Professor of Kuala Lumpur Infrastructure University and a doctoral supervisor at the Hong Kong Polytechnic University, a member of RILEM78MCA, and a member of the Architectural Society of Japan.

In China, he served as director of Diatomite Association, deputy director of Concrete and Cement Products Committee of Silicate Society, member of Building Materials Academic Committee of Architectural Society, member of Concrete Durability Committee, technical consultant of Beijing Municipal Government, and technical consultant of Beijing Building Materials Bureau,

He has published more than 10 books such as: (Construction materials), (Flowable Concrete), (High Strength Concrete Technologies), (High Performance Concrete), (Mineral Admixture for Cement and Concrete), (Testing and Quality Control in Cement Industry) and more than 260 technical papers in Chinese, English and Japanese.

Mr Lu Jin Ping is currently the President of the American Concrete Institute - Singapore Chapter and the Managing Director of Hitchins International Pte Ltd, Singapore. Mr Lu has more than 30 years of experience working in areas of research & development, testing and technical consultancy for construction materials. He was a Lecturer in the department of construction materials at Tongji University China from 1988-1994. Mr Lu serves as Advisory Committee member of Temasek Polytechnic, School of Applied Science and Member of Board of Directors, International Congress on Polymers in Concrete (ICPIC). Mr Lu has also presented more than 50 papers at various international conferences in the region and has published articles on testing, performance and research on construction materials. He is the lead auditor for certification of Ready-Mix Concrete Products by Singapore Accreditation Council. He received five years' service award of dedicated voluntary service to the community from People's Association and Merit Award from Spring Singapore for Meritorious service and contribution to the Singapore National Standardisation Programme.

Gai-Fei Peng is a professor of civil engineering materials at the Beijing Jiaotong University of China and, a vice chairman of the high-performance concrete committee in China Ceramic Society. He received his Ph.D. from the Hong Kong Polytechnic University and master degree from the Tsinghua University at Beijing, respectively. As an author of more than 100 research papers, Prof. Peng's research interests include high performance concrete, ultra-high-performance concrete and recycled aggregate concrete.

www.ingramcontent.com/pod-product-compliance
Lightning Source LLC
Chambersburg PA
CBHW071317210326
41597CB00015B/1257